(continued inside back cover)

Basic Statistics in Business and Economics

George W. Summers
University of Arizona

William S. Peters
University of New Mexico

Wadsworth Publishing Company, Inc.
Belmont, California

ISBN-0-534-00198-X

L. C. Cat. Card No. 72-93197
Printed in the United States of America

1 2 3 4 5 6 7 8 9 10—77 76 75 74 73

Preface

Basic Statistics in Business and Economics is the core of an integrated instructional system that has been under continuous development and class testing for nearly five years. While it can stand alone as a basic text, experience has shown that students more readily grasp the subject matter when supplemental material is provided. Therefore, references to the *Self-Instructional Supplement* and the *Self-Correcting Exercises* are included in the text.

Both organization and content of the text are generally in keeping with modern practice. Although some knowledge of finite mathematics is helpful to students, no mathematics beyond algebra is used. The text is extensive enough for courses covering a full academic year, but it is organized to accommodate courses of shorter duration as well.

The basic sequence in the first nine chapters progresses from descriptive statistics to probability to statistical inference. In the descriptive statistics of the first three chapters, however, we extend the usual coverage of central tendency and dispersion in the context of populations and parameters to include association and the measurement of relationships. We have found that early exposure to these areas in a nonsampling context helps to maintain the student's focus on the objectives of inference after he moves into the chapters on that subject.

Elementary decision making is introduced in Chapter 7, after the chapters on probability and discrete sampling distributions and before those on continuous probability distributions and elementary inference. This sequence allows the student an earlier and longer exposure to the important elements of Bayesian decision procedures, which are first discussed in a discrete-state context. Bayesian statistics for continuous-state variables are taken up in Chapter 18.

Chapters 9 through 12 discuss the topics in univariate statistical inference that are customary in an introductory course. After three chapters covering estimation and hypothesis tests for means and variances of observations made on a continuous variable, Chapter 12 discusses proportions and chi-square tests of independence. The final six chapters, concerned with other topics in inference and analysis that have important applications in economics and administration, are, for the most part, independent of one another. The instructor may wish to select among these later chapters in accordance with his course objectives. We cover nonparametric statistics in Chapter 15, and Chapter 16 covers sample survey methods in somewhat more depth than many books. In Chapter 17, we return to a descriptive parameter context for the discussion of time series and index numbers. To introduce inferential time series models in an introductory text would detract from adequate coverage of the basic uses of time series decomposition and index numbers.

Readability without sacrifice of precision has been a major goal, and the book has benefitted immeasurably from student and instructor evaluations. Their comments led us, for example, to explain fully each appendix table when it is introduced in the text. Class testing was conducted continuously at the University of Arizona and for shorter periods at the University of Rhode Island and the University of New Mexico.

The typical chapter is divided into two study units, each dealing with two major topics and concluding with a summary and a set of exercises. At the end of each chapter is a glossary of the most important equations introduced.

Students are referred to the *Self-Instructional Supplement* by marginal notations giving the section number of the corresponding programmed set. A bar over the number $(\overline{2.4})$ indicates the beginning of text material for which programming is provided. A bar under the number $(\underline{2.4})$ correspondingly indicates the end. The *Self-Correcting Exercises* are referenced at the end of each study unit. In this manual the step-by-step solutions reinforce the student's learning and help him discover misconceptions before they become fixed. Special aids for the instructor in course planning and testing are also available.

We wish to express our sincere appreciation to Donald L. Harnett, Manfred W. Hopfe, Joseph G. Monks, and Charles Warnock for their reviews during the development of this book and to Charles P. Armstrong, John Maier, James Lee, Robert Zmud, and other instructors who helped in class testing and offered many valuable suggestions. We also thank Eleanore de Gennaro for her patient typing of several versions of the text and exercises. We are grateful to the College of Business and Public Administration at the University of Arizona and to the School of Business and Administrative Sciences at the University of New Mexico for providing time and facilities. Finally, we are indebted to the Literary Executor of the late Sir Ronald A. Fisher, F.R.S., and to Oliver & Boyd, Edinburgh, for their permission to reprint Table IV from their book *Statistical Methods for Research Workers*.

George W. Summers
William S. Peters

Contents

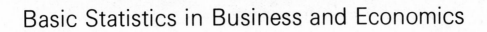

Basic Statistics in Business and Economics

To the Student

Two supplements based on this text are available: The *Self-Instructional Supplement,* by Elzey and Armstrong, contains a programmed expansion of the more difficult concepts. The *Self-Correcting Exercises,* by Summers and Peters, contains complete solutions that allow you to check your comprehension of statistical procedures.

Introduction

To most people statistics are collections of numerical facts. For example, one can find in the 1971 edition of The World Almanac that Bob Gibson of the St. Louis Cardinals set a National League Record earned run average in 1968 with an ERA of 1.12. One can also learn that the birth rate per thousand population in the United States went down from 23.7 in 1960 to 17.8 in 1969 and that General Motors Corporation in 1969 made a net profit of $1,710.7 million on total revenues of $24,295.1 millions. Thus, we see that numbers describing events of interest are collected and recorded in many fields of endeavor. The advantage of numerical description is well reflected in a statement of Lord Kelvin, a prominent British physicist (1824–1907): "When you can measure what you are speaking about, and express it in numbers, you know something about it; but when you cannot measure it, when you cannot express it in numbers, your knowledge is of a meager and unsatisfactory kind."

While statistics means numerical facts in everyday language, *statistics* as a field of study deals with collecting, summarizing, and drawing conclusions from numerical information about groups of events. For reasons that Lord Kelvin emphasized, a wide variety of events and problems in business and economics are described in quantitative terms. What information, or data, to collect about a problem, how much to collect, how to describe the information, and how to draw conclusions from it are the concerns of statistics.

A particularly important problem involves drawing conclusions from incomplete, or sample, data. It usually is impractical to collect information on an entire universe of events of interest. For example, testing the wearing quality of an automobile tire uses up the product being tested; and it would be virtually impossible to check every home in America to determine the audience for Monday Night Football. Therefore, since conclusions must be drawn from incomplete data, measurement of sampling error is a central task in statistics.

In order to give you an idea how this course in statistics applies to the operations of business, we ask you now to imagine that you are a member of a firm that manufactures a line of hardware and small hand tools for both consumer and industrial users in the United States. Here are some problem situations that your firm might encounter.

Kinds of measurement   The president and chairman of the Board of Directors are asked by a management consulting firm to rate numerically ten senior executives on several aspects of long-run leadership potential. How should the numbers (ratings) be treated (Chapter 1) and how can consistency of the judgments be measured (Chapter 15)?

Frequency distribution analysis   The Company considers offering a quantity discount to dealers on large orders of particular items. How can historical data be organized to provide information on the possible effects of different discount schedules (Chapter 2)?

Measurement of relationships   How can the production manager investigate what effect an increase in the number of days of training will have on the efficiency of employees assigned to certain assembly tasks (Chapters 3, 14)?

Sampling and quality control   The production manager complains to the purchasing manager about the quality of materials entering certain processes. How can purchasing set up a sample inspection plan that will provide a known degree of assurance that batches of substandard raw material will not be accepted (Chapters 6, 10)?

Analysis of experimental data   The advertising department seeks to expand the firm's market, by creating goodwill toward the company's brands among persons from a variety of cultural and ethnic groups. How can the effects of alternative themes and copy be tested (Chapters 12, 13)?

Design of sample surveys   The firm is concerned about the effects of marketing practices—pricing, service, and so forth—employed by middlemen handling certain of its industrial lines. It is proposed that detailed information be obtained from a sample of recent transactions. How can the sample be effectively designed (Chapter 16)?

Analysis of time series data   The treasurer believes that short-term bank loans can be reduced by more fully utilizing current receipts to meet accounts payable. How can a study of monthly cash flows be organized to reveal systematic patterns that can be used in forecasting cash needs (Chapter 17)?

Bayesian statistics   The sales manager cites instances of the firm's losing customers because items ordered were out of stock. He asks for a stock-level policy that balances the cost of losing customers and sales against the costs of holding larger stocks (Chapters 7, 18).

These problems illustrate where statistical methods can be useful in just one abbreviated context. Examples from economics or public administration could be given as well. Of course, many problems in human affairs defy quantitative description and analysis, and even statisticians are happy for that. But the range of fruitful applications is wide and growing all the time. The methodology of statistics is general—an integral part of what many call the scientific method. Knowledge of it will increase your ability to understand and work toward effective solutions of problems.

1

Fundamental Data Structures

Numbers are used in statistics in much the same way that they are used in beauty contests and sports. In fact, numbers are the raw materials for statistics. Since the term "statistics" identifies this entire branch of knowledge, however, the numbers themselves are called "data." The first part of this chapter describes how data arise from observations of people or other subjects in a study. This description involves, among other things, the types of measuring scales that produce data. Then, some important basic operations with data are discussed.

Just as baseball managers place more emphasis on the number of strikes a pitcher throws in an entire season than on his strikes in a particular game, we usually are more interested in an entire set of data in statistics than just a few numbers in the set. The second half of this chapter describes some of the most useful characteristics of sets of data. For the baseball example, one such characteristic would be the average number of strikes per game. Finally, algebraic operations with these characteristics are discussed.

Data

Early in a typical statistical study, a *unit of observation* must be selected. For instance, in studies of consumer purchasing patterns, the household often is selected because for such purchases as refrigerators the entire household is treated as a unit. In other studies, events such as traffic accidents or final examinations are the units to be observed.

The particular property to be observed in each unit is the second choice that must be made. The beef consumed by a household in a week may be such a property. Other examples are the relative success experienced on an examination by each student in a class, and a light bulb's capability to illuminate when proper voltage is applied.

After the unit of observation and the property to be observed are selected, a decision must be made on how to measure the varying amounts of the property possessed by the units—that is, the measurement scale to be used. Beef consumption can be measured in dollars per week or pounds per week, for instance. A student's relative success on a test can be measured by his rank on number of items correct or by his rank on number of items correct less some fraction of the number incorrect. A light bulb's ability to illuminate can be measured as "one" if it lights and "zero" if it doesn't, or the intensity of the light emitted can be metered.

Usually several choices for the unit of observation, the property, and the measurement scale are available. The decisions made almost always affect the time and money used for the study as well as the type of conclusion and the level of precision that can be realized.

Having explained the basic terms used in gathering data, we are now ready to define data:

Definition	**Data** are the numbers that result when a designated measuring scale is applied to a specified property of each unit of observation.

Data describe the quantities of a property possessed by every unit observed.

Measurement

Scales

Four types of measurement scale can be used for assigning numerical values to varying amounts of a property. They are, in order of increasing precision, nominal, ordinal, interval, and ratio scales. The characteristics of a scale that determine its level of sophistication are *order* (of numbers), *distance* (between numbers), and *origin* (a zero point). As we shall see, the most primitive scale, the nominal scale, does not use any of these properties. The most sophisticated, the ratio scale, uses all three.

The results of applying a selected scale to the amounts of a property possessed by given units of observation are numbers. We have already defined such numbers as data. Other terms used are *measurements* and *observations*. Unless we state otherwise, you are to treat *measurements*, *observations*, and *data* as synonyms.

Nominal Scales   When we use numbers merely to classify an observational unit with respect to a property, we are employing a *nominal* scale. If, for example, we assign the number "1" to the red pencils in a drawer, "2" to black pencils, and so forth, we are using a nominal scale. Note that the numbers can be chosen arbitrarily. We could just as well use any other numbers, making no use at all of the principle of *order* in the real-number system. We simply substitute numbers for class names in this case.

A typical situation in which nominal scales are useful is the *two-state*, or *binary*, situation. Consider, as an example, the light bulbs to which the numbers " 1 " and " 0 " are assigned, according to their ability to illuminate or not. As bulbs pass an inspection station, they produce a stream of numbers that might look like this:

$$0, 1, 1, 0, 1, 1, 1, 1, 0, 0, 1, \ldots.$$

Many statistical techniques are designed to deal with data of this sort.

Classification systems having more than two categories also may use the nominal scale. Educational status and colors of automobiles are examples. Once again, the numbers chosen to identify the classes make no use of the *order* property of the real-number system.

Ordinal Scales   A scale that uses the order principle in numbers but does not use the distance and origin principles is an *ordinal* scale. When a judge uses numbers to rank contestants in order of spelling skill, for instance, he applies an ordinal scale. Any number can be chosen to represent the amount of skill possessed by either the best or the worst speller, although, in practice, most investigators start with number " 1." The natural numbers that follow the initial number can be used to rank the others. Successively larger numbers can represent either increasingly competent or increasingly incompetent spellers. Grades of meat, military ranks, and morale are examples of other properties to which ordinal scales can be applied.

In using an ordinal scale to rank spelling contestants, the judge makes no use of the distance principle in the sequence of natural numbers. Assume contestants have been ranked from best to worst, and consider those who rank 2, 5, and 8, for instance. The distance from 2 to 5 is 3, and the distance from 5 to 8 is also 3. Yet the ordinal scale does *not* imply that the contestant who ranked second is as much better than the one who ranked fifth as the contestant who ranked fifth is better than the one who ranked eighth.

Interval Scales   An interval scale makes use of both the order and distance properties of numbers but does not use the origin property. The origins of interval scales are arbitrary. For instance, the zero point on the centigrade temperature scale is arbitrarily set at the freezing point of water. Other interval scales are Fahrenheit temperature, clock time, longitude, and test scores. The origins for clock time, for example, are the middle of the night and the middle of the day. Since these points in the daily cycle have no particular advantage over any other points, they represent arbitrary or conventional origins.

When two interval scales are used to measure the amount of change in the same property, the proportionality of differences (intervals) is preserved from one scale to the other. For example, suppose three pans of oil register 50, 100, and 120 degrees on the centigrade scale. The oil which registers 50 on the centigrade scale registers 122 on the Fahrenheit scale. The other two Fahrenheit readings are 212 and 248 degrees, respectively. As an example, for the centigrade readings the proportion of differences between the intervals 100 to 120 and 50 to 120 is

$$\frac{120 - 100}{120 - 50} = \frac{2}{7}.$$

On the Fahrenheit scale the proportion of the respective differences is

$$\frac{248 - 212}{248 - 122} = \frac{2}{7}.$$

Thus, the proportionality of differences is preserved as we change from a centigrade to a Fahrenheit scale, or *vice versa*.

Because origins are arbitrary, it is almost always misleading to state that a value on an interval scale is some multiple of another value. Suppose one student scores 16 correct on a test and another student scores 8. It seldom follows that the first student knows twice as much as the second, because a score of 0 seldom means "no knowledge."

Ratio Scales   For the majority of statistical techniques, measurement of properties on nominal, ordinal, or interval scales is satisfactory. On the other hand, most techniques that require interval-scale measurements·can also make use of ratio-scale measurements.

Ratio scales use the order, distance, and origin properties of numbers. Such properties as length and weight are suitable for measurement on ratio scales, because the concepts of "zero length" and "zero weight" are meaningful. In business and economics, properties such as profit and income usually are measured on ratio scales.

As with interval scales, proportionality of intervals is preserved as one shifts from one suitable ratio scale to another for a given property. Also, it is meaningful to state that one number on a ratio scale is some multiple of another. For example, it is proper to say that a firm which received $2000 of profit in a certain period had *one-half* the profit of another firm that received $4000 of profit.

Historically, the majority of statistical techniques were developed for data from interval or ratio scales, but many of the more recently developed techniques use ordinal and nominal scales. Typically, even though they are less powerful, these latter techniques are less restrictive and simpler.

Variables

Whatever the scale being used in a study, a large set of numbers can be expected when the selected scale is applied to all of the units of observation. In addition, there often are several properties being investigated in the same study and each property gives rise to another such set of numbers. For ease in referring to such a set, the term "variable" is used.

A variable usually is designated by a letter near the end of the alphabet. The letters x, y, or z, either lowercase or capital, are used most often.

Definition	A **variable** is a letter that denotes the set of all possible numerical values a given measuring scale can have in a given application.

We can, for instance, define the variable X to be the set of profits earned by firms in the drug industry last year. Or X could be the set of ranks of contestants in a beauty pageant.

A variable may be either discrete or continuous. It is said to be *continuous* if, theoretically, it can assume any value between two given values on the measuring scale used. A variable which cannot do so is *discrete*. Weight in pounds and portions of pounds is a continuous variable. Theoretically, a person can weigh 112 pounds, 112.3 pounds, 112.3278 pounds, and so forth, to any desired degree of precision. On the other hand, the number of coins in a cash drawer on each of several days is a discrete variable.

In practice, of course, there is a limit to the accuracy with which any continuous variable can be measured. Length can be measured only to the nearest eighth of an inch with most yardsticks, for example. When a piece of cloth is measured by such a yardstick as being eighteen inches long, it may, in fact, have any length from $17\frac{15}{16}$ inches to $18\frac{1}{16}$ inches.

In many instances, there is no need to carry measurements of continuous variables to the limit of accuracy possible with the measuring device. As an example, measuring weights to the nearest tenth of a pound may be sufficient for a given purpose. Actual measurements are *rounded* to this level of accuracy. A package that weighs 10.247 pounds is recorded as 10.2 pounds, and one weighing 10.281 pounds is recorded as 10.3 pounds. For measurements that fall exactly halfway between values on the scale, we shall use the *round-even* rule:

Rule	Original measurements that lie exactly halfway between values on the rounded scale must end in an even number after rounding.

Under this rule, if we round to tenths of a pound, weights measured as 10.250 and 10.350 pounds are recorded as 10.2 and 10.4 pounds, respectively. The rule tends to average out errors produced by rounding. Incidentally, we can say that these two observations have been rounded to three *significant digits*. The original observations were made to five significant digits; note that the final zero is relevant in both cases.

Operations with Statistical Data

Since several basic mathematical operations are used repeatedly throughout statistics, the symbols used to denote these operations are the same ones you have encountered in mathematics. Many basic algebraic formulas also simplify computations in statistics. This section offers a review of some symbols and algebraic formulas for basic numerical relationships.

1.1

Compact Notation

Sometimes the most direct way to describe how to perform a statistical procedure upon a collection of measured observations is with the language of *sets*. Therefore, we shall now review some basic terms of this language. Then we will show how some special symbols can be used to denote certain basic operations with sets.

Any clearly defined collection of objects or symbols is called a *set*. The objects or symbols that constitute a given set are its *elements*, or *members*. If and only if every element of one set is also an element of another set, then the first set is a *subset* of the second. For example, the students in a certain room at a certain time can be considered as a set. The individual students are the elements, and the male students in the room make up a subset of the set of students.

One way we shall define a set is by listing its elements in a horizontal row and enclosing them in braces. For example, the set 1, 8, 5, 12 can be written

$$\{1, 8, 5, 12\}.$$

Usually, the order in which the numbers appear in the list has no significance. On the other hand, if we state that we are defining the *ordered set* to be

$$\{1, 8, 5, 12\},$$

then we mean the numbers are to be listed only in the order given.

Rather than *list* its elements, we can define a set by stating a *rule*. We can, for example, write

$$A = \{x : x = \text{positive, even, real number}\},$$

which is to be read, "Let A be the set of all elements x such that x is a positive, even, real number." The set A, therefore, represents the infinite series 2, 4, 6, ...

Summation   The summation symbol is one of the most pervasive in statistics. In accordance with general practice among statisticians, we shall let $\sum$ (uppercase Greek letter sigma) represent the summing operation. Then, if we let y represent elements in the set of real, nonnegative integers,

$$\sum_{y=1}^{4} y$$

is read "summation of y with y running from 1 through 4." It is interpreted to mean "substitute 1, 2, 3 and 4, in turn, for y and sum the series." Thus,

$$\sum_{y=1}^{4} y = 1 + 2 + 3 + 4 = 10.$$

Similarly,

$$\sum_{y=3}^{5} y = 3 + 4 + 5 = 12$$

and

$$\sum_{y=2}^{4} y^2 = 2^2 + 3^2 + 4^2$$
$$= 4 + 9 + 16 = 29.$$

Indexes   Suppose that we are given the *ordered* set

$$\{x : x = 7, 4, 10, 2, 1, 8\}.$$

Further, we shall define i to be the *index* that specifies the ordinal number for any value of x in the set. For example, in the foregoing set, $x_1 = 7$ and $x_2 = 4$, because 7 is the first and 4 is the second value in the ordered set. With i as the index we can indicate that certain numbers in the set are to be summed; thus

$$\sum_{i=1}^{6} x_i$$

is interpreted as follows:

$$\sum_{i=1}^{6} x_i = x_1 + x_2 + x_3 + x_4 + x_5 + x_6$$
$$= 7 + 4 + 10 + 2 + 1 + 8 = 32.$$

Similarly, for the same set,

$$\sum_{i=3}^{4} \frac{5}{x_i^2} = \frac{5}{10^2} + \frac{5}{2^2} = \frac{5}{100} + \frac{5}{4} = \frac{130}{100},$$

where the summation is performed on only the third and fourth elements of the set.

When an entire set is to be summed (as in the first example above), we typically substitute "N" for the upper number used with the summation sign. If we were to write $\sum_{i=1}^{N} x_i$ for the set just defined, the interpretation would be exactly the same as that for $\sum_{i=1}^{6} x_i$.

It is apparent that the use of index notation permits us to describe summation for any set of numbers, so long as we consider the members of the set to be ordered as stated. This notation is much more versatile than that which does not use indexes.

Parentheses   When required for clarity, parentheses are used to designate the portion of a statement to which an operation such as summation applies. For example, we can define the ordered set

$$\{t : t = 4, 11, 6\}.$$

Then

$$\sum_{i=1}^{N} (t_i - 2) = (4 - 2) + (11 - 2) + (6 - 2) = 15,$$

but

$$\left(\sum_{i=1}^{N} t_i \right) - 2 = (4 + 11 + 6) - 2 = 19.$$

Care must be exercised in the placement of parentheses. For instance,

$$\sum_{i=1}^{N} (t_i) - 2$$

is ambiguous. Does it mean that we are to proceed as in the first or as in the second case above? Similarly, omitting the parentheses would also cause ambiguity.

1.1

Formulas

There are two formulas for obtaining sums that can save a considerable amount of time. We first present them as statements, and then give an example of each. Finally, the proofs are given so that you will not have to take the statements on faith. In connection with both statements, we shall let w, x, y, and z be variables and a be a constant (an unchanging numerical value).

Statement

Let a constant (a) be added to each of the N elements in a set of numbers. The sum of the elements in the new set is the sum of the elements in the original set and N times the constant. In symbols,

$$\sum_{i=1}^{N} (y_i + a) = \left(\sum_{i=1}^{N} y_i \right) + Na.$$

For example, suppose we already know that the sum is 22 for $\{y : y = 2, 7, 4, 9\}$. What will be the sum if 5 is added to each of the four elements in this set? The statement tells us the sum of the new set will be $22 + 4(5)$, or 42. We can state the example in symbolic form as follows.

Given:

$$a = 5 \qquad \{y : y = 2, 7, 4, 9\}$$

Then

$$\sum_{i=1}^{N} (y_i + 5) = (2 + 7 + 4 + 9) + 4(5)$$
$$= 22 + 20 = 42.$$

The proof is carried out by expanding the left side of the general symbolic form of the statement, regrouping the resulting elements, and then expressing the result in compact notation.

$$\sum_{i=1}^{N} (y_i + a) = (y_1 + a) + (y_2 + a) + \cdots + (y_N + a)$$
$$= (y_1 + y_2 + \cdots + y_N) + (a + a + \cdots + a)$$
$$= \sum_{i=1}^{N} y_i + Na.$$

The next statement concerns multiplication by a constant rather than addition of a constant.

Statement

Let each of the N elements in a set of numbers be multiplied by a constant to form a new set. The sum of the elements in the new set is the product of the sum of the old set and the constant. In symbols,

$$\sum_{i=1}^{N} (ay_i) = a \cdot \sum_{i=1}^{N} y_i.$$

The example is based upon the same set used in connection with the previous statement. The sum of that set was 22. If each of the four elements is multiplied by -5, the sum of the elements of the new set should be -110.

Given:

$$a = -5 \qquad \{y : y = 2, 7, 4, 9\}$$

Then

$$\sum_{i=1}^{N} (-5y_i) = -5 \left(\sum_{i=1}^{N} y_i \right)$$

$$= -5(22) = -110.$$

The proof uses the same approach used for the earlier statement.

$$\sum_{i=1}^{N} (ay_i) = ay_1 + ay_2 + \cdots + ay_N$$

$$= a(y_1 + y_2 + \cdots + y_N)$$

$$= a \sum_{i=1}^{N} y_i.$$

We have taken care to explain the preceding two concepts verbally as well as symbolically for the purpose of clarity. Study the symbolic statements until you can read them easily.†

Summary

The raw materials of statistical analysis are data—that is, observations, or measurements, stated numerically. These are synonyms for the numbers that describe the varying amounts of some property of interest possessed by each element (member) in a set of units of observation. The amounts are expressed as values on an appropriate measuring scale, which may be of the nominal, ordinal, interval, or ratio type. Characteristics of order, distance, and origin in the real-number system form the basis for distinguishing among the four scales. The entire set of values which it is possible for a scale to contain is called a variable. Variables are either discrete or continuous.

A summation symbol, an index, and parentheses can be used with ordered sets of measurements to achieve compact notation. A number of basic formulas provide helpful shortcuts when sums of some special types of numerical sets are required. These relationships are most readily understood when expressed in the compact notation developed earlier.

See Self-Correcting Exercises 1.A

---

† Practice helps. The *Self-Correcting Exercises*, referred to just above, and the *Self-Instructional Supplement*, referenced in the margins of this book (see Preface), provide efficient practice for mastering the concepts of statistics.

Exercises

1. A highway safety agency is studying factors that possibly contribute to single-car accidents. An expert mechanic judges what the condition of each automobile was just prior to its accident. Using a numerical rating scale with nine values running from "8: Perfect mechanical order" to "0: Extremely dangerous to operate," he records one value from this scale for each vehicle.

   a. What constitutes the unit of observation for this study?

   b. What is the common property of each unit to be observed?

   c. Name the variable.

   d. Is the variable discrete or continuous?

2. A quality control inspector wishes to classify shipments of replacement automobile bumpers. If a shipment is found to have no scratched bumpers, he records a 2. If any bumper is scratched, he records a 1.

   a. Is the variable discrete or continuous?

   b. What constitutes the unit of observation?

   c. What type of measurement scale is being used?

3. A tractor manufacturer, about to introduce a new model, has no indication of the price elasticity in the potential market for this model. He therefore chooses a random sample of his dealers and asks, "Now that you have examined this new model, how many units (n) would you order if the wholesale price were K dollars?" By varying the value of K, the manufacturer feels that he can obtain a rough estimate of the price elasticity.

   a. Is the variable n discrete or continuous?

   b. What type of measurement scale is being used for the variable n?

4. A personnel worker in a certain plant is collecting data on employee attitudes. She presents the following statement to employees and asks them to choose the appropriate response:

   I feel that I am performing a valuable service for society when I do my job well.

1	2	3	4	5
strongly agree	*agree*	*no opinion*	*disagree*	*strongly disagree*

   a. Name the variable under study.

   b. What type of measurement scale is being used?

   c. What constitutes the unit of observation?

   d. What name is given to the set of numbers resulting from this procedure?

5. The following measurements were taken by five different researchers. Round each of the measurements to four significant digits.

   a. 170.5532   b. 0.0032571   c. 25099.36   d. 1.234501   e. 1.23450

6. Given the ordered set $\{x : x = 2, 8, 1, 4\}$, evaluate

$$\sum_{i=2}^{4} (ix_i)^2.$$

7. Represent $100 + 300 + 500 + 700 + 900$ in compact notation. Use an index, parentheses, and the summation operator.

8. Using the relationships for obtaining sums, write the following expression in as simple a form as is possible:

$$(aw_1 + b) + (aw_2 + b) + (aw_3 + b) + \cdots + (aw_N + b).$$

Properties of Data Sets

Populations and Samples

In a statistical study, the set of collected data may be the complete set of interest or it may be only a portion of the complete set. A credit manager may be interested in the balances owed by all credit customers. He may, however, have immediately available only the accounts of every tenth customer in the file.

| Definition | The complete set of data of interest in a given study is called the **statistical population**, or the **statistical universe**. Any subset of the statistical universe is called a **sample**. |

In the foregoing example, balances owed by all credit customers constitute the statistical population, or universe, and balances owed by every tenth credit customer in the file constitute a sample from that population.

The observational units in the credit-customer example are the customer account cards, or similar unit records of account. The elements of both the statistical population and the sample mentioned are the *numerical credit balances* on the cards, but not the cards themselves nor the people whose names appear on the cards. Statistical populations and samples always are *sets of measurements*.

Some additional examples of statistical populations and samples can be mentioned. Household incomes in the United States for a certain year make up an important statistical population for many studies. Household incomes in any one state would constitute a sample, though not a very representative one, from that statistical population. The number of children that each living adult female in New York City has borne up to a certain date could be a population under investigation. The number of children borne by every fourth such female selected from a complete list would be a sample.

We have seen that sets of data can be classified as populations or samples. Statistical data also may be classified as belonging to a *finite* or an *infinite* set. The lengths of a certain number of copies of a particular part produced by a lathe on a given day will serve as an example. If this set of lengths is viewed as the entire set of interest, the statistical population is finite. On the other hand, this set of measurements can be viewed as a subset of all measurements that might conceivably be made on all output produced by all possible lathes of this type. For all practical purposes, this latter statistical population has an infinite number of elements in it. In the first three chapters we will be concerned with finite populations of measurements. Later chapters will consider infinite populations.

The primary interest of investigators seldom centers on the individual numbers in a set of data. Once the accuracy and relevance of the individual numbers have been established, the main concern becomes the entire set considered as an entity in itself. The credit manager referred to earlier probably is not interested in each credit balance. Rather, he might want to know how the average amount currently owed compares with the average amount owed a year ago. He would, therefore, be interested in a *property* of the entire statistical population (the average) rather than in the individual numbers which make up that population.

To illustrate our discussion of properties of sets of data, we can suppose that we have two sets of X values:

$$\{X: X = 3, 3, 4, 4, 4, 6\}, \quad \text{and} \quad \{X: X = 5, 7, 9, 11, 11, 12, 12, 12, 14\}.$$

The first set might be the number of automobiles passing a certain point on a street between 3:00 and 3:15 P.M. on six successive work days. The second set might be the data for the hours 5:00 to 5:15 P.M. on nine successive work days. We can plot both sets on the same X axis. In Figure 1.1, numbers in the first set are represented by dots. Crosses are used for the second set.

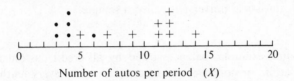

Number of autos per period (X)

Figure 1.1   Two Sets of Traffic Data

The first visual impression that emerges from the figure concerns the relative locations of the two sets. The general location of the first set is to the left of the second. Generally, there seems to be more traffic in late afternoon than there is earlier in the afternoon. This first impression gives rise to the concept of *central location*.

The second impression gained from Figure 1.1 is that the set of crosses spreads over a greater portion of the X scale than does the set of dots. In other words, there seems to be more variability in the amount of traffic during the late afternoon. This second impression gives rise to the concept of *dispersion*.

Central location and dispersion are two of the most useful basic properties of data sets. They are of prime importance throughout the entire field of statistics. Later on, in Chapter 2, we will discuss skewness, a third important property. The remainder of this chapter is concerned with measures of central location and dispersion.

Measuring Central Location

1.2

The Mean   This is the measure familiarly known as the "average." Because there are several measures of central location, or "average," however, each must have its own name.

The full name of the measure we are about to describe is the *arithmetic mean*, but we usually refer to it simply as the *mean*. This is the most important measure of central location. The symbol commonly used to represent the mean of a statistical population is the Greek letter μ (lowercase mu, pronounced "mew").

If we let $\{X\}$ be a statistical population of N measurements, then the **arithmetic mean** is the sum of the elements in the set divided by the number of elements in that set. In symbols,

Definition

$$\mu = \frac{\sum_{i=1}^{N} X_i}{N},$$

which we usually state more simply as

$$\blacksquare \quad \mu = \frac{\sum X}{N}. \quad \blacksquare \tag{1.1}$$

As an example of the computation of the mean, again consider the two data sets in Figure 1.1. If μ_1 and μ_2 are the means of these two sets, then

$$\mu_1 = \frac{3 + 3 + 4 + 4 + 4 + 6}{6} = 4,$$

and

$$\mu_2 = \frac{5 + 7 + 9 + 11 + 11 + 12 + 12 + 12 + 14}{9} = 10.33.$$

The two means are at the centers of "mass" of their respective data sets and satisfactorily describe the impression that the central location of the smaller set is well below that of the larger set.

1.2

Algebraic Operations with Means   Four algebraic operations with means are of major importance for future developments. Therefore, it is also important for you to understand *why* they are correct. For this reason, we give an example and a general proof.

Statement

Let a constant, a, be added to every value in a statistical population. The mean of the newly formed population, μ_2, will be the sum of the mean of the original population, μ_1, and the constant. In symbols,

$$\mu_2 = \mu_1 + a.$$

As an example, consider the set $\{2, 7, 9\}$, for which the arithmetic mean is 6. If we add 5 to each of the three numbers in this set, the new set will be $\{7, 12, 14\}$. The above statement tells us that the mean of this new set will be $6 + 5$, or 11. If we take the mean of 7, 12, and 14 we see that the result is also 11.

The proof is carried out as follows:

$$\mu_1 = \frac{\sum X}{N},$$

$$\mu_2 = \frac{\sum (X + a)}{N} = \frac{1}{N} [(X_1 + a) + \cdots + (X_N + a)]$$

$$= \frac{1}{N} (\sum X + Na) = \frac{\sum X}{N} + a,$$

or

$$\mu_2 = \mu_1 + a.$$

The next statement covers the result of multiplying every number in a set by a constant, rather than adding a constant to the numbers.

Statement | Let every measurement in a statistical population be multiplied by a constant, a. The mean of the resulting population, μ_2, will be the product of the constant and the original mean, μ_1. In symbols,

$$\mu_2 = a \cdot \mu_1.$$

Once again our example is the set $\{2, 7, 9\}$, for which the arithmetic mean is 6. If we multiply each of the three numbers in this set by 5, the new set will be $\{10, 35, 45\}$. The foregoing statement tells us that the mean of this new set will be 6 times 5, or 30. If we take the mean of 10, 35, and 45, we find that 30 is the correct value.

The proof proceeds as follows:

$$\mu_1 = \frac{\sum X}{N}$$

$$\mu_2 = \frac{\sum (aX)}{N} = \frac{1}{N} (aX_1 + \cdots + aX_N)$$

$$= \frac{a}{N} (\sum X) = a \frac{\sum X}{N},$$

or

$$\mu_2 = a\mu_1.$$

The third statement concerns the mean of a population composed of the elements of two other populations for which we know the means. In set language, we want to find the mean of a set composed of two subsets of numbers. We want to find this mean in terms of the means of the two subsets.

Statement

Let one statistical population, $\{X\}$, be composed of N_1 measurements and a second population, $\{Y\}$, be composed of N_2 measurements. Let the mean of $\{X\}$ be μ_1 and the mean of $\{Y\}$ be μ_2. Consider the mean, μ_3, of the set $\{Z\}$ composed of the N_3 measurements in $\{X\}$ and $\{Y\}$, where $N_3 = N_1 + N_2$. The mean, μ_3, is the sum of the *weighted* means of the original sets with weights (N_1/N_3) for μ_1 and (N_2/N_3) for μ_2. Stated symbolically,

$$\mu_3 = \frac{N_1}{N_3}\mu_1 + \frac{N_2}{N_3}\mu_2 .$$

The two initial sets for our example are $\{2, 7, 9\}$ and $\{4, 18\}$. The mean of the first set is 6 and the mean of the second set is 11. If we combine these two sets into a single set, the result is $\{2, 7, 9, 4, 18\}$, which has five elements. Since there are three elements in the first set and two in the second, the statement says that we can find the mean of the combined set by adding 3/5 of the first mean to 2/5 of the second. The result is $0.6(6) + 0.4(11)$, which is 8. Using the basic definition of the mean, we find that the mean of the combined set is, in fact, 8.

The proof follows.

$$\mu_1 = \frac{\sum X}{N_1} \quad \text{and} \quad \mu_2 = \frac{\sum Y}{N_2} .$$

When we multiply both sides of the first equation by N_1 and the second by N_2, we get

$$\sum X = N_1 \mu_1 \quad \text{and} \quad \sum Y = N_2 \mu_2 .$$

We defined N_3 in the statement such that

$$N_3 = N_1 + N_2 .$$

Then

$$\mu_3 = \frac{\sum X + \sum Y}{N_1 + N_2} = \frac{N_1 \mu_1 + N_2 \mu_2}{N_3} ,$$

or

$$\mu_3 = \frac{N_1}{N_3}\mu_1 + \frac{N_2}{N_3}\mu_2 .$$

It follows that, for K original sets with means μ_1, μ_2, ..., μ_K and with elements $N_1, N_2, \ldots, N_K$,

$$\mu_C = \frac{N_1}{N}\mu_1 + \frac{N_2}{N}\mu_2 + \cdots + \frac{N_K}{N}\mu_K ,$$

where μ_C is the mean of the combined sets and

$$N = N_1 + N_2 + \cdots + N_K .$$

The fourth statement is concerned with the mean of a population, the elements of which are the ordered sums of the elements in two other populations. We want to find the mean of a set composed of the ordered sums of the measurements in two ordered subsets in terms of the means of these subsets.

Statement	Let one statistical population be the ordered set $\{X\}$ with N elements and mean μ_1. Let a second statistical population be the ordered set $\{Y\}$ also with N elements and mean μ_2. Finally, let $\{Z\}$ be the set having as its ith element the sum of the ith elements in $\{X\}$ and $\{Y\}$. The mean of $\{Z\}$ is defined to be μ_3 and is equal to the sum of μ_1 and μ_2. In symbols,

$$\mu_3 = \mu_1 + \mu_2.$$

The two initial ordered sets for our example must contain the same number of elements. These sets are $\{1, 5, 3\}$ and $\{7, 6, 8\}$. The means are 3 and 7. The combined set is formed by adding 1 to 7, 5 to 6, and 3 to 8. The result is $\{8, 11, 11\}$. The statement says that the mean of this combined set should be $3 + 7$, or 10. Indeed, the mean of 8, 11, and 11 is 10.

The proof can be carried out by expanding and rearranging the sums. Since this is the same approach used in the preceding cases, the proof for this case is omitted. From the proof, it follows that for K original ordered sets of N elements each with means $\mu_1, \mu_2, \ldots, \mu_K$,

$$\mu_T = \mu_1 + \mu_2 + \cdots + \mu_K,$$

where μ_T is the mean of the ordered set formed by summing the ith elements of the K sets.

The Median   For a finite statistical population, the median, φ (the Greek letter phi), is loosely defined as the middle value when a data set is arranged in order of magnitude. This definition is all right when there is an odd number of values. It is sometimes ambiguous when there is an even number of values. In a set composed of the three values 7, 4, and 8, the middle value in order of magnitude is 7. Therefore, $\varphi = 7$. Similarly, in four families composed of 2, 3, 3, and 5 persons each, $\varphi = 3$ persons per family. But, for four families composed of 2, 3, 6, and 10 persons each, what is the median family size? Any value between 3 and 6 can be the middle value. By convention, the median in such a situation is arbitrarily defined to be the mean of the two values at the ends of the interval of ambiguity. For the last example, φ is the mean of 3 and 6, or 4.5 persons per family.

These examples illustrate that the median, as compared with the mean, is a less precisely defined measure of central location. On the other hand, for a set with a few outliers—that is, numbers widely separated from the majority of values—the median usually is considered a better descriptive measure than the mean. For example, in the set

$$\{X: X = 1, 2, 2, 3, 3, 3, 4, 4, 32\},$$

$\mu = 6$ and $\varphi = 3$. The latter measure is less influenced by the single extreme value.

Another important measure of central location is the mode, the most frequently recurring value in a set. In the foregoing set, the mode is 3. We shall defer further discussion of this measure until we consider grouped data in the next chapter.

Several other measures of central location that have rather specialized uses are described in references 1 and 2 in the list at the end of the book.

Measuring Dispersion

1.3

Recall Figure 1.1, in which the larger data set is more widely dispersed than the smaller. Having found ways to measure the property of central location, we now want to find a suitable measure for the *dispersion* of data within sets.

Initially, it may seem reasonable to attempt to measure dispersion by finding the amount by which every value in a set differs from some central reference value and averaging these deviations. We shall use the mean as the reference value. For the population {3, 4, 4, 9}, the mean is 5. The set of *differences* from the mean is

$$\{-2, -1, -1, 4\}.$$

In statistics, these differences are called *deviations from the mean*.

Definition	In general, for the statistical population $\{X\}$ with mean μ, the set of **deviations** from the mean is $\{x : x = X - \mu\}$.

For our example, given $\{X : X = 3, 4, 4, 9\}$, then $\{x : x = -2, -1, -1, 4\}$ is the set of deviations from the mean. Note that a deviation from the mean is represented by a lowercase "x," rather than capital "X."

We now want to average the deviations from the mean. But when we sum these deviations, we find that $\sum x = (-2) + (-1) + (-1) + 4 = 0$. Indeed, in general,

$$\sum x = 0;$$

the sum of the deviations of the measurements in a statistical population from the mean is always zero. This conclusion is reached by noting that

$$\sum x = \sum (X - \mu),$$

or

$$\sum x = (X_1 - \mu) + (X_2 - \mu) + \cdots + (X_N - \mu),$$

or

$$\sum x = \sum X - N\mu.$$

But, since $\mu = \sum X/N$, it follows that $N\mu = \sum X$. When we substitute $N\mu$ for $\sum X$ in our last expression for $\sum x$ above, we get

$$\sum x = N\mu - N\mu = 0.$$

| Statement | The algebraic sum of deviations from the arithmetic mean is always zero. |

Positive deviations are cancelled by negative deviations. Whether the values in a set are widely scattered or concentrated, the mean of the *deviations* will always be zero, because the sum of the deviations is zero. This is a consequence of the mean being the center of mass, or the balance point, of a data set. Our initial attempt to develop a measure of dispersion is not a success.

The Variance   We can keep negative deviations from cancelling positive deviations by squaring all deviations from the mean and then averaging these squares. The result is the most pervasive measure of dispersion in the field of statistics, the *variance*.

| Definition | The **variance**, σ^2 (lowercase sigma), of a statistical population is the arithmetic mean of the squared deviations from μ. In symbols, $$\blacksquare \quad \sigma^2 = \frac{\sum [(X - \mu)^2]}{N}, \quad \text{or} \quad \sigma^2 = \frac{\sum (x^2)}{N}, \quad \blacksquare \quad (1.2)$$ where $x = X - \mu$ and N is the number of observations in the population. |

As an example of the computation of the variance, consider the statistical population {4, 10, 16}. For this set, the mean (μ) is 10 and deviations of the observations in the set from the mean are {−6, 0, +6}. The squares of these deviations are {+36, 0, +36}, and the mean of these squared deviations is 72/3, or 24. Hence, the variance of the set {4, 10, 16} is 24.

In addition to its central role in many other branches of statistics, the variance is a reasonably satisfactory measure of dispersion for descriptive purposes. For instance, consider a set of data in which all of the values are the same. This set will have no dispersion. When we calculate the variance, none of the values in the set will be different from the mean. Consequently, all will have deviations of 0 and the variance will also be 0, as we would wish. Furthermore, when the values in one set of data are more widely scattered than those in another set measured on the same scale, the value of the variance for the first set will be larger than the value of the variance for the second. The value of the variance increases as does the amount of scatter or dispersion.

One of two techniques for calculating the variance will usually prove to be effective for small sets of data. The first is used when the mean and the deviations from the mean turn out to be whole numbers. In this case we use the equation

$$\sigma^2 = \frac{\sum (x^2)}{N}$$

to find the variance of the set. To illustrate this procedure we can use the smaller set of data illustrated in Figure 1.1. This set is listed in the left column in Table 1.1.

Table 1.1   Calculating Variance from Deviations

X	x	x^2
3	−1	1
3	−1	1
4	0	0
4	0	0
4	0	0
6	2	4

$$\Sigma X = 24 \qquad \Sigma x = 0 \qquad \Sigma x^2 = 6$$

$$\mu = \frac{\Sigma X}{N} = \frac{24}{6} = 4 \qquad \sigma^2 = \frac{\Sigma x^2}{N} = \frac{6}{6} = 1$$

The first step is to find the mean, μ. Having found the mean, we find and record in the middle column of Table 1.1 the deviations of each of the observations from the mean. As we expect from our earlier proof, the sum of these deviations is 0. We took this step only as a check; we don't need to do it to find the variance. The next step is to square the deviations and write the results in the right column. Appendix J lists the squares of numbers from 1 to 9.99. Finally, we sum these squared deviations and average them to find the variance, which is 1.

The second technique for calculating the variance is used when we can see that the deviations from the mean will not be whole numbers. To illustrate this procedure we will use the larger of the two data sets in Figure 1.1. This set appears in the left column of Table 1.2.

1.3

Table 1.2   Calculating Variance Directly

X	X^2
5	25
7	49
9	81
11	121
11	121
12	144
12	144
12	144
14	196

$$\Sigma X = 93 \qquad \Sigma (X^2) = 1025$$

$$\mu = \frac{93}{9} = 10.333 \cdots$$

Again we begin by finding the mean, which is found to be $10\frac{1}{3}$. This makes it apparent that deviations will not be whole numbers, so we abandon the earlier approach. Instead, we square the original observations, record the squares in the right column, and find the sum of the squares. By doing a little algebra we can show that the sum of squared deviations can be expressed in terms of the sum of the squares of the original observations and the sum of the original observations squared. In symbols,

$$\blacksquare \qquad \sum [(X - \mu)^2] = \sum (X^2) - \frac{(\sum X)^2}{N}. \qquad \blacksquare \qquad (1.3)$$

For our example,

$$\sum [(X - \mu)^2] = 1025 - \frac{93^2}{9}$$

$$= \frac{9225 - 8649}{9}$$

$$= \frac{576}{9},$$

or, simply,

$$\sum [(X - \mu)^2] = 64.$$

This is the numerator of Equation (1.2), which defines the variance. To complete the calculation, we need only divide by N:

$$\sigma^2 = \frac{\sum [(X - \mu)^2]}{N} = \frac{64}{9} = 7.11 \cdots.$$

The variance of the larger set of data in Figure 1.1 is 7.11, to two decimal places, while the variance of the smaller set is only 1. This shows that the variance of a widely scattered set is greater than the variance of a set that is more concentrated about the mean.

1.4   *The Standard Deviation*   There are several reasons why another measure of dispersion based on the variance is preferable to the variance. Among them is the fact that the dimension of the variance is awkward. For example, the dimension of the values in the two data sets in Figure 1.1 is *automobiles*. Since we *square* the deviations of these numbers from their mean, the dimension of the variance of either set is *automobiles squared*. For descriptive purposes, we would prefer a measure of dispersion stated in the original dimension of the data. Such a measure can be had simply by taking the square root of the variance. The square root of a number is another number which, when multiplied by itself, gives the original number. For example, the square root of 9 is 3, because 3 times 3 is 9.

The positive square root of the variance is the *standard deviation*, σ, where

$$\blacksquare \qquad \sigma = + \sqrt{\frac{\sum x^2}{N}}. \qquad \blacksquare \qquad (1.4)$$

The standard deviation is probably the most widely used measure of dispersion for descriptive purposes.

To obtain the standard deviation, we first calculate the variance and then find the square root of the variance. Square roots can be found in Appendix J. For the smaller set of data in Figure 1.1 we found that the variance is 1. Hence, the standard deviation will be $\sqrt{1}$, which is also 1. But now the dimension will be "auto" instead of "auto squared." For the larger set, we know that σ^2 is 7.11 autos squared. In Appendix J we look under N until we find 7.11. Under $\sqrt{N}$ is a number opposite 7.11 that begins with a 2. Under $\sqrt{10N}$ is another number which begins with an 8. Since 7.11 lies between 4 and 9 (that is, between 2^2 and 3^2), we know that $\sqrt{7.11}$ lies between 2 and 3. Consequently we choose the digits under $\sqrt{N}$ and fix the decimal point properly to obtain, to two decimal places,

$$\sigma = \sqrt{7.11} = 2.67 \text{ autos.}$$

Complete instructions for Appendix J immediately precede it.

Insight into the usefulness of the standard deviation for interpreting the degree of concentration in a set of data is gained from a theorem developed by the Russian mathematician Chebyshev. One statement of Chebyshev's theorem follows.

Statement	The proportion of measurements in a set of data that lie within K standard deviations of the mean is not less than $1 - (1/K^2)$, where K is at least 1.

The theorem is perfectly general. It applies to any set of statistical data. To illustrate, suppose we form an interval the lower end of which is two standard deviations less than the mean and the upper end of which is two standard deviations greater than the mean. Then K is 2 and K^2 is 4, and no fewer than $1 - 1/4$, or 3/4, of the observations must be within this interval. Similarly, at least $1 - 1/9$, or 8/9, of the observations must lie within three standard deviations of the mean. For the set of data in Table 1.2, we know that the standard deviation is 2.67 and the mean is 10.33. Since two times 2.67 is 5.34, we should find no fewer than three-fourths of the nine observations within the interval from 5 autos $(10.33 - 5.34)$ to 15 autos $(10.33 + 5.34)$ inclusive. Indeed, all nine of the observations are within this interval.

1.4

Summary

When considering a finite statistical population, our interest usually is focused on one or more properties of the entire population, such as its central location or its dispersion. Two measures of central location are the arithmetic mean and the median. The mean is the center of mass, or the balance point, of the set of data. When the data are arranged from smallest to largest, the median is either the middle value or the mean of the two values that bracket the middle. Two measures of dispersion are the variance and the standard deviation. The variance is the mean of the squared deviations from the arithmetic mean of the observations. The standard deviation, which is the square root of the variance, has the same dimension as the original observations. Fewer than $1/K^2$ of the observations in a set of data are farther than K standard deviations from the mean $(K \geq 1)$.

See Self-Correcting Exercises 1.B

Exercises

1. The daily numbers of sick employees in a small job shop are given by
$$\{X: 0, 2, 5, 7, 4, 1, 0, 4, 2, 0\}.$$
Find the value of each of the following measures of central location for this set.
   a. The mean,   b. The median,   c. The mode.

2. An investment portfolio consisting of 200 stocks is worth \$4000.00. If each stock increases \$1.00 in value, what is the mean stock price for the portfolio?

3. An across-the-board 25% reduction in fee must be put into effect by a consultant firm consisting of 10 consultants. Before the reduction, six of the consultants charged \$20/hour and the others charged \$30/hour. What will be the mean fee per consultant after the reduction?

4. Although the 20 management personnel in a purchasing department have a mean vacation period of 4 weeks per year, the mean vacation length per employee for all 120 employees in the department is 1.5 weeks per year. What is the mean vacation period for the non-management personnel in the department?

5. An automobile dealer sold the following numbers of cars each day in a six-day period: 14, 8, 8, 14, 14, 8. What is the variance in the cars sold per day?

6. The variance in the weight of steel beams in a certain shipment is 784 pounds. If the mean weight of a beam is 1500 pounds, at least what percentage of the beams weigh within the range 1416–1584 pounds?

7. At least 75% of a bakery's daily sales are in the range \$1000–\$1500. The mean daily sales total for the bakery is \$1250. What is the maximum value of the variance in sales for the firm?

Glossary of Equations

$$\sum_{i=1}^{N} x_i = x_1 + \cdots + x_N$$

Capital sigma is defined to be the summation operator and i, the index.

$$\sum_{i=1}^{N} (y_i + a) = \sum_{i=1}^{N} y_i + Na$$

If a constant is added to every value in a data set, the sum will be the sum of the original set plus the product of the constant and the number of observations.

$$\sum_{i=1}^{N} (ay_i) = a \sum_{i=1}^{N} y_i$$

If every value in a data set is multiplied by a constant, the sum will be the sum of the original set multiplied by the constant.

$$\mu = \frac{\sum X}{N}$$

The arithmetic mean of a statistical population is the sum of the observations in the population divided by the number of observations in the set.

$$\mu_2 = \mu_1 + a$$

If a constant is added to every number in a population, the mean will be the mean of the original set plus the value of the constant.

$$\mu_2 = a \cdot \mu_1$$

If every number in a population is multiplied by a constant, the mean will be the mean of the original set multiplied by the constant.

$$\mu_C = \frac{N_1}{N} \mu_1 + \cdots + \frac{N_K}{N} \mu_K$$

The mean of a population composed of K populations is the sum of the weighted means of these subsets. The weight for a given subset is its proportion of the total number of observations in all K sets.

$$\mu_T = \mu_1 + \cdots + \mu_K$$

The mean of a population composed of the ordered sums of the elements in K ordered subsets of data is the sum of the means of these subsets.

$$\sigma^2 = \frac{\sum [(X - \mu)^2]}{N} \quad \text{or} \quad \sigma^2 = \frac{\sum (x^2)}{N}$$

The variance of a population is the arithmetic mean of the squared deviations of the data from their arithmetic mean.

$$\sigma = +\sqrt{\frac{\sum x^2}{N}}$$

The standard deviation of a statistical population is the positive square root of its variance.

2

Description of Large Sets

In the last chapter, concerned with small sets of data, graphical description and measured properties of these sets were based upon the *individual* observations. When the number of observations is small, these techniques for *ungrouped* data are satisfactory. However, when we are concerned with a set composed of a large number of observations, the data frequently must be *grouped* into classes.

Consider the problem of describing the ages of heads of households in any large city. If one lists the ages, the mass of detail will obscure any general pattern. Impressions about the average age, the variation of ages, and the general shape of the pattern are nearly unobtainable. Grouping, however, permits visual displays that bring out these factors. Particularly if one does not have access to a computer, grouping also makes it easier to compute measures of the desired properties.

Frequency Distributions

2.1   The frequency distribution is the device generally used to describe a large set of data. It uses a graph or table or both to show the frequency of occurrence of observations by groups, or classes. In a table, the values of observations are shown in one column and the frequency of occurrence for these values in a second column. Preparation of a frequency distribution can be described in four steps:

Rules

1. Subdivide the range of values into convenient classes.
2. Sort the observations into these classes.
3. Count and record the number of observations in each class.
4. Summarize the results in a table, a graph, or both.

Discrete Variables

Single-Value Classes   When values can occur only at isolated points on the scale measuring a variable, the variable is said to be *discrete*. For instance, a quality inspector has counted the number of flaws on each of 50 sheets of galvanized steel. In the results listed below, note that only integral, nonnegative values of this variable can occur.

1	2	1	0	3	0	2	4	2	1
3	4	2	0	2	1	3	1	3	2
1	0	3	2	2	0	2	0	0	1
5	1	2	1	3	1	0	5	3	1
0	1	2	4	2	1	4	1	0	3

Even for this modest number of observations it is difficult to get any idea of pattern from the list itself.

We can begin constructing a frequency distribution by noting that the variable ranges from 0 through 5, with several instances of each value. These six values will make convenient classes. Formally, we can define the variable X as the number of flaws per sheet and let it assume the integral values 0, 1, 2, 3, 4, 5.

We now must sort the 50 observations into the six classes and count the number of observations in each class. When sorting is done by hand, tallying is perhaps the easiest approach. The following table illustrates tallying and counting.

X	Tally	Count
0	ﬀ ﬀ	10
1	ﬀ ﬀ ||||	14
2	ﬀ ﬀ ||	12
3	ﬀ |||	8
4	||||	4
5	||	2

The classes are described in the left column. As we come to each observation in the list, a tally mark is made to the right of the appropriate X value, as shown. The final column is the result of counting the tally marks in each class.

A frequency distribution table is one form of summary for presentation of grouped data. When the table is intended for general audiences, presumably not familiar with the subject matter, table headings must be fully descriptive. One possible display for our example is shown in Table 2.1.

For statisticians in the steel plant the table may be simplified as shown in Table 2.2. Among statisticians, f is customarily used to denote the frequency of occurrence for each value of the variable, and N (where $N = \sum f$) is used to represent the total number of observations in the set.

In our current example, every possible value of the variable within the range of observations constitutes a class. For frequency distributions of this type, a *frequency diagram* is suitable for graphic display. As shown in Figure 2.1, the frequencies (f)

Table 2.1   Frequency Distribution of Flaws in Galvanized Steel Sheets

Number of flaws per sheet	Number of sheets
0	10
1	14
2	12
3	8
4	4
5	2
	Total 50

Table 2.2   Frequency Distribution of Flaws per Sheet

X	f
0	10
1	14
2	12
3	8
4	4
5	2
	$N = 50$

are plotted on the vertical scale (the ordinate). The number of flaws per sheet (X) are plotted on the abscissa. When the context is clear, the axes can be labeled f and X. Ordinates equal to the proper frequencies are constructed for every relevant value of the variable.

Notice how grouping the data brings out the pattern in the distribution. In Figure 2.1, the average appears to be between one and two flaws per sheet. The major portion of the observations is concentrated in the range 0 to 3 flaws. The distribution is not symmetrical with respect to the average; it tails off toward higher values of the variable. The facility it offers for displaying pattern illustrates one important benefit to be gained from grouping large sets of data.

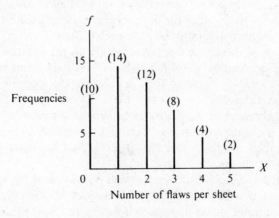

Figure 2.1   Frequency Diagram

Another tabular device for summarizing grouped data is the *relative frequency* table (see Table 2.3). Instead of the frequencies for each class being listed, the *proportion* of the entire set is listed. The symbol f' is often used to denote proportion. *Percentages* for each class can be obtained by multiplying the proportions by 100. The relative frequency approach is particularly useful when two distributions with widely different numbers of observations are to be compared.

Table 2.3   Relative Frequency Distribution of Flaws
Per Sheet

X	f'
0	0.20
1	0.28
2	0.24
3	0.16
4	0.08
5	0.04
	1.00

2.1

Many-Value Classes   When a set of observations falls along a wide range of the variable and when the density of observations is low, a class should not be defined for every possible value of the variable in that range. Rather, a *class interval*, which includes several possible values of the variable in every class, should be defined. As an example, look at Table 2.4. The variable is the number of automobiles passing a

2.2

Table 2.4   Distribution with Single-Value Classes

No. of autos (X)	No. of periods (f)	No. of autos (X)	No. of periods (f)
0	2	12	0
1	0	13	1
2	1	14	2
3	1	15	0
4	0	16	3
5	3	17	0
6	2	18	0
7	1	19	0
8	0	20	1
9	4	21	0
10	1	22	0
11	3	23	1

certain point during three-minute time periods. The frequency is the number of periods during which 0, 1, ..., or 23 automobiles were counted. For example, 5 autos passed the observer in each of three different periods.

Imagine the result if we had used the same graphic device for these observations that we used in Figure 2.1. The pattern would be analogous to irregularly spaced saw teeth, with rather wide gaps between some teeth.

By including a number of values of the variable in every class, we usually can bring out the pattern in such data. But we must be careful. If we include too few values within each class interval, the sawtooth effect will not be corrected sufficiently. Alternatively, the more values per class, the closer we approach putting all values in one class. Too few classes would disguise any underlying pattern just as effectively as would too many. There is no generally accepted formula for choosing the proper length of a class interval. Sometimes several schemes must be tried before an effective compromise between too much detail and too much summarization can be found.

For the traffic data of Table 2.4, class intervals of length 5 were selected for Table 2.5. A major difference between the two tables is noted in the following statement.

Statement | When several values of the variable are grouped into a class, the degree of precision present in the raw data is reduced.

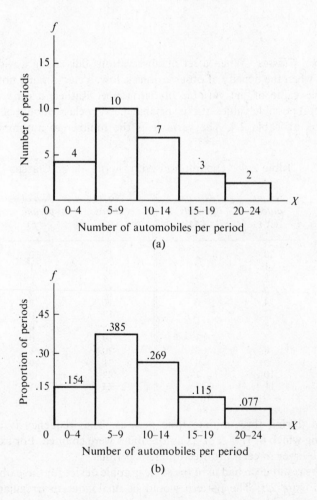

Figure 2.2   (a) Histogram, (b) Relative Frequency Histogram

Table 2.5   Frequency Distribution Using Class Intervals

Number of autos (X)	Number of periods (f)
0– 4	4
5– 9	10
10–14	7
15–19	3
20–24	2
	26

The exact locations of observations within any one class are ignored. Typically, the midvalue of a class is substituted to represent every value in the class. *The midvalue is the value at the center of a class.* For the traffic example, the midvalues are X values of 2, 7, 12, 17, and 22 automobiles. This use of the midvalue lends support to employing as many classes in a frequency distribution as the data will conveniently support. The more classes used, the more accurately can their midvalues be expected to represent the original values.

A graphic device often used to represent frequency distributions with data sorted into intervals is the *histogram*. Figure 2.2 shows two histograms based on Table 2.5. For classes of equal length, adjacent vertical bars of equal width are erected above the horizontal axis. Thus, the range of values is shown on the X axis. Frequencies or relative frequencies are shown on the vertical axis.

2.2

Continuous Variables

As we stated earlier, when a variable theoretically can assume an infinite number of values between any two given values on a measuring scale, it is a *continuous* variable. For example, consider a bag of sugar which, according to the label, weighs 5 pounds. If the bag were weighed on instruments of increasing sensitivity, it might show weights of 4.9, 4.93, 4.927, 4.9274 pounds, and so forth, to the limit of accuracy for the particular instrument.

In practice, observations of continuous variables are recorded to the limit of precision possible with a given measuring instrument or they are rounded to some lesser degree of precision. Suppose a bag of sugar is weighed on an instrument capable of weighing to the nearest ten thousandth of a pound and suppose we read 4.9274 pounds as the weight. For the purpose at hand, weights to the nearest hundredth of a pound may be sufficient. We would round 4.9274 pounds to 4.93 pounds and record this latter number as our observation. To the nearest hundredth of a pound, Table 2.6 lists the net content weights of 120 bags labeled as containing 5 pounds.

Class Length and Class Limits   Since the volume of data in Table 2.6 obscures any pattern that may be present, we want to prepare a frequency distribution for these data. The data run from a low of 4.84 pounds to a high of 5.11. This means that they have a range of 0.27 pounds (5.11 − 4.84). If we allow each class a length of 0.03 pounds, we will end up with about 9 classes. Since there are 120 observations, there will be an average of slightly more than 13 observations per class. This grouping should prevent a sawtooth effect and still give us enough classes to see the pattern.

Table 2.6   Content Weights of 120 Bags of Sugar (pounds)

4.98	5.08	5.00	5.04	4.92
4.99	4.96	5.00	5.02	5.05
5.01	4.98	5.05	4.94	5.11
4.95	5.08	4.95	5.01	5.07
5.02	5.01	5.00	5.00	5.00
5.04	5.03	5.01	5.01	5.08
4.98	5.07	5.03	4.97	4.98
5.01	4.99	5.02	5.00	5.06
4.96	4.98	5.01	4.95	5.00
5.01	5.01	4.91	5.01	4.97
4.99	4.97	4.98	5.00	5.08
4.92	4.89	4.95	4.93	4.90
5.05	4.99	5.06	4.98	4.84
4.96	5.03	5.04	4.92	5.02
5.00	4.90	5.02	5.02	4.95
5.01	4.95	4.97	4.94	4.98
4.99	5.03	5.06	4.96	4.98
5.06	5.06	4.97	5.00	5.05
5.08	5.00	5.07	4.98	5.02
5.03	5.01	5.00	4.98	5.05
5.00	4.96	4.93	5.05	4.93
5.03	5.00	5.03	5.00	5.04
5.07	4.99	4.96	4.98	4.95
5.03	5.06	5.06	4.92	4.96

We must now consider how to describe the upper and lower limits of each class in such a way that every observation can be correctly assigned to one and only one class. Suppose we begin with 4.84 pounds, the smallest value in Table 2.6, and use a class length of 0.03 pounds. We must realize that the *exact* weights of bags that were recorded as 4.84 pounds actually ranged from 4.835000 ··· to 4.844999 ··· pounds (4.835 to 4.845 rounded to three decimal places). The *exact* lower limit of our first class is 4.835, and by adding the class length of 0.03 pounds to this value, we obtain the exact upper limit, 4.865. Observations recorded as 4.84, 4.85, and 4.86 will be tallied into this class.

One way to find the midvalue of this class is by taking the mean of the exact class limits. The mean of 4.835 and 4.865 is 4.85, the class midvalue.

Now, let us consider the next class. The next possible recorded observation larger than 4.86 is 4.87. The exact lower limit of a weight recorded as 4.87 is 4.865. Hence, we can use 4.865 as both the exact *upper* limit of the first class and the exact *lower* limit of the next class. A class length of 0.03 will result in exact limits of 4.865 and 4.895 for this second class. The next class has limits of 4.895 and 4.925, and the remaining exact limits are found by continuing the process. Observations recorded as 4.87, 4.88, and 4.89 will go into the second class. Values of 4.90, 4.91, and 4.92 will go into the third class, and so on through the remaining classes. No possible observation is omitted and every possible observation has just one class to which it can be assigned.

The procedure just described is summarized in Table 2.7, where the left column presents the exact class limits, and the center column shows the class midvalues. The result of tallying the 120 observations from Table 2.6 into these classes and counting the frequencies appears in the right-hand column of the table.

The pattern of frequencies in Table 2.7 is well behaved, in that it exhibits no severe sawtooth effect and there are enough classes to bring out the shape of the distribution. Consequently, our decision to use a class length of 0.03 pounds was well founded.

Table 2.7   Class Limits, Midvalues, Frequencies

Exact class limits	Midvalues	Number of bags
4.835–4.865	4.85	1
4.865–4.895	4.88	1
4.895–4.925	4.91	7
4.925–4.955	4.94	12
4.955–4.985	4.97	24
4.985–5.015	5.00	33
5.015–5.045	5.03	19
5.045–5.075	5.06	17
5.075–5.105	5.09	5
5.105–5.135	5.12	1
	Total	120

Note that the class midvalues are stated *to the same level of precision* as the original observations (hundredths of a pound). This result is a consequence of our making the class length an *odd* number (0.03). Midvalues stated to the same level of precision as the observations are convenient for calculations based on grouped continuous data. Our choice of class length (0.03 pounds) and exact lower limit of the initial class (4.835 pounds) guaranteed such midvalues.

Three criteria governed our choice of the class length. It has the same level of precision as the recorded observations (hundredths of a pound). It has an odd number as its final digit (3). When divided into the range of the data (0.27), it produced an estimated 9 classes. We see that this will result in 120/9, or about 13, observations per class. An average of over 5 per class should prevent a sawtooth effect in most instances.

Our choice for the exact lower limit of the initial class was the exact lower limit of the smallest observation (4.835). Actually, any exact lower limit which is less than 4.835 by an integral multiple of a hundredth of a pound (4.835 − 0.01, 4.835 − 0.02, etc.), also will produce midvalues to the desired level of precision. Sometimes, lowering the starting point slightly in this manner is desirable. For example, it happens that our arrangement of class limits resulted in a midvalue of 5.00 pounds for one of the classes. Since this is the advertised weight, it would be desirable to find out if the data appear to be clustered around this value. Had the smallest observation been 4.85 pounds rather than 4.84, we still could have realized the 5.00 pound midvalue by beginning with 4.835 pounds as our initial exact lower class limit.

Given a set of data that constitute observations of a continuous variable, we can summarize the selection of classes for a frequency distribution.

Rules

1. Select a class length that
   (a) is stated to the same level of precision as the data,
   (b) has an odd number as its final digit,
   (c) divides the range of observations into as many classes as can be had without a sawtooth effect. Usually from five to 15 classes with an average of at least five observations per class will do.

2. Select the exact lower limit of the smallest observation as the exact lower limit of the initial class. Add the class length to this lower limit to find the exact upper limit of the initial class.

3. Set the exact lower limit of the next class equal to the exact upper limit of the initial class. Add the class length to find the exact upper limit of the next class. Continue the process until there is a class for every observation.

These rules assure us of having class midvalues that are stated to the same level of precision as are the original data. This same set of rules is appropriate for forming grouped-value classes for observations of a discrete variable if we use the artifice of pretending that the observations come from a continuous variable. Table 2.5 was formed from Table 2.4 in this manner. The exact lower limit of the first class was taken to be -0.5. An observation of 0 autos, treated as a continuous variable, runs from -0.5 to $+0.5$. Then a class length of 5 was applied to exact lower limits successively, as required by the rules just stated.

The rules for selecting classes for a frequency distribution will suffice for most large sets of data. There are alternative approaches to setting class limits for continuous data. References 1 and 2 describe several of these alternatives.

Frequency Distribution   The frequency distribution for the net weights of the 120 five-pound bags of sugar, listed individually in Table 2.6, appears in Table 2.8. As is

Table 2.8   Frequency Distribution

Weights in pounds (X)	Number of bags (f)
4.84–4.86	1
4.87–4.89	1
4.90–4.92	7
4.93–4.95	12
4.96–4.98	24
4.99–5.01	33
5.02–5.04	19
5.05–5.07	17
5.08–5.10	5
5.11–5.13	1
Total	120

customary, the *stated* class limits have the same level of precision as the observations. Exact class limits will be called *class boundaries* to distinguish them from stated class limits. The class length can be found by taking the difference between the lower (or upper) boundaries of any two adjacent classes. A class midvalue can be found by taking the mean of the lower and upper boundaries. Alternatively, the stated limits can be averaged to find the midvalue.

Graphic Representation   The histogram based on Table 2.8 appears in (a) of Figure 2.3. Class midvalues, rather than boundaries, are used on the horizontal axis.

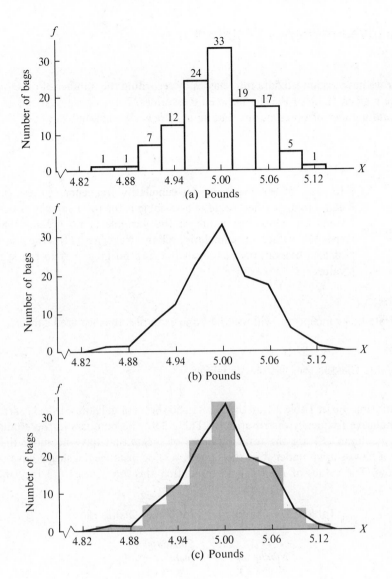

Figure 2.3 (a) Histogram, (b) Frequency Polygon, (c) Polygon Superimposed on Histogram

This choice prevents clutter and possible confusion. Breaking the scale at the left of the horizontal axis simply makes it possible to avoid a large, empty space.

Another graphic device for presenting a frequency distribution is the *frequency polygon*. The frequency polygon based on Table 2.8 appears in (b) of Figure 2.3, and its relationship to the histogram for the same table appears in (c) of that figure. To construct a frequency polygon, we begin by plotting a point vertically above the midvalue of each class at a height equal to the class frequency. These points are then connected with straight lines. Finally, at either end of the distribution, straight lines connect the midvalue of the last occupied class with the midvalue of the adjacent empty class, to close the figure.

Cumulative Frequency Distributions

So far we have organized data into classes by recording the number of observations that have a given value or that lie between two values. Now we want to classify data by the *total* number of observations that lie *at or below* designated values of the variable.

Definition	In a set of observations, the **cumulative frequency** (F) less than or equal to a specified value of a variable is the total *number* of observations for which the value of the variable is no greater than the specified value. The **cumulative relative frequency** (F') is the *proportion* of all observations in the set that are no greater than the specified value.

To illustrate this concept, we will use the same examples that we used earlier.

Single-Value Classes for Discrete Variables

The information in Table 2.1 on the flaws in 50 sheets of galvanized steel is arranged as a cumulative frequency distribution in Table 2.9. The numbers in the cumulative frequency column (F) are the total numbers of sheets that have no more than the number of flaws shown under X in the same row. For instance, 36 sheets have two or fewer flaws. The values of F are found by adding the frequencies (f) in Table 2.1

Table 2.9   Cumulative Frequency Distribution

Number of flaws (X)	Cumulative number of sheets (F)
0	10
1	24
2	36
3	44
4	48
5	50

for all classes with X values equal to or less than the given value of X in Table 2.9. Hence, 36 is the sum of the 10 sheets with no flaws, the 14 with one flaw apiece, and the 12 with two flaws. In other words, a cumulative frequency distribution table is constructed from a frequency distribution table. By the same token, given a cumulative frequency distribution table, we can construct a frequency distribution table by finding the differences in successive values of the cumulative frequency column

(F). The difference between the 36 sheets that have two flaws or fewer per sheet and the 24 sheets that have no more than one flaw is 12, the number of sheets that have exactly 2 flaws per sheet.

The cumulative frequency distribution of Table 2.9 is changed to a cumulative *relative* frequency distribution in Table 2.10. This is accomplished by dividing each of

Table 2.10   Cumulative Relative Frequency Distribution

Number of flaws (X)	Cumulative proportion of sheets (F')
0	0.20
1	0.48
2	0.72
3	0.88
4	0.96
5	1.00

the entries in the F column of Table 2.9 by 50, the total number of observations. The values of F' tell us the proportion of sheets in the set that have the stated number of flaws (X) or fewer per sheet. For instance, 0.72 is the proportion that have two flaws or fewer per sheet. By subtracting 0.48 from 0.72, we find that 0.24 is the proportion with exactly two flaws per sheet. This is the value of the relative frequency (f') associated with two flaws (X) in Table 2.3.

The *step graph* in Figure 2.4 is the proper graphical device for cumulative frequency

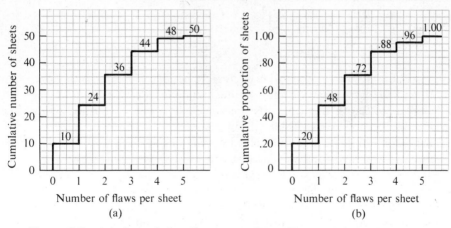

Figure 2.4   (a) Cumulative Frequency Step Graph; (b) Cumulative Relative Frequency Step Graph

distributions and cumulative relative frequency distributions for observations of discrete variables when values of the variable (X) have not been grouped into class intervals. The step graph for the cumulative *number* of sheets appears in (a), and the cumulative *relative* frequency graph appears in (b). From (a) we can see, for instance, that the first 24 sheets (F) have no more than one flaw per sheet (X). Or we can ask how many sheets have three flaws or fewer (X) and read 44 (F) at the top of the vertical step above 3. The heights of the steps are equal to the individual class frequencies (f) in Table 2.1. From (b) we can see that 0.88 (F') have three (X) or fewer flaws each—that is, 88 percent have no more than three flaws apiece. Or we may want to know what proportion have two flaws or fewer (X) and find that 0.72 (F') is the answer. The heights of steps are equal to individual class relative frequencies (f') in Table 2.3.

Many-Value Classes

When several values of a variable (either discrete or continuous) are grouped into each class, the result is a frequency distribution such as that in Table 2.5. A cumulative frequency distribution based on this information appears in Table 2.11. We see that

Table 2.11   Cumulative Frequency Distribution

Number of autos per observation period (X)	Cumulative number of periods (F)
0– 4	4
5– 9	14
10–14	21
15–19	24
20–24	26

14 or fewer autos (X) were observed in each of 21 periods (F). By subtracting 14 (F) from 21, we find that between 10 and 14 autos per period were observed in 7 periods. The cumulative relative frequency distribution table would be formed by dividing each of the entries under F in Table 2.11 by 26.

A step graph cannot be used to display information for the type of data illustrated in Table 2.11. For example, if we placed the entire vertical rise of 4 for the first class above 0 on the X scale, this would erroneously indicate that no autos were observed in 4 periods. Alternatively, placing the vertical rise of 4 above 4 on the X scale would also create the wrong impression. Rather, we want to indicate that the data are scattered throughout the class, as we did with the histograms in Figure 2.2 for these same data.

Another complication must be resolved before we can graph cumulative frequency distributions for many-value classes. The graph should indicate that the cumulative frequencies shown in a table such as Table 2.11 are not all accounted for until we reach the upper boundary of the related class of X values. But the second class in that table, for instance, ends on 9 autos and the third class begins on 10, an ambiguous situation. We resolve this dilemma by forming the following rule.

Rule | Many-value classes in frequency distributions for discrete variables will be given *exact* class limits by using the same rules that were used for continuous variables.

Hence, the exact limits for the classes in Table 2.11 are −0.5 to 4.5, 4.5 to 9.5, 9.5 to 14.5, and so forth. Even though we understand that negative values of the variable cannot occur, we make the lower boundary of the first class (−0.05) conform with the rule in order that class lengths remain equal on the graph.

The graphical display for many-value classes is called a *cumulative frequency ogive*. The ogive for Table 2.11 is shown in Figure 2.5. Cumulative frequencies (F) of 0, 4, 14,

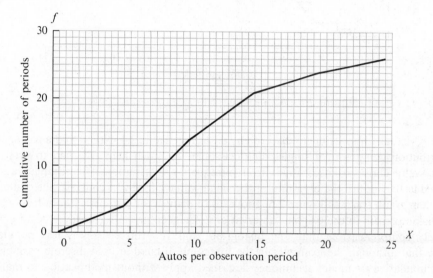

Figure 2.5   Cumulative Frequency Ogive

21, 24, and 26 are plotted vertically above X values of −0.5, 4.5, 9.5, 14.5, 19.5, and 24.5. The six points are then connected with straight line segments. As in the case of the histogram, these straight lines imply that the data in a given class are scattered uniformly throughout the class. This implication is at least approximately true.

Preparation of a cumulative *relative* frequency ogive for many-value classes offers no new difficulty. We can simply convert the F (cumulative frequency) scale to an F' (cumulative relative frequency) scale as we did in Figure 2.4.

A useful characteristic of cumulative frequency graphs can be illustrated with the traffic data we have been discussing. We do not have to form many-value classes to see the pattern in data plotted as a cumulative step graph. For instance, all we must do prior to plotting a cumulative step graph is to order the data from the smallest value of the variable to the largest. The result is called an *array*. Then, if desired, the data can further be arranged in a frequency distribution composed of single-value classes, as are the traffic data in Table 2.4. The cumulative step graph shown by the solid lines in Figure 2.6 is plotted directly from Table 2.4. The cumulative frequency ogive from Figure 2.5 appears as a thin dashed line superimposed on the step graph. Most viewers see essentially the same pattern in both types of graph.

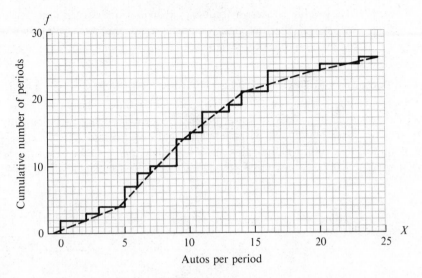

Figure 2.6   Cumulative Step Graph and Ogive

It is virtually impossible to discern any pattern in Table 2.4. Nor would a frequency distribution composed of single-value classes be of much help. We had to prepare many-value classes to discern pattern in a frequency distribution. This summarization forced us to sacrifice the precision of the single-value classes. In contrast, the cumulative step graph (such as Figure 2.6) allows us to discern the pattern without losing the precision present in the original observations. The step graph is, however, less immediately comprehensible than the histogram, and for this reason is less popular. Note that although the examples in the foregoing discussion used discrete variables, the cumulative frequency techniques described apply without modification to many-value classes of continuous variables.

Summary

The frequency distribution is a device used to bring out the pattern in a large set of data. The range of values in the set is divided into enough intervals of equal length to produce an average of five or more observations per class. The number of observations in each class is counted and recorded.

If the density of observations is high, the frequency distribution for a discrete variable is based on classes, each of which consists of a single value of the variable. If the density of observations along a discrete scale is low, each class contains several values of the variable. These are called many-value classes. Frequency diagrams are appropriate graphical representations of frequency distributions based on single-value classes of discrete variables. Histograms are appropriate for many-value classes.

Observations of a continuous variable are always rounded. To form a frequency distribution from such data, a class length is selected which is stated to the same level of precision as the data and has an odd number as its final digit. Using class lengths formed in this manner in conjunction with exact class limits makes convenient class midvalues. This same rule also is appropriate for finding many-value class lengths for a discrete variable. Histograms and frequency polygons can be used for graphic representation of frequency distributions based on many-value classes of continuous variables.

Cumulative frequency distributions are another device for bringing out patterns in large sets of data. Frequencies listed opposite classes in a cumulative frequency distribution show the total number of observations for which the value of the variable is less than or equal to the upper boundary of the class. The step graph and the ogive are used for graphic representations of cumulative distributions. A special advantage of the step graph is that it will display pattern without destroying the precision of the original data on which it is based.

See Self-Correcting Exercises 2.A

Exercises

1. The following ordered set of data shows the number of days per month in a twenty-month period that production of light bulbs was below quota:

$$\{3, 5, 0, 1, 4, 2, 2, 1, 4, 3, 0, 2, 1, 2, 1, 3, 0, 1, 2, 1\}.$$

   a. Prepare a frequency distribution table with single-value classes.

   b. Display the distribution in part (a) by means of the appropriate graphical device.

   c. Using two values per class, prepare a frequency distribution table.

   d. Display the distribution from part (c) by means of an appropriate graphical device.

2. The lengths (Y) of 47 pieces of scrap lumber were measured to the nearest foot and the results are given in the following frequency distribution.

Y:	1	2	3	4	5	6	7	8	9	10	11	12	13	14	15	16	17	18	19	20
f:	0	1	0	6	1	5	2	5	8	6	2	3	4	2	0	2	1	0	2	0

Prepare a frequency distribution table for these data that satisfies the criteria for continuous variables.

3. Data set X is a set of observations of a continuous variable precise to integers.

Data set X

X	f	X	f
0	3	10	6
1	0	11	2
2	1	12	3
3	0	13	4
4	6	14	2
5	0	15	0
6	5	16	2
7	0	17	1
8	5	18	0
9	8	19	2

The data were grouped as follows:

X	f
1– 8	17
8–15	30
16–23	3

List at least three errors that were made in grouping this set.

4. For the situation described in Exercise 1:

   a. Prepare a cumulative frequency distribution table.

   b. Prepare a cumulative relative frequency distribution table.

   c. Display the table from part (b) of this exercise in an appropriate graph.

5. Change the frequency distribution you prepared in Exercise 2 to a cumulative frequency distribution and graph it.

Measures of Properties

The concept of useful properties of data sets was introduced in the latter portion of Chapter 1, where central location and dispersion were defined. The mean and the median were developed as measures of location. The variance and standard deviation were developed as measures of dispersion. We are now ready to discuss calculation techniques applicable to grouped data for all of these meaures. Additional measures of location also will be presented. Finally, a new property, *skewness*, will be introduced and a measure of this property will be described.

Location

2.3

The Mean   Consider again the frequency distribution for the number of flaws in each of 50 sheets of galvanized steel given in Table 2.2. Using this table as a description of the statistical population of interest, we can find the mean number of flaws per sheet by making use of the following statement.

Statement

The basic formula for finding the mean (μ) of any finite statistical population from its frequency distribution is

$$\blacksquare \quad \mu = \frac{\sum (fX)}{N}, \quad \blacksquare \qquad (2.1)$$

where X is the value representative of every observation in a given class, f is the number of observations in that class, and N is the total number of observations in the population. The representative value for each class (X) is multiplied by the number of times it occurs (f). These products, one for each class, are summed and the total is divided by the number of observations in the population (N).

The frequency distribution from Table 2.2 is repeated in the two columns to the left in Table 2.12. Recall that X is the number of flaws per sheet and f is the number of sheets with the stated number of flaws per sheet. The third column is the product of each value of X and the number of times that value occurs (f). The final calculation gives a mean value of 1.76 flaws per sheet. The process represented by $\sum (fX)$, which gave us 88, is just a rapid way to find the sum of the ten 0's, fourteen 1's, . . ., and two 5's listed as the 50 original observations on the second page of this chapter.

Table 2.12   The Mean from Single-Value Classes

X	f	fX
0	10	0
1	14	14
2	12	24
3	8	24
4	4	16
5	2	10
	$N = 50$	$88 = \sum (fX)$

$$\mu = \frac{\sum (fX)}{N} = \frac{88}{50}, \text{ or } 1.76.$$

Table 2.4 provides another example of a frequency distribution for a discrete variable in which single-value classes are employed. We can use the basic formula in Equation (2.1) to find the mean (μ) number of automobiles passing the observation point in a typical three-minute period. Since the procedure is identical to that described in connection with Table 2.12, we shall not repeat it. The result is 9.61 autos per period.

With only a minor modification we can use the same basic formula [Equation (2.1)] to find the means of frequency distributions with many-value classes when the variable observed is either discrete or continuous. The modification is to *use the class midvalues as the values of X in the basic formula.* For example, the left column and the column of frequencies in Table 2.13 repeat the information from the automobile traffic distribution in Table 2.5. We want to find the mean number of autos that passed the observer in a typical three-minute period. The second column contains the values of X to use in the basic formula. These are the midvalues of each class.

Table 2.13   The Mean from Many-Value Classes

Number of autos	Midvalues (X)	Frequency (f)	Class total (fX)
0– 4	2	4	8
5– 9	7	10	70
10–14	12	7	84
15–19	17	3	51
20–24	22	2	44
		$N = 26$	$\sum (fX) = 257$

$$\mu = 257/26 = 9.88 \text{ autos per period}$$

As compared with Table 2.5, we have moved the symbol X in Table 2.13 from the first column to the second. We are now assigning specific values to X, whereas previously we only defined the variable X as the number of autos. After we complete the process described by the basic formula, the result is a mean (μ) of 9.88 autos per period. When we found the mean of the distribution based on single-value classes, the result was 9.61 autos per period. The 9.88 autos per period found in Table 2.13 reflects the slight inaccuracy that results from using midvalues to represent all the observations in each class.

Exactly the same procedure as that just described can be used to find the mean of a frequency distribution where the variable is continuous. Midvalues are found for the classes. These can be viewed as approximations for the means of each class. Then the frequency counts can be viewed as weights, and the basic formula becomes a way to find the weighted mean, as described on pages 16 and 17 of Chapter 1.

Several procedures developed to simplify calculations for finding the mean of grouped data are discussed in reference $\bar{3}$. These formulas are less generally applicable than the basic formula, and their major computational advantages apply to large distributions with many classes. We omit them here because computer processing for this latter type of distribution has reduced the need for such special formulas.

2.3

The Median   We considered a brief description of this measure of location in the previous chapter. Now we must be more thorough.

Definition	The **median**, φ, of a set of numbers is a number greater than not more than half the numbers in the set and less than not more than half.

The relative frequency step graph provides a way to illustrate the concept of the median. Figure 2.7 repeats the relative frequency step graph of Figure 2.4 for the data on galvanized steel sheets summarized in Table 2.10.

To find the median we draw a horizontal line in Figure 2.7 through the point 0.50 on the cumulative relative frequency (F') scale and continue the line to the right until it intersects the step graph. Then we drop a vertical line from the point of intersection to the X scale. The X value at which the vertical line intersects the X scale is the median. In this case, the median (φ) is two flaws per sheet. Since the cumulative relative frequency (F') is 0.50 for $X = 2$, we reason that 2 should be greater than no more than half the observations and less than no more than half. As we see in Table 2.10, the number of flaws (X) is either 0 or 1 for 48 percent of the observations. Further, X is 3, 4, or 5 for 28 percent. Hence the value 2 satisfies the definition of the median.

Note that in Figure 2.7 the horizontal projection from 0.50 on the cumulative relative frequency (F') scale intersects the step graph on a vertical portion of a step. This is the typical case. Sometimes, however, it coincides exactly with the horizontal portion of a step. In this case, any value of the variable (X) which lies below any point

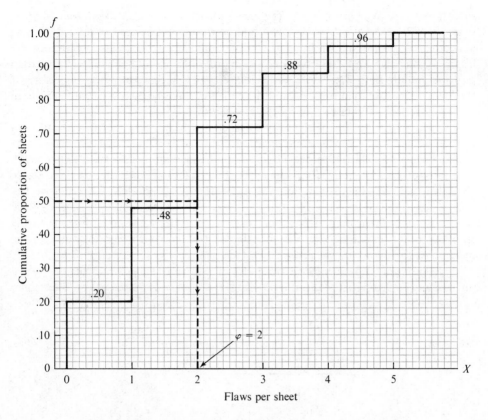

Figure 2.7   Cumulative Relative Frequency Step Graph

on this horizontal portion of the step satisfies the definition of the median. There is a range of ambiguity in this instance, and the median is arbitrarily defined to be the midvalue of this range. For example, suppose the dashed horizontal line in Figure 2.7 had fallen exactly on the horizontal segment of the step graph above X values running from 1 to 2, instead of just above that horizontal segment. Then the median would be 1.5 flaws per sheet.

If the step graph is not available, we can find the median (φ) of a distribution with single-value classes directly from the cumulative frequency distribution. In the table below for the sheet-steel example, we see in the cumulative frequency column (F)

X	F
0	10
1	24
2	36
3	44
4	48
5	50

that there are 50 (N) observations. The median (φ) must be a value such that no more than 25 ($N/2$) observations are below it and no more than 25 are above it. Beginning at the top of the cumulative frequency (F) column, we see that 24 observations are 0's or 1's. The 25th and 26th observations, in order of magnitude, bracket the median position and are both 2's. Hence, the median (φ) is two flaws per sheet, as before. If the 25th observation had been a 1 and the 26th a 2, then the median would have been 1.5, as explained in connection with the step graph.

Frequency distributions with many-value classes based on either discrete or continuous variables will now be considered. If a cumulative frequency ogive such as the one in Figure 2.5 is available, we can estimate the median from it accurately enough for most purposes. Reading across from the top of the ogive, we see that there are 26 observations (N) in the distribution. It follows that 13 ($N/2$) on the cumulative frequency scale (F) will correspond with 0.5 on the cumulative relative frequency scale (F'). We move horizontally from 13 on the F scale to the ogive, then drop vertically from this point on the ogive to the X axis. On the X axis we find the value of the median (φ) to be 9 autos per period.

If we have only a frequency distribution table, such as Table 2.5, available, we can find the median from it. The information from that table is repeated below. Since there

Number of autos (X)	Number of periods (f)
0– 4	4
5– 9	10
10–14	7
15–19	3
20–24	2
	26

are 26 observations (N), there can be no more than 13 ($N/2$) observations smaller in value than the median, and no more than 13 larger in value. Beginning at the top of the frequency (f) column, we see that four observations are less than 4.5 (X), where 4.5 is the exact upper limit of the first class. The following class has 10 observations. Hence, there are 14 observations less than 9.5 (X), the exact upper limit of the second class. But we want the value of X, below which lie only 13 observations. Assuming that the 10 observations in the second class are distributed uniformly throughout the class, we must go only part way through this class to find the median. Proceeding from the small end of the X scale toward greater values of X, we have accumulated four frequencies (f) when we reach an X value of 4.5 and 14 frequencies when we reach an X value of 9.5. Since we need only 13 frequencies, we can use only nine of the 10 frequencies in the second class. In other words, we need only go 9/10 of the way through this class on the X scale. The class begins with an X value of 4.5 and the class length is 5. Hence the median (φ) is the sum of 4.5 and 0.9(5), or $4.5 + 4.5 = 9$, autos per period, the same value we estimated from the ogive.

We can summarize the foregoing discussion in the following statement:

The median (φ) of a frequency distribution with grouped-value classes is

Statement

$$\blacksquare \qquad \varphi = X_e + \frac{f_r}{f_m} \cdot c, \qquad \blacksquare \qquad (2.2)$$

where X_e is the exact lower limit of the class in which the median lies (the median class), f_r is the number of frequencies required in the median class, f_m is the number of frequencies in the median class, and c is the class length. In this equation,

$$f_r = \frac{N}{2} - F_e,$$

where N is the total number of observations and F_e is the cumulative number of observations less than the exact lower limit of the median class (X_e).

For the data on content weights of sugar sacks in Table 2.8, $N/2$ is 60. Then

$$f_r = \frac{N}{2} - F_e$$

$$= 60 - 45, \quad \text{or } 15,$$

and

$$\varphi = X_e + \frac{f_r}{f_m} \cdot c$$

$$= 4.985 + \frac{15}{33}(0.03), \quad \text{or } 4.999 \text{ pounds.}$$

Fractiles   By definition, the median is greater than not more than half and less than not more than half of the values in a data set. We can generalize the definition to an entire class of descriptive measures called *fractiles*. We do this by specifying fractional parts other than *one-half* in the definitions. For example, the *first quartile* is a value of the variable greater than not more than 1/4 and less than not more than 3/4 of the values in the data set. The *third decile* is a value greater than not more than 3/10 and less than not more than 7/10 of the values in the data set. The *sixtieth percentile* is a value greater than not more than 60/100 and less than not more than 40/100 of the values in the data set.

To find fractiles for distributions based on grouped-value classes, we can use Equation (2.2) if we modify the definitions. Suppose we want to find the third quartile (Q_3) for the weight distribution in Table 2.8. We change $N/2$ in the definition of f_r to $3N/4$. Then, $3N/4$ is $3(120)/4$, or 90, and F_e becomes 78, the cumulative frequency at the exact lower limit of the class in which the third quartile lies. Continuing,

$$f_r = \frac{3N}{4} - F_e$$

$$= \frac{3(120)}{4} - 78, \quad \text{or} \quad 12,$$

and

$$Q_3 = X_e + \frac{f_r}{f_m} \cdot c$$

$$= 5.015 + \frac{12}{19}(0.03), \quad \text{or} \quad 5.034 \text{ pounds.}$$

The Mode   The *mode* is that value that occurs the most often in a set of values. For example, in Table 2.2, there are 14 steel sheets with one flaw apiece. This is the most frequently recurring value of X. Hence, the mode of this distribution is 1 flaw per sheet.

For distributions with more than one value per class, the same basic concept may be used to find the *modal class*. In other words, the class with the most frequencies in it is the modal class. For the weight distribution in Table 2.8, there are 33 frequencies in the class 4.985 to 5.015 pounds. Since 33 is the greatest class frequency, the modal class is 4.985 to 5.015 pounds. For frequency distributions, *we will define the **mode** as the midvalue of the modal class*. Hence for the example just cited, the mode is 5.00 pounds. References 4 and 5 discuss another common definition of the mode for frequency distributions.

The mode is the least frequently used measure of location of the three we have discussed, primarily because the modal value is not precisely defined for many data sets or even for some frequency distributions. For example, in the set of data

$$\{0, 2, 3, 5, 8\},$$

every value satisfies the definition of the mode. In a frequency distribution, two or more classes that are not adjacent may be equally popular and more popular than any other class. Again, there would be more than one mode.

2.4   Dispersion

Since the variance and the standard deviation are by far the most frequently used measures of dispersion for frequency distributions, they are the only ones we will consider.

As in the case illustrated in Table 1.1, the variance (σ^2) calculation for frequency distributions can sometimes be based on deviations from the mean ($x = X - \mu$). This will be the case when the deviations are whole numbers. In this case, we use the equation

$$\blacksquare \qquad \sigma^2 = \frac{\sum [f(X - \mu)^2]}{N}, \quad \text{or} \quad \sigma^2 = \frac{\sum fx^2}{N}, \qquad \blacksquare \qquad (2.3)$$

to find the variance. For example, in Table 2.14, we see that the distribution is symmetrical about the central value of the distribution ($X = 72$ inches). This will be the balance point, and, therefore, the arithmetic mean (μ) of the distribution. The deviations of the class midvalues from the mean are recorded in a column labeled x, where

Table 2.14 The Variance from a Frequency Distribution

Length of pipe (inches)	Midvalues (X)	f	x	x²	fx²
60–64	62	4	−10	100	400
65–69	67	6	−5	25	150
70–74	72	10	0	0	0
75–79	77	6	5	25	150
80–84	82	4	10	100	400
		$N = 30$			1100

$$\mu = 72 \text{ inches}; \quad \sigma^2 = \frac{\Sigma fx^2}{N} = \frac{1100}{30} = 36\frac{2}{3} \text{ inches squared};$$

$$\sigma = \sqrt{36.7} = 6.058 \text{ inches.}$$

$x = X - \mu$. These values are squared and recorded under the heading x^2. There are four observations (f) in the class for which 62 inches is the representative value. Each of these observations has 100 as the square of its deviation from the mean (x^2). This class contributes 400 (fx^2) to the sum of squared deviations for all 30 observations, which is 1100. The mean of this sum is the variance (σ^2), which has the value $36\frac{2}{3}$ inches squared. The square root of 36.7 is the standard deviation (σ), which has the value 6.058 inches, as found from Appendix J.

When the value of the mean cannot be ascertained easily or when deviations from the mean are not small whole numbers that are easily squared, another procedure for finding the variance of a frequency distribution is preferred. Again, this procedure is an extension of one discussed in Chapter 1 [see Table 1.2 and Equation (1.3)]. If we cannot determine the mean of a frequency distribution by inspection, we find the variance from the equation

$$\sigma^2 = \frac{\Sigma (fX^2)}{N} - \left[\frac{\Sigma (fX)}{N}\right]^2, \tag{2.4}$$

which is the algebraic equivalent of Equation (2.3). We illustrate its use in Table 2.15.

Table 2.15 An Alternate Procedure for the Variance

Number of autos	Midvalues X	X²	f	fX	fX²
0– 4	2	4	4	8	16
5– 9	7	49	10	70	490
10–14	12	144	7	84	1008
15–19	17	289	3	51	867
20–24	22	484	2	44	968
		Totals	26	257	3349

The quantities to substitute in Equation (2.4) are found in the bottom row of totals in Table 2.15. The substitutions are:

$$\sigma^2 = \frac{\sum (fX^2)}{N} - \left[\frac{\sum (fX)}{N}\right]^2$$

$$= \frac{3349}{26} - \left[\frac{257}{26}\right]^2$$

$$= 128.808 - 9.88^2,$$

or

$$\sigma^2 = 31.1.$$

2.4

It follows immediately that the standard deviation (σ) is $\sqrt{31.1}$, or 5.58 autos per period.

Skewness

A frequency distribution that is symmetrical with respect to its median (φ) has no skewness. A frequency distribution is said to possess *positive* skewness when its graph has a long tail extending toward larger values of the variable. A frequency distribution that is unsymmetrical because of a long tail extending toward smaller values of the variable possesses *negative* skewness.

In typical symmetrical distributions, the values of the mean, the median, and the mode coincide. For skewed distributions the values of these three measures of location seldom coincide. Generally, the mean is most affected by extreme values and is pulled farthest in the direction of the long tail in skewed distributions. The median is the second most affected. The mode, the least affected, remains closest to the major concentration of the data.

Histograms for three frequency distributions are shown in Figure 2.8. The distribution in (a) is symmetrical with respect to the median because the side to the left of the median is a mirror image of the side to the right. The distribution in (b) exhibits positive skewness and in (c) the distribution exhibits negative skewness. Arrows point to the mean, median, and mode of each distribution. With positive skewness, the mean is greater than the median and the median is greater than the mode. With negative skewness, the relationships among the mean, the median, and the mode are reversed.

The measure of skewness we will present is known as *Pearson's coefficient of skewness*. This measure will be represented by the symbol κ (the Greek letter kappa). The Pearson coefficient of skewness (κ) is based on the tendency of the mean and median to differ in the manner illustrated in Figure 2.8 when skewness is present in a set of data.

Definition	For a statistical population, Pearson's coefficient of skewness (κ) is defined as $$\blacksquare \quad \kappa = \frac{3(\mu - \varphi)}{\sigma}, \quad \blacksquare \qquad (2.5)$$ where μ is the mean of the set, φ is the median, and σ is the standard deviation.

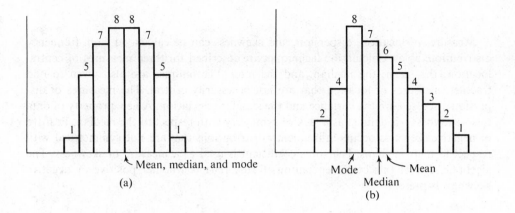

(a)

(b)

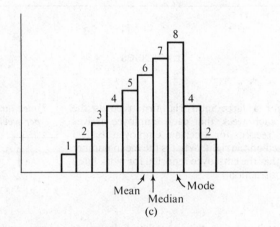

(c)

Figure 2.8   (a) Symmetrical Distribution, (b) Positive Skewness, (c) Negative Skewness

For the distribution in Table 2.15 we have already found that μ is 9.88, φ is 9.00, and σ is 5.58 autos per period. From this information, we find that

$$\kappa = \frac{3(9.88 - 9.00)}{5.58} = 0.47.$$

Since both numerator and denominator have the same dimensions, κ is a dimensionless number, a coefficient in the pure sense of the word. Note also that because the median is always subtracted from the mean, the algebraic sign of Pearson's coefficient always will reflect the direction of skewness automatically. When the mean is less than the median, the value of κ will be negative, as it should be to reflect negative skewness.

Summary

Measures of location, dispersion, and skewness can be calculated from frequency distribution tables. Calculation techniques are described for three measures of central location: the mean, the median, and the mode. Techniques are also given to find fractiles, measures of location that are not necessarily central. The measures of dispersion discussed are the variance and the standard deviation. A new property of data sets—skewness—is defined as a lack of symmetry with respect to the median. Positive or negative skewness occurs in frequency distributions that are not symmetrical with respect to their medians. Pearson's coefficient is used as a measure of skewness. The algebraic sign of this coefficient automatically reflects whether positive or negative skewness is present.

See Self-Correcting Exercises 2.B

Exercises

1. The timekeeper for a large industrial firm records the number of times each week that each employee arrives late for work. His records for a certain employee for last year are summarized on the right. What is the mean number of times per week that the employee reported for work late? (Leave answer as a fraction.)

Times tardy per week	Number of weeks
0	32
1	10
2	8
3	2
	52

2. The cumulative distribution on the right shows the gallons of gas used daily (X) by a groundskeeper. The length of the study was ten days. What is the mean daily use of gasoline by the groundskeeper?

X	F
0	2
1	6
2	6
3	7
4	9
5	10

3. A local junior college holds evening classes in real estate management. The director of evening studies, wishing to know something about the distribution of the ages of students in the class, obtains the data shown on the right. Ages are recorded to the nearest year. What is the mean age of the students in the class?

Age	Frequency
25–29	4
30–34	6
35–39	6
40–44	2
45–49	2
	20

4. The relative frequency diagram showing the number of sales per week for a mutual fund salesman is:

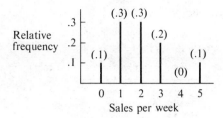

Sales per week

What is the mean number of sales per week?

5. An agronomist wishes to study the effect of the potassium content in the soil on tomato yield. He obtains the distribution below for one level of potassium content.

Number of tomatoes per plant	Number of plants
0– 2	4
3– 5	8
6– 8	8
9–11	20
12–14	10
	50

What is the median number of tomatoes per plant for this type of soil?

6. The balances for 200 accounts receivable, recorded to the nearest dollar, for a local retail store are summarized in the distribution below. Find the 25th percentile of the distribution.

Balance (Y)	Frequency (f)
0–$14	20
$15–$29	45
$30–$44	55
$45–$59	40
$60–$74	40
	200

7. The distribution of grades for a certain statistics class is shown below. Calculate the mode of the distribution.

Grade	Frequency
60–64	12
65–69	14
70–74	16
75–79	20
80–84	18
85–89	16
90–94	14
95–99	4
	114

8. The daily sales, recorded to the nearest dollar, for a corner newspaper stand can be summarized in tabular form as follows:

Daily sales (Z)	Number of days (f)
0– $2	10
$3– $5	5
$6– $8	10
$9–$11	5
	30

a. If the mean of daily sales is $5.00, what is the variance of daily sales?

b. What is the standard deviation?

9. The distribution of defective items in 20 production runs from a certain machine is shown below. (Note that single-value classes are shown.) If the mean of the distribution is 12 and its standard deviation is 4.6, find the value of Pearson's coefficient of skewness.

Number of defectives	Number of production runs
5	4
10	6
15	8
20	2
	20

Glossary of Equations

$$\mu = \frac{\sum (fX)}{N}$$

The arithmetic mean of a frequency distribution is the sum of the products of the class midvalues and frequencies divided by the number of observations.

$$\varphi = X_e + \frac{f_r}{f_m} \cdot c,$$

where $f_r = \frac{N}{2} - F_e$

The median of a frequency distribution with grouped-value classes is the sum of the exact lower limit of the median class and a portion of the median class length. The portion is the ratio of (1) the frequencies required from the median class to bring the cumulative frequency count up to half of the total number of observations and (2) the number of observations in the median class.

$$Q_3 = X_e + \frac{f_r}{f_m} \cdot c,$$

where $f_r = \frac{3N}{4} - F_e$

The third quartile of a frequency distribution with grouped-value classes is the sum of the exact lower limit and a portion of the length of the class in which the third quartile lies. The numerator of the portion is the number of observations required from the class in which Q_3 lies to bring the cumulative frequency up to 3/4 of the total observations. The denominator is the number of observations in the class in which Q_3 lies. The formula can also be adapted for other fractiles.

$$\sigma^2 = \frac{\sum [f(X - \mu)^2]}{N}, \quad \text{or}$$

$$\sigma^2 = \frac{\sum (fx^2)}{N}$$

The variance of a frequency distribution is found by summing the products of the squared class midvalue deviations from the mean and the frequencies associated with them and then dividing this sum by the total number of observations. This equation is useful when the mean is obvious and the deviations are small whole numbers.

$$\sigma^2 = \frac{\sum (fX^2)}{N} - \left[\frac{\sum (fX)}{N}\right]^2$$

The variance of a frequency distribution is the difference between two quantities. The sum of the products of class frequencies and squared midvalues divided by the total number of observations is the first quantity. From this quantity is subtracted the square of the arithmetic mean of the distribution.

$$\kappa = \frac{3(\mu - \varphi)}{\sigma}$$

Pearson's coefficient of skewness is 3 times the difference between the mean and the median, divided by the standard deviation.

3

Association

So far we have considered observations of only a single variable. In other words, an observation consisted of a single number that represents a value on only one measurement scale—for example, the weight of sugar in a particular sack. We discussed ways to describe entire sets composed of such *univariate observations*.

In this chapter we shall work with sets of *bivariate observations*, in which an observation will consist of a *pair* of numbers—that is, a number for each of two variables observed simultaneously. The weight of sugar in a particular sack and the age of the machine that filled the sack provide an example of a bivariate observation. The content weight of one sack and the age of the machine that filled a different sack is *not* a bivariate observation. The two numbers that constitute the observation must describe two characteristics of the *same unit of observation*, the same sugar sack in our example.

A set of bivariate observations can disclose a highly useful relationship, or association, between the two variables. A relationship between sales of a product and amount spent on television spot advertising or a relationship between employee morale and number of lost-time accidents would constitute valuable knowledge for most firms.

Basic Concepts

Before discussing ways to find association and to describe relationships between two variables, we must consider two preliminary topics. These are the nature of mathematical functions and a discussion of what we mean by search techniques for association.

Functions

A good way to describe the association between two variables is to find a function that relates the two.

Definition	**A function** is a rule that assigns to a given element in one set exactly one element in another set.

Consider three squares. The area and length of a side are given for each square in Table 3.1. To the element 3 in the set of side lengths is assigned the element 9 in the set

Table 3.1   Association between Square's Side Length and Area

Square	Length of side	Area
a. ☐	3←	———— 9
b. ☐	2←	———— 4
c. ☐	7←	————49

of areas. Similarly, to a side of length 2 is assigned an area of 4, and to a side of length 7 is assigned an area of 49. These assignments are indicated by arrows. Suppose we let s be the length of the side of any given square and A be the area of that square. Then we see that the function by which an element in $\{A\}$ is assigned to an element in $\{s\}$ is

$$A = s^2. \tag{3.1}$$

The units of observation in Table 3.1 are the squares a, b, and c. There are three bivariate observations—the number pairs (3, 9), (2, 4), and (7, 49).

As another example of a function, Table 3.2 shows the sales for a certain firm as a

Table 3.2   Salesmen and Sales for Four Weeks

Week	Number of salesmen (X)	Sales (Y)
1	4←	$1800
2	2←	900
3	3←	1600
4	5←	

function of the number of salesmen working in each of four weeks. Here the functional relationship is not so easy to determine as the relationship between the side of a square and its area. We can let the variable X represent the number of salesmen and the variable Y represent sales, but no functional relationship between the two variables is apparent. In this and other instances, a general notation which signifies that one variable (Y) is a function of another variable (X) will be useful. If we simply wish to make the general statement that Y is a function of X, we can write

$$Y = f(X) \tag{3.2}$$

and read this as "Y is a function of X." X is called the *independent* variable and Y is the *dependent* variable. In other words, the sales for a given week (Y) *depend* upon the number of salesmen working (X). Similarly, the area of a square (A) *depends* upon the length of its side (s), the independent variable. The use of f to denote *function* is completely unrelated to the use of f to mean *frequency*. Furthermore, in Equation (3.2), $f(X)$ does not mean the product of f and X.

Search Techniques for Association

When examining two or more variables simultaneously, we need some way to organize our observations, the objective being to get an indication as to whether one variable is functionally dependent on another. If we can find an association, we will be able to make useful predictions of the value of the dependent variable that will result when the independent variable is set at a given value. For example, given a square with a side (s) of 5 inches we can use Equation (3.1) to predict that the area (A) will be 25 square inches.

We will consider two techniques for seeking association among variables, *tabular analysis* and *linear regression*. As we shall see, tabular techniques emphasize the search for the *existence* of association among variables more than they do the *description* of any functional relationship that may exist. Regression techniques place more emphasis on establishing an empirical description of the relationship.

Tabular Analysis

Suppose that a mail order firm studying the effectiveness of its catalog wants to determine whether or not the position of an item's description on the catalog page affects sales of the item. One executive thinks the upper right corner of a page commands more attention. Other executives champion other positions.

In the study, the units of observation are people selected as representative of potential customers. Several different versions of the catalog have been prepared to show to these people. In one version the description of a certain test item is printed in the upper left quadrant; in another version it is in the lower right quadrant, and in other versions it is located in the two other quadrants. Since the ad's position on the page can be varied by the firm, it is *the independent variable*.

After being shown a test ad page, along with several other pages, a subject is asked whether or not he recalls seeing ads describing each of several items, among which are some of the test items. He is then asked questions about the specific content of any ads he remembers, to verify the recall. The subject's recall is the second variable in this study. The objective of the study is to see whether this second variable is dependent upon, or associated with, changes in the independent variable.

To facilitate punched card processing, each of the four quadrants on a page has been numbered. The upper left corner is designated number 1 for the position variable, the upper right is 2, the lower left is 3, and the lower right is 4. Similarly, if a subject recalls a test ad, the recall variable is given the value " 1 " and, if he does not, it is given the value "2." Then, for these two nominal scales, the set of original observations for a given ad could be shown as follows

$$\{(3, 2), (1, 1), (4, 1), \ldots, (2, 2)\}.$$

In this set of ordered pairs, the first value in each pair is X_i, the value of the independent variable (position) and the second value is Y_i, the value of the dependent variable (subject's recall). This is the customary order. Hence, (4, 1) is (X_3, Y_3) for a person who recalls the test ad in the lower right corner of the page.

Cross Classification

After being tallied and counted, the usable responses are tabulated, as shown in Table 3.3. The totals in the bottom row and in the right column of Table 3.3 tell us

Table 3.3   Responses to Catalog Survey

| | *Ad position on page* | | | | |
| | *Upper half* | | *Lower half* | | |
Recall	*Left*	*Right*	*Left*	*Right*	*Total*
Yes	23	25	10	6	64
No	51	67	43	61	222
Total	74	92	53	67	286

how responses are distributed over the recall scale and, independently, the position scale. It is in the individual cells of the table, simultaneously *cross-classified* on both variables, that we will find data for determining whether ad position is associated with recall. Even here, however, the differing frequencies in the four column totals tend to conceal any pattern. This leads us to consider stating the cell values as percentages of some total.

In Table 3.3, three choices are available as bases for cell percentages. We can state each cell frequency count as a percentage of the grand total (286). Second, we can state each cell value in the top row as a percentage of all the " Yes " responses (64) and all values in the bottom row as percentages of the " No " responses (222). Third, we can state each cell value as a percentage of its column total. If we accept the first choice described, we will simply scale down all the values in Table 3.3 proportionately. This would leave us with our original problem unsolved.

If we take the second choice, the result is shown in Table 3.4. Erroneously, one

Table 3.4   Percentages of Recall Totals

Ad position on page

| | Upper half | | Lower half | | |
| | Left | Right | Left | Right | |
Recall	Left	Right	Left	Right	Total
Yes	36%	39%	16%	9%	100%
No	23	30	19	28	100%

might now infer that the upper right position is the most effective because it has the highest percentage of "Yes" responses. It has, but only because a larger group of people (92) were exposed to ads in this position than in any other. Thus, we see that taking percentages within classes of the *dependent* variable, as in Table 3.4, sometimes leads to the wrong conclusion.

For the third choice, we state each value in the body of the table as a percentage of its column total. In Table 3.5 we take percentages within each class of the *independent*

Table 3.5   Percentages of Position Totals

Ad position on page

| | Upper half | | Lower half | |
| | Left | Right | Left | Right |
Recall	Left	Right	Left	Right
Yes	31%	27%	19%	9%
No	69	73	81	91
Total	100%	100%	100%	100%

variable. This has the effect of subjecting a group of 100 people to each ad position so that we can see the influence of ad position upon recall separately from the irrelevant effect of differences in audience sizes. We can see which ad treatment produces the highest percentage recall. We can infer correctly that ad positions *in order of effectiveness* are top left, top right, bottom left, and bottom right for this group of 286 people. The recall variable appears to be dependent upon ad position, the independent variable.

The general conclusion we have illustrated can be stated as follows:

Rule | In cross-classified tables, percentages within classes of the *independent* variable expose any relationship between the independent variable and the dependent variable.

Chance Effects

Were we interested in only this one group of 286 people, our task would be finished. The table showing percentages of each group's response within ad position class has established the relationship. In this instance, however, the people were selected only as being representative of the entire set of potential customers. These 286 people are only a sample. The sample could be unrepresentative to some extent, or the tabled relationship might have come from a special or temporary effect that does not apply to all potential customers. These and other possible *chance effects* must be taken into account before we can trust the sample evidence and go on to estimate the relationship for the entire statistical population. Techniques for dealing with samples arranged in cross-classified tables constitute the subject matter in Chapter 12.

Spurious Association

Sometimes, apart from chance effects, the association between two variables is not what it may appear to be because an intervening variable has not been taken into account. When this happens, we have an instance of *spurious association*.

Suppose that on a certain university campus, Table 3.6 is presented as a partial

Table 3.6   Society Membership by Residence

	Permanent residence	
Status	*In state*	*Out of state*
Greek	62 (47%)	48 (55%)
Independent	70 (53%)	39 (45%)
Total	132 (100)	87 (100)

justification for funding a sociological study to determine what special appeal Greek fraternities and sororities hold for out-of-state students.

When family income, as well as residence, is taken into account, Table 3.6 is revised as shown in Table 3.7. Note in the table that within a given income group,

Table 3.7   Society Membership by Residence and Family Income

	Permanent residence			
	In state		Out of state	
Family income	*Under $15,000*	*$15,000 or more*	*Under $15,000*	*$15,000 or more*
Greek	43 (42%)	19 (61%)	10 (42%)	38 (60%)
Independent	58 (58%)	12 (39%)	14 (58%)	25 (40%)
	101 (100%)	31 (100%)	24 (100%)	63 (100%)

residence had no effect upon membership in Greek societies. Rather, the initial apparent relationship between residence and membership status was entirely *spurious*. Family income is an intervening variable that was not taken into account. Tuition may very well be considerably higher for out-of-state residents. As a result, a higher percentage of students from families with larger incomes are found among out-of-state students than among in-state students. Differences in income, rather than differences in residence, may account for differences in membership.

The foregoing example serves to make the point that *detecting* association and *explaining* association are separate processes. Statistical techniques are available to detect association, but they are completely mute when it comes to explaining any association detected. Only *substantiated theory* from the subject matter field that gave rise to the data can supply an explanation. Nonetheless, statistical techniques are of central importance for checking the many ramifications that convert pure speculation into substantiated theory.

Linear Regression

Whenever we are concerned with detecting association between two variables, both of which are measured on interval or ratio scales, we may be able to go beyond tabular analysis. We may be able to go as far as describing the association with a functional relationship between the two variables that is expressed in the form of an algebraic equation. A widely used technique for finding a straight-line relationship from a set of bivariate observations is called *linear regression*. In addition to providing a way to

Table 3.8 Observations of Calls and Sales

Agent	Number of calls (X)	Number of sales (Y)
1	26	11
2	13	7
3	21	8
4	37	20
5	17	9
6	20	12
7	17	4
8	28	16
9	28	11
10	6	2
11	23	11
12	25	7
13	38	18
14	33	14
15	12	2
16	30	15
17	30	10
18	10	4
19	18	7
20	21	10

find a relationship, linear regression also includes procedures to measure the adequacy of the relationship. These latter procedures are discussed in the last half of this chapter.

Suppose that an analyst for an insurance company with 20 agents seeks a relationship between the number of customer calls, or visits, an agent makes in a week and the number of customers to whom the agent makes a sale. The analyst reasons that the number of sales should depend to some degree on the number of calls made. This leads him to designate sales as the dependent variable (Y) and calls as the independent variable (X).

In this chapter, we shall treat the 20 paired observations as if they are the only data set of interest to the analyst. We will treat them as a *statistical population* of 20 bivariate observations. In Chapter 14 we will discuss how to deal with samples from bivariate populations. The 20 observations are listed in Table 3.8. Note that a single observation consists of a pair of numbers that describe two characteristics of a particular agent.

The Scattergram

It is usually worth while to begin a linear regression analysis with a search for a functional relationship between two variables by plotting a *scattergram*. It is customary to plot the dependent variable on the vertical axis and the independent variable on the horizontal axis. A point is placed at the intersection of each of the observed values. The scattergram for our insurance sales example appears in Figure 3.1. Note that a given point represents the simultaneous observation of two characteristics of a particular agent. Thus the point (10, 4), where X is 10 and Y is 4, shows the numbers of calls and sales for the eighteenth agent on the list.

In the scattergram in Figure 3.1, the points appear to fall reasonably close to an

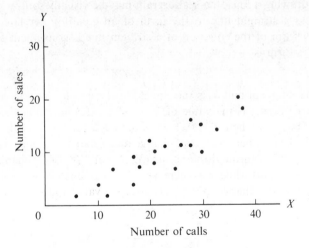

Figure 3.1   Scattergram for Insurance Data

imaginary straight line running from the lower left to the upper right. Their failure to fall even closer may be the net effect of many other factors that we could reasonably expect to exert some influence on sales.

One way to find a straight line to describe the relationship between calls and sales in our example is simply to draw one that we judge to be a good fit by its visual appearance on the scattergram. This technique does not constitute part of linear regression but is a useful point of departure. One such line is shown in Figure 3.2. This line cuts

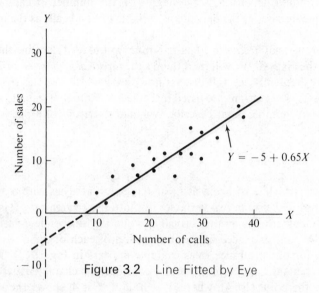

Figure 3.2   Line Fitted by Eye

the extended Y axis at *minus 5* and cuts the X axis at 8. The coordinates of these points are $(0, -5)$ and $(8, 0)$, respectively.

Although drawing a line on a scattergram may be visually satisfactory, we must be able to express a straight line in the form of an equation for later efficiency. The *slope-intercept* form of the equation of a straight line is convenient for this purpose. In general, this form is

$$Y = a' + b'X,$$

where a' is the intercept and b' is the slope. On the graph, the intercept (the Y intercept in the strict sense) is the value of Y at which the line cuts the Y axis. In the above figure, Y is -5 when X is 0. Hence, $a' = -5$ for our specific line.

The slope is the number of units of change in Y relative to a unit increase in X. For example, the line goes through the points $(39, 20)$ and $(8, 0)$. Y changes by a positive 20 points $(20 - 0 = 20)$ while X is increasing by 31 points $(39 - 8 = 31)$. The ratio of the change in Y to the change in X yields both the value and the algebraic sign of the slope; that is,

$$b' = \frac{20 - 0}{39 - 8} = \frac{+20}{+31} = +0.65.$$

We now know the values of both the Y intercept and the slope for the straight line in Figure 3.2. The value of a' is -5 and the value of b' is 0.65.

The resulting equation of the line is

$$Y = -5 + 0.65X. \tag{3.3}$$

This equation summarizes the data in the table and can be used to approximate the number of sales resulting from any number of calls within the range of the data. For instance, if an agent is known to be one of those who made 30 calls during the week, the approximation would be

$$Y = -5 + 0.65(30) = 14.5,$$

which rounds to 14 sales by the round-even rule.

The line seems to indicate that if a salesman made no calls during the week, he could expect his sales to be *negative* 5, a meaningless concept. Similarly, the line shows negative values of Y for all values of X less than 8. In fact, the true relationship would never drop below zero by the logic of the situation. We have extended the line so that we can use the convenient slope-intercept form of its equation. This illustrates the danger in using a functional relationship outside the range of observations on which the relationship is based.

The Least Squares Line

There are major shortcomings in fitting a straight line to data by eye, as was done above. Probably, no two people will produce exactly the same line under these circumstances. Furthermore, there is no agreed-upon standard with which to compare the lines drawn by eye.

We need a procedure that will produce only one line for a given set of data. We also want that line to be, in some sense, the *best fitting* line for the given set of points. The *least squares* technique produces such a line and constitutes a part of standard linear regression procedures.

In Figure 3.3, we repeat the scattergram for the insurance example and show the line for which the least squares regression equation is

$$Y_C = -1.69 + 0.51X. \tag{3.4}$$

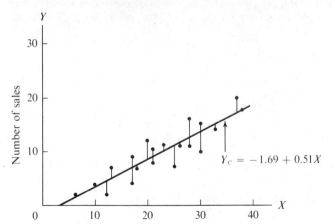

Figure 3.3   Least Squares Line and Deviations

Also shown in Figure 3.3 are the deviations of the observations from the line in the Y direction (vertical). These deviations are the errors that result when we use the least squares equation to approximate the number of sales (Y_C) that will result from a given number of calls (X). For example, the agent who made 20 calls actually made 12 sales. In this observation, the value of Y is 12 for an X value of 20. If we substitute an X value of 20 in Equation (3.4), we find that Y_C is only 8.51. Hence, the deviation ($Y - Y_C$) is $12 - 8.51$, or 3.49. The approximation based on 20 calls falls 3.49 short of the actual sales for this observation.

Together, the deviations for the 20 observations shown in Figure 3.3 represent the *dispersion* of the data about the line shown. Any other line would have a similar set of deviations. We can build a measure of dispersion upon these deviations. Then it would seem reasonable to define the best-fitting line as that line for which the dispersion of the data set is a minimum. The least squares line is just such a line.

Definition | Of all lines that can represent the functional relationship for a given set of bivariate observations, the **least squares line** is the one for which the sum of the squared deviations of the data from the line is a minimum. In this sense, the least squares line is the line of best fit.

In symbols, suppose we let the equation of the least squares straight line be

$$Y_C = \alpha + \beta X, \tag{3.5}$$

where α is the intercept and β is the slope. Then for a given set of observations,

$$\sum [(Y - Y_C)^2] \tag{3.6}$$

is the sum of the squared Y deviations. In order for Equation (3.5) to be the *least* squares line for a given set of data points, the sum of squared Y deviations must be smaller than is the comparable sum for any other line.

Differential calculus is used to find the two conditions which α and β in Equation (3.5) must satisfy to minimize Expression (3.6) for a given set of data:

Statement | The numerical values of α and β found from a simultaneous solution of the equations

$$\blacksquare \qquad \sum Y = \alpha \cdot N + \beta \cdot \sum X, \qquad \blacksquare \tag{3.7a}$$

$$\blacksquare \qquad \sum XY = \alpha \cdot \sum X + \beta \cdot \sum X^2 \qquad \blacksquare \tag{3.7b}$$

will be the intercept (α) and slope (β) of the least squares line for a given set of bivariate observations. In these equations all values except α and β are obtained from the data.

For our insurance example, the 20 pairs of numbers that constitute our observations appear under X (calls) and Y (sales) in Table 3.9. In the column headed by X^2, the

Table 3.9   Calculations Required for Least Squares Procedure

Observation	X	Y	X^2	XY
1	26	11	676	286
2	13	7	169	91
3	21	8	441	168
4	37	20	1,369	740
5	17	9	289	153
6	20	12	400	240
7	17	4	289	68
8	28	16	784	448
9	28	11	784	308
10	6	2	36	12
11	23	11	529	253
12	25	7	625	175
13	38	18	1,444	684
14	33	14	1,089	462
15	12	2	144	24
16	30	15	900	450
17	30	10	900	300
18	10	4	100	40
19	18	7	324	126
20	21	10	441	210
Totals	453	198	11,733	5,238

squared value of X for every observation is listed. These squares can be obtained from Appendix Table J. In the column headed by XY, the product of X and Y for each observation is shown. Then the columns are totaled to give us

$$\sum X = 453, \qquad \sum X^2 = 11{,}733,$$

$$\sum Y = 198, \qquad \sum XY = 5{,}238.$$

Finally, since there are 20 observations, $N = 20$. The given set of observations has been used to find every value we need to substitute in Equations (3.7) in order to find the particular values of α and β that define the least squares line.

We next substitute the sums from Table 3.9 for our 20 observations into Equations (3.7) and solve them simultaneously. In Table 3.10, the two top rows show these substitutions. Row (3) is row (1) divided through by the coefficient of α, which is 20. Row (4) is row (2) divided through by 453, the coefficient of α in the second equation. Five-place accuracy is required for a subsequent use. Row (5) is the difference between

Table 3.10   Solution of Simultaneous Equations

$$
\begin{array}{rl}
(1) & 20\alpha + 453\beta = 198 \\
(2) & 453\alpha + 11733\beta = 5238 \\
\hline
(3) & \alpha + 22.65000\beta = 9.90000 \\
(4) & \alpha + 25.90066\beta = 11.56291 \\
\hline
(5) & 3.25066\beta = 1.66291 \\
(6) & \beta = 0.51156 \\
(3) & \alpha + 22.65(0.51156) = 9.90000 \\
(7) & \alpha = -1.68683
\end{array}
$$

rows (4) and (3). Row (6) is row (5) divided through by the coefficient of β to give the value of β for the least squares line. This value is substituted in row (3), which is then solved for the value of α for the least squares line. To two-place accuracy, the least squares line for the insurance example is, then,

$$Y_C = -1.69 + 0.51X,$$

which is the same as Equation (3.4). Because -1.69 and 0.51 satisfy Equations (3.7) for our data, we know that no other straight line can fit the data as well. We know that these two values for the intercept and slope produce the smallest value of Expression (3.6) possible for any straight line fitted to the 20 observations.

The line $Y_C = \alpha + \beta X$, where α and β are defined by Equations (3.7), is called either the least squares line or the least squares regression line.

For practical calculation, it may be somewhat simpler at times to solve Equations (3.7) for β and use β to get α. Then numerical substitutions can be made in these general solutions. The general solutions are

$$\beta = \frac{N \cdot \sum XY - \sum X \cdot \sum Y}{N \cdot \sum X^2 - (\sum X)^2} \qquad (3.8a)$$

and

$$\alpha = \frac{\sum Y - \beta \cdot \sum X}{N}. \qquad (3.8b)$$

The proper substitutions are:

$$\beta = \frac{20(5238) - 453(198)}{20(11,733) - (453)^2} = 0.51156$$

and

$$\alpha = \frac{198 - 0.51156(453)}{20} = -1.68683.$$

As we expect, these are the same values we got earlier.

An algebraically equivalent alternative for Equation (3.8b) is

$$\alpha = \mu_Y - \beta \cdot \mu_X,$$

which shows that the least squares line passes through the point (μ_X, μ_Y), the point where the means of the Y values and the X values intersect on the scattergram.

3.1

Other Relationships

The relationship between Y and X in our insurance example is called a *positive linear* relationship, because a straight line appears to be the proper shape of the function and because that line has a positive slope (the value of β is positive). Not all linear relationships are positive, however. Studies have shown that as family income increases, the consumption of bread tends to decrease. A scattergram for these two variables might be like the one in Figure 3.4(a). Here we have a *negative linear* relationship and the value of β will be negative.

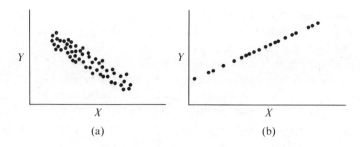

(a)   (b)

Figure 3.4   Linear Relationships

Figure 3.4 (b) shows an example of a *perfect positive linear* relationship. It is perfect because no data point deviates from the straight regression line. Such a relationship would be very nearly realized by simultaneously reading the temperatures of several different containers of water on well-calibrated Fahrenheit and centigrade thermometers.

Equations (3.7) are proper for finding the least squares regression lines for any scattergram to which a linear function applies. These formulas would apply to all three of the scattergrams we have discussed so far.

The two scattergrams in Figure 3.5 are examples of *curvilinear* relationships. The

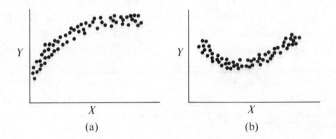

(a)   (b)

Figure 3.5   Curvilinear Relationships

one in (a) might represent sales as increasing amounts of advertising are applied, for instance. The underlying relationships appear to be represented by curved lines, rather than straight ones. In neither case do all the points fall exactly on the curved line. Consequently, neither one is a perfect curvilinear relationship. Algebraic techniques for fitting curves to such scattergrams can be found in Reference 1.

Even when a relationship is known to be curvilinear, straight-line techniques may be useful. For instance, in Figure 3.5(a), suppose we were interested in only the left half of the X values in the scattergram. In other words, suppose we are interested in only a restricted range of the independent variable. Then, by covering the right half of the data, you can see that a straight line would do. The same would be true if we were interested only in the right half of the X scale or in any other similarly restricted range.

Just as tabular analysis has the sole function of detecting association, regression techniques only *describe* association between variables. They have nothing whatsoever to do with *explaining* that association. The variable chosen to be independent for a particular analysis may, indeed, be all or part of the "cause" of variation in the dependent variable. Or both may vary in response to changes in another variable. Or the apparent relationship may represent only an isolated accident. Only sound theory from the subject matter area can provide the explanation.

With regression techniques we use nothing other than the appearance of the scatter-gram to decide the *algebraic form* of the functional relationship. This procedure is entirely *empirical*, in that it is based on no theory from the subject matter. Scientifically, it is always more satisfying if *analytical* reasons can be found to explain why a partic-ular function should have a particular form. Regression techniques can describe associ-ation and provide an interim functional relationship to serve utilitarian prediction purposes while satisfactory theory is being developed. But such relationships should be considered only as interim replacements.

Summary

A bivariate observation is the observation of the values of two variables that apply to a single unit of observation. A person's weight and age is an example. Tabular analysis and linear regression are two techniques for finding and describing a relation-ship between two variables for bivariate observations. In tabular analysis, per-centages of the dependent variable occurring within each class of the independent variable are used to detect any relationship between variables.

When bivariate observations come from interval or ratio scales, an algebraic equa-tion is often useful to describe the functional relationship between the two variables. A straight line is a satisfactory form for many such relationships. The least squares criterion is the major one used to define the slope and intercept of the best-fitting line. For this line, the sum of squared deviations of the dependent variable from the line is a minimum. Fitting least squares relationships and measuring the adequacy of these relationships is called regression analysis. When the relationship is to be a straight line, linear regression analysis is the proper term.

See Self-Correcting Exercises 3.A

Exercises

1. The marketing staff of a large firm was investigating the packaging of a new product. The particular aspect being studied was customer response to color. The following data were accumulated:

Response	Color	
	Red	Blue
Good	18	30
Bad	6	18

a. Utilizing the same form, make the data more understandable by taking percentages.

b. Interpret the result of part (a).

2. A production engineer is testing two processes in order to determine which to use on an assembly line. He will use the process that produces better quality output. His research is summarized as follows:

| | Process | |
Units	A	B
Above standard	175	525
Below standard	75	225

Which process should he choose based on this information alone?

3. A stock analyst, trying to find a way to predict small changes in the Dow-Jones index, noticed the following facts.

When automobile production (X) did not change, the Dow-Jones index (Y) rose 2%;

when automobile production rose 1%, the Dow-Jones index rose 6%;

when automobile production dropped 1%, the Dow-Jones index dropped 2%.

From this information alone, determine the change in the Dow-Jones index per unit change in automobile production.

4. By using a forecast of daily high temperatures, the owner of an ice cream stand wants to estimate how many gallons of ice cream he will sell. For a two-week period, he has kept track of the amount of ice cream sold each day and the temperature that was forecasted on the previous day. He has arrived at the following relationship:

$$Y = -25 + 0.5X,$$

where

$$Y = \text{gallons/day of ice cream sold,}$$
$$X = \text{the day's forecasted temperature.}$$

During the two-week period, sales ranged from 13 to 38 gallons per day and high temperatures ranged from 85 to 98 degrees.

a. If tomorrow's forecast is for a temperature of 96 degrees, how much ice cream will be sold according to the estimating procedure?

b. If tomorrow's forecast is for a temperature of 108 degrees, how much ice cream will be sold?

5. A moving firm wishes to find a relation between estimated weight of an upcoming job and the time it will take to load the goods on a truck. Data kept on 10 jobs resulted in the following figures, where X is weight in tons and Y is hours:

$$\sum X = 20, \qquad \sum Y = 30, \qquad \sum XY = 70;$$
$$\sum X^2 = 50, \qquad \sum Y^2 = 85.$$

Determine the least squares linear relationship.

6. A major appliance company was trying to develop a relationship between per capita income of a household and the number of appliances in that household. The company surveyed 50 households and arrived at the following set of data (Y = number of appliances per household; X = income per household/1000):

$$\sum X = 90, \qquad \sum Y = 120, \qquad \sum X^2 = 170, \qquad \sum XY = 220.$$

a. Find the equation of the least squares line.

b. Find the estimated number of appliances owned in a household that has a per capita income of $8000.

Prediction

We want to show how a least squares linear relationship between two variables helps us predict a value of the dependent variable. We also want to show how much better this type of prediction can be than one based only on knowledge concerning the dependent variable. We will begin with the latter case by assuming that we know only the sales results for our 20 insurance agents. We will describe the best we can do in this univariate case. Then we will develop a technique applicable to the bivariate situation and compare the results of the two approaches.

Univariate Prediction

Suppose that Table 3.8 (page 62) shows only the 20 sales values in the third column and we know only that each of these numbers represents the result for some one of the 20 agents in our study. We do not know either the identity of the agent who produced any given sales result or how many calls were made for any given sale.

Our task is to make the best sales prediction that we can for one agent selected at random from the 20 in the study. From the set of Y values in Table 3.8 we must somehow predict sales for a given agent in the group of 20.

One interpretation of our task is that we are to find a single value of Y, the sales variable, which represents our best guess as to how many sales a given agent made. In this circumstance it seems reasonable to select that Y value which is typical of the set. This suggests that we choose a measure of central location for the set and use it as our prediction. We select the mean because it is well defined and because, as we will see shortly, it can serve a further purpose. We found in Table 3.9 that the sum of the Y values is 198 for our insurance example. Consequently,

$$\mu_Y = \frac{\sum Y}{N} = \frac{198}{20}, \quad \text{or } 9.9,$$

sales constitutes our *point prediction* of sales for an agent selected at random from the set.

A second interpretation of our task is that we are to find an interval on the Y scale within which we predict a given agent's sales will fall. Furthermore, if we center our interval on the mean (μ_Y) and make use of Chebyshev's theorem (see page 23), we can state a level of assurance that our prediction will be correct. To make use of this theorem we need the standard deviation of the set of 20 sales (Y) values. From Equation (1.3),

$$\sum (Y - \mu_Y)^2 = \sum Y^2 - \frac{(\sum Y)^2}{N}$$

$$= 2440 - \frac{(198)^2}{20}, \quad \text{or } 479.8.$$

From Equation (1.2) we find the variance (σ_Y^2):

$$\sigma_Y^2 = \frac{479.8}{20} = 23.99;$$

and from Equation (1.4) we find the standard deviation (σ_Y):

$$\sigma_Y = \sqrt{23.99}, \quad \text{or } 4.9 \text{ sales.}$$

Suppose that we construct an interval, the lower limit of which is two standard deviations below the mean ($\mu_Y - 2\sigma_Y$) and the upper limit of which is two standard deviations above the mean ($\mu_Y + 2\sigma_Y$):

$$\mu_Y - 2\sigma_Y = 9.9 - 2(4.9) = 0.1 \quad \text{sales,}$$

$$\mu_Y + 2\sigma_Y = 9.9 + 2(4.9) = 19.7 \text{ sales.}$$

Chebyshev's theorem assures us that no less than $1 - 1/(2)^2$, or 3/4, of the Y values lie within this 2σ interval. *We predict that a given agent's number of sales falls in the interval from 0.1 to 19.7 with no less than 75 percent assurance.*

Since our prediction interval (0.1 to 19.7) includes all but one value in the distribution of Y values, we know that the assurance level in this case is actually 95 percent ($19 \cdot 100/20$) for this interval. Sometimes, however, we know the mean and standard deviation for a population, but the distribution itself is not available. In this case a general technique such as the one used in Chebyshev's theorem is necessary.

For univariate *prediction* we used the mean (μ) as our *point prediction* and an interval four standard deviations wide centered on the mean as our *interval prediction* for no less than 75 percent assurance ($\mu \pm 2\sigma$). This interval ranges all the way from 0.1 sales to 19.7 sales.

The Standard Error of Estimate

We now return to the bivariate situation we have been discussing in connection with our insurance example. In describing how to find the least squares regression line illustrated in Figure 3.3, we pointed out that this line minimizes the sum of squared deviations from the line as defined in Expression (3.6). We can use this same expression as the basis for a measure of dispersion of the data about the regression line.

Definition	The **standard error of estimate** ($\sigma_{Y \cdot X}$) is the standard deviation of the predictive errors measured with reference to the least squares regression line. In symbols, $$\blacksquare \qquad \sigma_{Y \cdot X} = +\sqrt{\frac{\sum [(Y - Y_C)^2]}{N}}. \qquad \blacksquare \qquad (3.9)$$

When we had only the 20 values of Y at our disposal, we measured deviations from the fixed mean ($\mu_Y = 9.9$) to get the standard deviation. Now, in Figure 3.3, we can consider the least squares regression line to be a *moving* mean, the value of which depends on the number of calls (X) an agent makes. When we calculate the root mean square of the sales deviations from this line, we get the standard error of estimate. The standard error of estimate is the special standard deviation that applies in least squares regression.

The purpose of Table 3.11 is to find the sum of squared deviations from the regression line $[\sum(Y - Y_c)^2]$ for our insurance example. The two columns on the left repeat the 20 bivariate observations. Each value in the third column is found by substituting the value of X for that observation into the least squares regression equation,

Table 3.11   Calculation of Sums Required for the Standard Error of Estimate

X	Y	Y_c	$(Y - Y_c)$	$(Y - Y_c)^2$
26	11	11.614	−0.614	0.377
13	7	4.963	2.037	4.149
21	8	9.056	−1.056	1.115
37	20	17.241	2.759	7.612
17	9	7.010	1.990	3.960
20	12	8.544	3.456	11.944
17	4	7.010	−3.010	9.060
28	16	12.637	3.363	11.310
28	11	12.637	−1.637	2.680
6	2	1.382	0.618	0.382
23	11	10.079	0.921	0.848
25	7	11.102	−4.102	16.826
38	18	17.752	0.248	0.062
33	14	15.195	−1.195	1.428
12	2	4.452	−2.452	6.012
30	15	13.660	1.340	1.796
30	10	13.660	−3.660	13.396
10	4	3.429	0.571	0.326
18	7	7.521	−0.521	0.271
21	10	9.056	0.944	0.891
453	198	198.000	0	94.444

Equation (3.4). For example, in the first observation the agent made 26 calls ($X = 26$). The regression line approximation is 11.614 sales [$Y_c = -1.68683 + 0.51156(26)$]. The fourth column shows the deviations of actual sales from regression line approximations, and the fifth column is the square of these deviations.

We can substitute 94.44 from Table 3.11 into Equation (3.9) to find the standard error of estimate.

$$\sigma_{Y \cdot X} = +\sqrt{\frac{94.44}{20}} = +\sqrt{4.7222},$$

or

$$\sigma_{Y \cdot X} = 2.17 \text{ sales.}$$

The approach taken in Table 3.11 to find the standard error of estimate works best when values of $(Y - Y_c)$ are small integers. Otherwise, as in the insurance example, we must carry along several decimal places to avoid serious rounding errors. In the latter situation, an alternative calculation technique is to find the sum of squared deviations in the numerator of Equation (3.9) from

$$\sum[(Y - Y_c)^2] = \sum(Y^2) - \alpha \cdot \sum Y - \beta \cdot \sum(XY), \tag{3.10}$$

or, for our example,

$$\sum [(Y - Y_C)^2] = 2440 - (-1.68683)(198) - 0.51156(5238)$$

$$= 94.44.$$

This avoids having to find the individual values of Y_C for each observation and keeps rounding errors to a minimum.

Bivariate Prediction

Suppose we are told that one of the 20 insurance agents in our statistical population will be selected at random to state the number of calls he made during the experiment. Given this information together with the least squares regression equation and the standard error of estimate, we must estimate the number of sales the agent made.

To make a point estimate for the selected agent's sales, we will substitute this number of calls (X) in the least squares regression equation $[Y_C = -1.68683 + 0.51156(X)]$ and use the resulting value of Y_C for our estimate. For example, if the agent made 30 calls, we find that the associated value of Y_C is 13.67 sales, which we would round to 14 for our prediction.

To make an interval estimate for which we can have at least 75 percent assurance of being correct, we would take advantage of the fact that Chebyshev's theorem also applies to the standard error of estimate for a bivariate population. We can therefore add and subtract two standard errors of estimate $(\sigma_{Y \cdot x})$ to and from our point estimation. For an agent who made 30 calls, the lower and upper limits would be

$$Y_C - 2\sigma_{Y \cdot \bar{x}} = 13.67 - 2(2.17), \quad \text{or } 9.33 \text{ sales}$$

and

$$Y_C + 2\sigma_{Y \cdot \bar{x}} = 13.67 + 4.34, \quad \text{or } 18.01 \text{ sales.}$$

This interval is $18.01 - 9.33$, or 8.68 sales. The lower and upper limits of this interval are plotted above an X value of 30 calls on the scattergram in Figure 3.6. The least

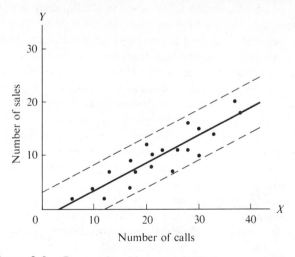

Figure 3.6   Regression Line and 0.75 Assurance Band

squares regression line and similar intervals for all other X values are also shown. The entire set of intervals forms the band shown between the dashed lines. Chebyshev's theorem assures us that no less than 75 percent of the observations must lie in this band. In fact, all the observations do.

The width of any interval in Figure 3.6 is only 8.7 sales, to one decimal place. The comparable interval in the univariate case discussed earlier had a width of 19.6 sales. Our knowledge of how the number of calls is related to sales permits us to estimate with much greater precision. The sales observations are concentrated about the regression line to a much greater extent than they are about their own mean. We can state the general conclusion as follows:

Statement	The dispersion of actual observations of a variable from predicted values can be reduced by finding another variable to which the original variable is functionally related. The stronger the relationship, the greater the reduction in dispersion that will result.

To illustrate the final point in the statement, we can refer to the perfect positive linear relationship in Figure 3.4(b). For a statistical population, a knowledge of the value of the independent variable (X) for any observation in the set would allow us to estimate the associated value of the dependent variable (Y) *without error*. On the other hand, if we had only the set of values of the dependent variable, dispersion would be present and we could not estimate without error.

3.2

3.3   Coefficients of Determination and Correlation

We have just seen that dispersion in the values of a variable causes trouble when we want to predict a value of that variable. The smaller the dispersion, the smaller we can expect the error in our prediction to be when it is compared with the actual value observed. Suppose that, in the extreme case, we know there is no dispersion in a certain statistical population and we know the value of the mean. Then we can predict that the value of a random observation from that population will be equal to the mean and we can be absolutely certain that there will be no error in our prediction.

Only on very rare occasions do we encounter a statistical population in which all values are the same. In the typical case there is enough dispersion to make our predictions far less reliable than we would wish. We sometimes can reduce the expected error appreciably by finding a least squares regression function which relates the variable (now dependent) to an independent variable. The regression relationship must satisfy two conditions to reduce the expected error in predictions. First, the dispersion of the dependent variable about the regression function must be appreciably less than its dispersion about its own mean. Second, we must be able to determine the value of the independent variable for a given unit of observation so that we can substitute it into the regression function to obtain our prediction. In the insurance example, we were able to predict a randomly selected agent's sales with much less expected error when we knew the number of calls he made.

A measure of the *relative* strength of any regression function can be very useful. Such a measure tells us how much better we can predict with the function than without it, and it helps us decide which of several alternative functions is preferable. As we shall see, a measure of relative strength also makes it possible to compare regression functions not measured in the same units. The remainder of this section describes two such measures: the coefficient of determination and the coefficient of correlation.

The Coefficient of Determination

Returning to our insurance example, we can begin by focusing on the variance in sales (σ_Y^2). This is a measure of the dispersion we must overcome in order to predict without error. If we can find the portion of this variance that is removed by a particular regression function, that portion will be a direct measure of the strength of the function. It happens that such a measure is readily available. It is called the *coefficient of determination*, for which the symbol is ρ^2 (the Greek letter rho).

We know that the variance in sales (σ_Y^2) for our example is based on deviations of individual sales from the mean $(Y - \mu_Y)$. We also know that after finding the least squares linear relationship between sales (Y) and number of calls (X), a measure of the dispersion remaining is the square of the standard error of estimate $(\sigma_{Y \cdot X}^2)$. This measure is based on deviations of individual sales from the least squares regression line $(Y - Y_C)$. It appears that the deviation of a given sales observation from the mean sales can be expressed as the sum of two other deviations, namely,

$$(Y - \mu_Y) = (Y - Y_C) + (Y_C - \mu_Y). \tag{3.11}$$

We already know that $(Y - Y_C)$ is the deviation from regression. The other component $(Y_C - \mu_Y)$ is nothing more than the deviation of the predicted number of sales (Y_C) from the mean number of sales (μ_Y). This deviation $(Y_C - \mu_Y)$ is the component *explained by linear regression*. The entire deviation of a Y value from the mean of the Y values can be expressed as the sum of a deviation explained by regression and of a remaining unexplained deviation from regression.

For our example, consider the observation (37, 20) in the fourth row from the top in Table 3.11. The value of Y is 20 sales. We know also that the mean number of sales (μ_Y) is 9.9. In Table 3.11 we see that the prediction (Y_C) based upon least squares regression with the number of calls (X) is 17.241 sales. When we substitute into Equation (3.11), the result is

$$(20 - 9.9) = (20 - 17.241) + (17.241 - 9.9),$$

or

$$10.1 = 2.759 + 7.341.$$

The entire deviation from the mean Y value is 10.1. Of this, 7.341 is accounted for by regression with sales, and 2.759 remains unexplained. These three types of deviation for this observation and for all 20 observations in our example are illustrated in Figure 3.7.

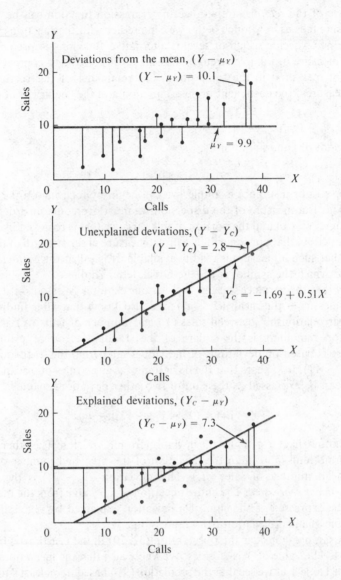

Figure 3.7   Components of Variation

Table 3.12 presents the numerical values of the three types of deviation described in Equation (3.11). The two columns to the left in Table 3.12 repeat the 20 observations for our example. The third column shows the sales predictions. The next three columns show the total deviations, the unexplained deviations, and the explained deviations. We want to concentrate upon the three columns on the right in Table 3.12. They contain the squares of the three types of deviation described in Equation (3.11). Now look at the totals of these columns. Note that

$$479.8 = 94.4 + 385.4.$$

Table 3.12   Components of Variation and Their Squares

X	Y	Y_c	$Y - \mu_Y$	$Y - Y_c$	$Y_c - \mu_Y$	$(Y - \mu_Y)^2$	$(Y - Y_c)^2$	$(Y_c - \mu_Y)^2$
26	11	11.614	1.1	−0.614	1.714	1.21	0.377	2.938
13	7	4.963	−2.9	2.037	−4.937	8.41	4.149	24.374
21	8	9.056	−1.9	−1.056	−0.844	3.61	1.115	0.712
37	20	17.241	10.1	2.759	7.341	102.01	7.612	53.890
17	9	7.010	−0.9	1.990	−2.890	0.81	3.960	8.352
20	12	8.544	2.1	3.456	−1.356	4.41	11.944	1.839
17	4	7.010	−5.9	−3.010	−2.890	34.81	9.060	8.352
28	16	12.637	6.1	3.363	2.737	37.21	11.310	7.491
28	11	12.637	1.1	−1.637	2.737	1.21	2.680	7.491
6	2	1.382	−7.9	0.618	−8.518	62.41	0.382	72.556
23	11	10.079	1.1	0.921	0.179	1.21	0.848	0.032
25	7	11.102	−2.9	−4.102	1.202	8.41	16.826	1.445
38	18	17.752	8.1	0.248	7.852	65.61	0.062	61.654
33	14	15.195	4.1	−1.195	5.295	16.81	1.428	28.037
12	2	4.452	−7.9	−2.452	−5.448	62.41	6.012	29.681
30	15	13.660	5.1	1.340	3.760	26.01	1.796	14.138
30	10	13.660	0.1	−3.660	3.760	0.01	13.396	14.138
10	4	3.429	−5.9	0.571	−6.471	34.81	0.326	41.874
18	7	7.521	−2.9	−0.521	−2.379	8.41	0.271	5.660
21	10	9.056	0.1	0.944	−0.844	0.01	0.891	0.712
453	198	198.000	0	0	0	479.80	94.444	385.366

$$\mu_Y = \frac{198}{20} = 9.90 \text{ sales} \qquad Y_c = -1.68683 + 0.51156X$$

That is, for this set of 20 observations, the sum of squared total deviations is equal to the sum of squared unexplained deviations added to the sum of squared explained deviations. The remarkable fact is:

Statement | It can be shown in general that
$$\sum [(Y - \mu_Y)^2] = \sum [(Y - Y_c)^2] + \sum [(Y_c - \mu_Y)^2]. \quad (3.12)$$
The sum of squared deviations from the mean is the sum of squared deviations from the least squares regression line added to the sum of squared deviations of predictions.

Suppose that we divide Equation (3.12) through by N, the number of observations. The quantity on the left of Equation (3.12) divided by N is the Y variance (σ_Y^2). The quantity just to the right of the equal sign divided by N is the squared standard error of estimate ($\sigma_{Y.X}^2$), or the variance of Y values about the regression line. This is the unexplained variance remaining in the set of Y values after regression. Finally, the quantity on the far right divided by N is the variance of the predictions about the mean. This quantity is, therefore, the variance explained by linear regression, and we will notate this variance as σ_R^2.

Statement

$$\frac{\sum (Y - \mu_Y)^2}{N} = \frac{\sum (Y - Y_C)^2}{N} + \frac{\sum (Y_C - \mu_Y)^2}{N},$$

or

$$\sigma_Y^2 = \sigma_{Y \cdot X}^2 + \sigma_R^2. \tag{3.13}$$

The total variance in the observations of a dependent variable is the sum of the unexplained variance and the variance explained by linear regression.

For our insurance example,

$$\sigma_Y^2 = \sigma_{Y \cdot X}^2 + \sigma_R^2,$$

or

$$\frac{479.8}{20} = \frac{94.444}{20} + \frac{385.366}{20},$$

or

$$23.99 = 4.722 + 19.268,$$

where 23.99 is the Y variance, 4.722 is the unexplained variance, and 19.268 is the explained variance. As a check, we see that the square root of 23.99 is 4.9 sales, the same value of the standard deviation we found on page 73, and the square root of 4.722 is 2.17, the standard error of estimate found on page 74.

Recall that our task in this section is to develop a measure of the relative strength of a regression function by finding the *portion* of total variance in observations of a dependent variable that is accounted for by regression. With Equation (3.13) we can complete this task. We can divide through by the entire Y variance (σ_Y^2) so that each quantity in the right member of the equation will be expressed as a portion of that total variance:

$$1 = \frac{\sigma_{Y \cdot X}^2}{\sigma_Y^2} + \frac{\sigma_R^2}{\sigma_Y^2}. \tag{3.14}$$

The fraction just to the right of the equal sign in Equation (3.14) is the portion of total variance left unexplained by regression, and the fraction to the extreme right is the portion of total variance explained by regression. It is this latter fraction we want, because it measures the strength of regression. This fraction is the coefficient of determination (ρ^2). If we substitute and then solve Equation (3.14), we have the following definition.

Definition

$$\rho^2 = \frac{\sigma_R^2}{\sigma_Y^2}, \tag{3.15a}$$

or

$$\rho^2 = 1 - \frac{\sigma_{Y \cdot X}^2}{\sigma_Y^2}. \tag{3.15b}$$

The **coefficient of determination** (ρ^2) is the portion of variance (σ_R^2/σ_Y^2) in the observations of the dependent variable (Y) that is explained by least squares linear regression with an independent variable. Alternatively, the coefficient of determination is the remainder when the portion of variance unexplained by linear regression ($\sigma_{Y \cdot X}^2/\sigma_Y^2$) is subtracted from the whole.

For our insurance example we know already that the total Y variance (σ_Y^2) is 23.99, the variance about the regression line ($\sigma_{Y \cdot X}^2$) is 4.722, and the explained variance (σ_R^2) is 19.268. Consequently, the portion of Y variance explained by linear regression is 19.268 divided by 23.99, or 0.803. This is the value of the coefficient of determination (ρ^2). It tells us that just over 80 percent of the original variance in sales is explained by the observed differences in number of calls made. We could also have arrived at this result as follows:

$$\rho^2 = 1 - \frac{\sigma_{Y \cdot X}^2}{\sigma_Y^2},$$

$$\rho^2 = 1 - \frac{4.722}{23.99}, \quad \text{or } 0.803.$$

As another example, for the perfect regression relationship in Figure 3.4(b), the coefficient of determination (ρ^2) must be 1. All the variance in observations of the dependent variable is accounted for by the relationship of these observations with observations of the independent variable.

In practice, we seldom use the procedure shown in Table 3.12 to find the coefficient of determination. This approach entails finding all the predicted values (Y_C) and all the deviations and squares of all deviations, as well as the sums of the latter. Usually calculations must be carried to many decimal places to avoid sizable errors from rounding. Instead, we can find the variance of the observations of the dependent variable (σ_Y^2) by using Equations (1.2) and (1.3). Then we can find the variance about regression ($\sigma_{Y \cdot X}^2$) by using Equations (3.7) and (3.10). Finally, we find the coefficient of determination by using Equation (3.15b).

The coefficient of determination is a pure ratio [see Equation (3.15a)], and as such it can be used to compare regression functions measured in different units. We chose to use the *number of sales* as the dependent variable for our insurance example, and found 80.3 percent of the variance in the observations of this variable to be accounted for by least squares regression with the number of calls. Had we chosen to measure the dependent variable in *dollars*, we might have found that only, say, 52 percent of the variance in sales *dollars* is accounted for by regression with the number of calls. We cannot use the standard error of estimate ($\sigma_{Y \cdot X}$) to compare results, because it is expressed in the same dimension as is the dependent variable. In the example we have been discussing, the standard error of estimate is 2.17 sales. If the dependent variable were measured in dollars, the standard error of estimate might be something like $337. There is no basis for comparing the two standard errors of estimate in such a circumstance. Instead, the coefficients of determination provide a proper comparison.

The Coefficient of Correlation

The coefficient of determination is a readily understood measure of the strength of regression. The coefficient of correlation can be defined in terms of the cofficient of determination:

Definition	The **Pearson coefficient of correlation** (ρ) is the square root of the coefficient of determination. For linear regression based on bivariate observations, the coefficient of correlation is given the same algebraic sign as the slope (β) of the regression line. In symbols, $$\rho = \pm\sqrt{\rho^2}, \tag{3.16a}$$ or $$\rho = \beta\frac{\sigma_X}{\sigma_Y}, \tag{3.16b}$$ where σ_X and σ_Y are the standard deviations of the observations of the independent and dependent variables.

To aid comprehension, we developed the coefficient of determination (ρ^2) first. In practice, Equation (3.16b) is the usual source of both the coefficients of correlation and determination. In addition, since ρ takes the sign of β, it provides information that ρ^2 does not. This coefficient is named in honor of Karl Pearson, and the name is retained to distinguish this coefficient from the other types of correlation coefficient.

For the insurance example, we found that the coefficient of determination (ρ^2) is 0.8032 and that the slope of the regression line (β) is $+0.51156$. From Equation (3.16a), then, the correlation coefficient is $\sqrt{0.8032}$, or $+0.896$, where the positive sign is chosen to match the sign of β. Alternatively, from the totals in Table 3.9 we find that $\sum X^2$ is 11,733 and $\sum X$ is 453. Then, from Equation (1.3),

$$\sum (X - \mu_X)^2 = 11,733 - \frac{453^2}{20}$$

and

$$\sigma_X^2 = 73.6275,$$

from which

$$\sigma_X = 8.581 \text{ calls.}$$

We have already found σ_Y to be $\sqrt{23.99}$, or 4.898. Then from Equation (3.16b),

$$\rho = +0.51156\frac{8.581}{4.898}, \quad \text{or} +0.896,$$

as before. Since σ_X and σ_Y can never be negative, it is apparent from Equation (3.16b) that ρ must have the same sign as β.

We can observe that the coefficient of determination (ρ^2) may take on values from 0 to 1 because the ratio of explained variance to total variance must lie in this range. As a consequence of Equation (3.16a), the absolute value of the correlation coefficient

(ρ) must also lie between 0 and 1, inclusive. But since the correlation coefficient can be negative, it can range from -1 through 0 to $+1$ on the real-number line. Values of either $+1$ or -1 for the correlation coefficient indicate a perfect linear relationship in the sense that all the variance in the dependent observations is explained by that relationship. In the former case the regression line has a positive slope, and in the latter case it has a negative slope. If and only if the coefficient of determination (ρ^2) is 0, then the correlation coefficient must also be 0. Only when the coefficient of determination (ρ^2) is either 1 or 0 will the coefficient of correlation (ρ) have the same absolute value. When ρ^2 is greater than 0 but less than 1, the absolute value of ρ will be greater than that of ρ^2. For instance, when ρ^2 is 0.64, then $|\rho|$ is $\sqrt{0.64}$, or $|0.8|$. Hence, a correlation of either $+0.8$ or -0.8 indicates that the accompanying linear relationship explains 64 per cent of the variance in the dependent observations.

3.3

Summary

The standard error of estimate ($\sigma_{Y \cdot X}$) is the special standard deviation that applies to least squares regression. It is the standard deviation of the predictive errors with respect to the regression line. When a variable is strongly dependent upon another variable, the standard error of estimate will be much smaller than the standard deviation of the observations of the dependent variable about their own mean. In this case, given knowledge of the independent variable, we can predict with much greater precision than we can without that knowledge.

The total variance in a set of observations of a variable is the sum of (1) the variance explained by the least squares linear relationship with another variable, and (2) the remaining variance about the relationship. The portion of total variance explained by the relationship is the coefficient of determination. The square root of the coefficient of determination is the Pearson coefficient of correlation. Both are relative measures and can be used to compare the effectiveness of regression relationships measured in different units.

See Self-Correcting Exercises 3.B

Exercises

1. An exclusive clothing store caters to only a few customers, each of whom is very rich. The owners of the store want a univariate prediction of the range in amount of their next sale. If the mean sale is $2000 and the standard deviation of sales is $500, within what interval will their next sale fall with at least an 8/9 level of assurance?

2. An accountant wants to estimate the total weekly operating cost of a machine from the number of items produced on it during the week. He calculates the following values:

   Slope of regression line $= 0.5$,
   Y intercept of regression line $= \$5500$.

   If the machine produces 2000 items in a week, what is the bivariate point estimate of total weekly cost of operation for the machine?

3. A stock market buff is trying to predict the price of Acme common stock by using quarterly dividend as an independent variable. He has found the least squares regression line and has used it to predict the five stock values given to the right below. Use this and the other data given to find the standard error of estimate.

Quarterly dividend	Actual value of Acme common	Predicted value of Acme common
$.20	$30	$31.20
.15	25	25.80
.22	35	33.40
.06	17	16.10
.10	20	20.50

4. A local restaurant owner wants to be able to estimate the cost per person when he is serving large groups of people. He searches his records and finds the number of people served (X) and cost per person (Y) for several large dinners. Next, he calculates the following values:

$$\text{Slope of least squares regression line} = -0.01,$$
$$Y \text{ intercept of least squares line} = \$4,$$
$$\text{Standard error of estimate} = \$0.20.$$

Find the limits of an interval prediction for the cost per person of serving 100 people. The owner requires at least 0.99 assurance that the cost figure lies within the interval.

5. The variance of the dependent variable, σ_Y^2, in a linear regression problem is found to be 16. If the sum of the squared deviations from the regression line, $\sum (Y_c - \mu_Y)^2$, is 2400 for the 200 observations, what is the coefficient of determination?

6. A researcher wants to know which of two straight lines is the better fit for the same set of data. All else being equal, we can determine which one is better by examining the values given below:

	Line A	Line B
$\sigma_{Y \cdot X}$	10 inches	200 pounds
σ_Y	20 inches	600 pounds

Which line is better and why?

7. A least squares straight line for a set of twenty observations passes through the points (80, 40) and (100, 20). The first number in each pair is the value of the independent variable, and the second is the value of the dependent variable. If the standard deviation of the dependent variable is 20 and the standard deviation of the independent variable is 18, what is the value of the Pearson coefficient of correlation?

Glossary of Equations

$Y = f(X)$

The dependent variable Y is a function (unspecified) of the independent variable X.

$Y_c = \alpha + \beta X$

The functional relationship between the dependent variable Y and the independent variable X is a straight line in which α is the Y intercept and β is the slope.

$$\sum Y = \alpha \cdot N + \beta \cdot \sum X$$
$$\sum XY = \alpha \cdot \sum X + \beta \cdot \sum X^2$$

The values of α and β which satisfy these equations are the intercept and slope of the least squares regression line for a given set of bivariate observations. These equations are known as the "normal" equations of the least squares line.

$$\beta = \frac{N \cdot \sum XY - \sum X \cdot \sum Y}{N \cdot \sum X^2 - (\sum X)^2}$$
$$\alpha = \frac{\sum Y - \beta \cdot \sum X}{N}$$

These are alternative equations for finding α and β when rounding errors would be troublesome in the preceding approach.

$$\sigma_{Y \cdot X} = +\sqrt{\frac{\sum [(Y - Y_c)^2]}{N}}$$

The standard error of estimate is the standard deviation of the observations of the dependent variable with respect to the least squares regression line.

$$\sum [(Y - Y_c)^2] = \sum Y^2 - \alpha \cdot \sum Y - \beta \cdot \sum XY$$

The sum of squared deviations of the observations of the dependent variable from the least squares regression line can be found without finding all the values of Y_c.

$$\sum [(Y - \mu_Y)^2] = \sum [(Y - Y_c)^2] + \sum [(Y_c - \mu_Y)^2]$$

The sum of squared deviations of the observations of the dependent variable from their mean is the sum of their squared deviations from the least squares regression line added to the sum of the squared deviations of predictions from the mean.

$$\sigma_Y^2 = \sigma_{Y \cdot X}^2 + \sigma_R^2$$

The variance in the observations of a dependent variable from their mean is the variance of these observations measured from the least squares regression line added to the variance explained by the regression.

$$\rho^2 = \frac{\sigma_R^2}{\sigma_Y^2}, \quad \text{or} \quad \rho^2 = 1 - \frac{\sigma_{Y \cdot X}^2}{\sigma_Y^2}$$

The coefficient of determination is the portion of the total variance in the observations of the dependent variable that is explained by least squares regression.

$$\rho = \pm\sqrt{\rho^2}, \quad \text{or} \quad \rho = \beta\frac{\sigma_X}{\sigma_Y}$$

The Pearson coefficient of correlation is the square root of the coefficient of determination. This coefficient of correlation has the same sign as does the slope of the least squares regression line and can also be expressed in terms of the slope.

Probability

Up to now we have made only brief mention of inferential procedures in statistics. Descriptive procedures concerning measurable properties of sets of observations—central tendency, variability, and association—have been our chief concern.

The next several chapters develop the foundations of inferential methods—formulating statements about *parameters* (measurable characteristics of populations) based on *statistics* (measurable characteristics of samples). Inferential methods depend heavily on *probability*, which, in turn, is based on the algebra of sets. In this chapter we begin with the language and logic of sets and go on to the meaning of probability and its relation to the important concept of a sample space. Finally, some basic probability theorems are introduced and some of their applications shown.

Basic Concepts of Sets

Definitions

1. A **set** is any well-defined collection of entities or elements within some frame of discourse.
2. The set comprising all elements within the frame of discourse is termed the **universal set**, denoted by the letter S.
3. A **subset** of S is a well-defined collection of elements, all of which belong to the universal set, S.
4. The **null** (or **empty**) **set** (denoted by $\emptyset$) is a set which contains no elements of S. It is a member of every subset in S.

In statistical work the *collections of entities*, or *sets*, we are concerned with are comprised of units of observation. These units may be persons, families, corporations, parts produced by a machine, farms, cities, city blocks, or dwelling units, and so forth. In other words, they are the objects that can be assessed quantitatively by our measurement scales.

A *universal set* appropriate to many studies might be all dwelling units within the incorporated limits of a specified city at a specified time. First, a specific definition of what is and what is not a "dwelling unit" would have to be developed. *Subsets* of interest might be dilapidated dwelling units, dwelling units with a market value exceeding $10,000, or dwelling units in the urban renewal area.

The idea of a *null*, or *empty, subset* is necessary for formal completeness. Also, to be complete, we should note that the possible subsets of S include the entire collection of S (reread Definition 3 above with this in mind).

Sets are frequently represented as shown in Figure 4.1. The elements are dots, or points. All the points within the bounded rectangle comprise S; and A and B, subsets of S, are defined by the points within the inscribed enclosures. For example, S might be the dwelling units in the 1100 block of Sesame Street,

$$S = \{a, b, c, d, \ldots, m, n, o, p\} = \{\text{all dwelling units}\},$$

A might be the subset of these that are for sale,

$$A = \{f, g, j, k\} = \{\text{dwelling units for sale}\},$$

and B might be the subset of units that are in sound condition,

$$B = \{i, m, o\} = \{\text{sound dwelling units}\}.$$

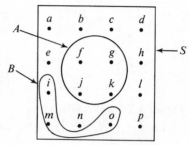

Figure 4.1   Representation of Sets

The conceivable subset of sound dwelling units for sale is, in this case, empty. There are no dwelling units in S that are both sound and for sale.

Basic Set Relationships

Three basic set relationships are union, intersection, and complement. A fairly standard symbolism is used to indicate these relationships.

Definitions

1. If A and B are subsets of a universal set, S, then the **union** of A and B includes all elements in S that are elements of A or B (or both). In symbols,

   $$A \cup B = \text{the union of } A \text{ and } B.$$

2. The **intersection** of two subsets, A and B, in S includes all the elements that are common to both A and B. In symbols,

   $$A \cap B = \text{the intersection of } A \text{ and } B.$$

3. The **complement** of a subset A in S is comprised of all elements in S that are not included in A. In symbols,

   $$A' = \text{complement of } A.$$

In the example of Figure 4.1, the union of A and B consists of all the dwelling units that are for sale or sound (or both). The intersection of A and B includes all the dwelling units that are both for sale and sound (an empty set in this case). The complement of A includes all the dwelling units not for sale. The set $(A \cup B)'$ includes the dwelling units that are neither for sale nor sound. This subset has no elements in either A or B. It could be written $A' \cap B'$.

Further Set Relations

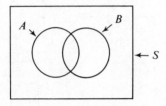

Figure 4.2   Venn Diagram for Two Subsets

A general way of representing subsets in a universal set is by means of a Venn diagram such as the one shown in Figure 4.2. This is a Venn diagram for two subsets. The circular regions representing subset A and subset B are drawn with an overlap, to suggest that the intersection, composed of elements in S that belong to both A and B, may not be empty. If $A \cap B = \emptyset$ (the empty set), then the number of elements in the union will be the sum of the number of elements in A and the number of elements in B. This leads to a definition.

Definition | If two subsets have no elements in common, they are said to be **mutually exclusive**. Subsets A and B are mutually exclusive if $A \cap B = \emptyset$.

We are now in a position to give two rules for counting elements in subsets.

Statements

1. The number of elements in the union of A and B is obtained by adding the number of elements in B to the number of elements in A and subtracting out the number of elements in the intersection of A and B.

    $$N(A \cup B) = N(A) + N(B) - N(A \cap B).$$

2. If A and B are mutually exclusive subsets, the number of elements in the union of A and B is the sum of the number of elements in A and the number of elements in B.

    $$N(A \cup B) = N(A) + N(B) \quad \text{if} \quad A \cap B = \emptyset.$$

Venn diagrams help us visualize relations among subsets. Figure 4.3 shows three subsets of a universal set. Let us return to our example of dwelling units and suppose that

$A = \{\text{sound dwelling units}\}$,

$B = \{\text{dwelling units for rent}\}$,

$C = \{\text{dwelling units for sale}\}$.

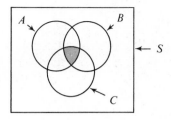

The union of these three sets can be written in the following three ways.

$$A \cup B \cup C = (A \cup B) \cup C = A \cup (B \cup C).$$

In other words, the union relationship is associative. Thus, it makes no difference whether we say

Figure 4.3   Venn Diagram for Three Subsets

(sound or for rent or for sale),

(sound or for rent) or (for sale),

or

(sound) or (for rent or for sale).

In any case, an element in S that is a member of any one (or more) of the subsets belongs to the union $A \cup B \cup C$. This union includes any element within the region circumscribed by the three overlapping circles.

Consider the intersection $A \cap B \cap C$. This is represented by the shaded region, which is within all three of the circles. In the Venn diagram one can see that

$$A \cap B \cap C = (A \cap B) \cap C = A \cap (B \cap C).$$

In words, sound units for rent that are also for sale are the same as sound units that are for rent as well as for sale, and both of these correspond to units that are sound and for rent and for sale. Thus the relation of intersection of sets is also associative.

Two other relations among subsets can be stated as follows.

$$A \cap (B \cup C) = (A \cap B) \cup (A \cap C),$$
$$A \cup (B \cap C) = (A \cup B) \cap (A \cup C).$$

These equations state two distributive laws that apply to set relations. In our example the left side of the first equation can be translated as "sound units that are for rent or for sale." The equation says this set is logically the same as "sound units for rent or sound units for sale." An urban renewal agency might say this subset of dwelling units constitutes the housing available for relocating families.

Now consider the subset

$$C \cup (A \cap B) = (C \cup A) \cap (C \cup B).$$

This is the subset of dwelling units that are for sale or that are sound and for rent. The equation says this is logically equivalent to units that are for sale or are sound and units that are for sale or for rent. This is not as immediately evident as the earlier relation, but Venn diagrams can show us the equivalence. In (a) of Figure 4.4 we see the region for subset C shaded horizontally and the region for $A \cap B$ shaded vertically. The subset $C \cup (A \cap B)$ corresponds to the entire shaded region. In (b) of the figure we see $C \cup A$ shaded horizontally and $C \cup B$ shaded vertically. The subset $(C \cup A) \cap (C \cup B)$ is represented by the intersection, which shows up as cross-hatched. This is clearly the same as the total shaded region in (a).

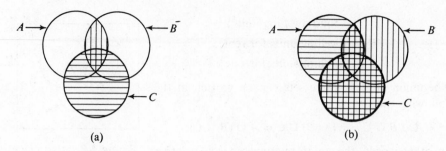

Figure 4.4   Equivalence of Two Set Statements

If the urban renewal agency could purchase and repair unsound units that are for sale, the subset $C \cup (A \cap B)$ might constitute the housing inventory that could be used for relocation.

Probability

4.1   We are now in a position to define probability.†

Definition

> **Probability** is an assignment of numbers to the subsets E_i (where $i = 1, 2, \ldots, j, \ldots, n$) of a universal set S satisfying the following conditions.
>
> 1. $0 \le P(E_i) \le 1$.
> 2. $P(S) = 1$.
> 3. $P(E_i \cup E_j) = P(E_i) + P(E_j) - P(E_i \cap E_j)$.

The first property says that each subset is assigned a number (probability) between zero and one. By the second property, the number (probability) assigned to the universal set (remember it is a subset of itself) is unity. The third property tells us that the total probability assigned to the union of two subsets is the sum of the probabilities assigned to the two subsets minus the probability assigned to the intersection of the subsets. This third property of probability is often called the *general additive property*.

Note that the definition of probability says nothing about how probabilities are used. Mathematically speaking, probability is nothing more nor less than a measure, associated with sets and subsets, that has the three stated properties. Probabilities are used in connection with events subject to uncertainty. Some examples are the outcome of a proposed coin toss (heads or tails), the weather tomorrow (sunshine, cloudy, or rain), the value of the Dow Jones closing stock average next week, the number of defective parts in a bin of untested parts.

When uncertainty prevails, we usually recognize a set of mutually exclusive and exhaustive outcomes that are of interest to us. Mutually exclusive outcomes are

† Many students find probability difficult. We remind you that the learning supplements may be helpful.

outcomes that cannot occur together. Exhaustive means that the set of outcomes is complete, in that one of the outcomes will occur. In an example of change of the price of a stock over the next week we might have

$$S = \{E_1, E_2, E_3\},$$

Change in price = {down, unchanged, up}.

Two probability assignments that might be made by two different security analysts are shown below.

Outcome (E_i)	Probability assigned by	
	Mr. A	Mr. B
Down (E_1)	0.3	0.1
Unchanged (E_2)	0.2	0.2
Up (E_3)	0.5	0.7
	1.0	1.0

Note that the E_i are exhaustive. Because they are mutually exclusive, their intersections $E_1 \cap E_2$, $E_2 \cap E_3$, and $E_1 \cap E_3$ are empty subsets and have zero probability. Consequently the additive relation

$$P(E_1 \cup E_2) = P(E_1) + P(E_2) - P(E_1 \cap E_2)$$

becomes

$$P(E_1 \cup E_2) = P(E_1) + P(E_2).$$

Because the subsets are exhaustive as well, $\sum P(E_i) = 1.0$.

Statement | If the subsets, E_i (where $i = 1, 2, \ldots, j, \ldots, n$) of a universal set, S, are defined in a mutually exclusive and exhaustive manner, then a proper assignment of probabilities to the subset will have the properties that

$$P(S) = \sum P(E_i) = 1.0,$$
$$P(E_i \cup E_j) = P(E_i) + P(E_j), \quad \text{for all } i \neq j.$$

Now that we have an assignment of numbers called probabilities to subsets in S having to do with an uncertain event, let us talk about what the probabilities are supposed to measure. There are two traditional answers that correspond to two schools of thought about what probability means. Those of one school look upon probability as the long-run relative frequency of a given outcome among a set of possible outcomes. Others view probability as a measure of degree of belief in outcomes. Each outcome among a set of possible outcomes can be assigned a degree of belief (probability) that the particular outcome will occur.

Relative Frequency

In some situations uncertainty involves a set of underlying circumstances that could be repeated over and over again. One can toss a coin or roll a die again and again in essentially the same manner. One can think of predicting the next day's weather on all days when meteorological conditions are essentially as they are today. One can think of repeatedly sampling a bin of parts and recording the number of defective parts out of each ten parts drawn. Such conceivably repeatable sets of circumstances are called *experiments*.

The probability of an event is viewed by one school as the long-run relative frequency of that outcome among the set of all possible outcomes of an experiment. For example, recall the frequency distribution of flaws per sheet in 50 galvanized steel sheets in Table 2.1, page 28. Ten sheets were found to have no flaws. Now, imagine the experiment *select a steel sheet from among the fifty* and ask what probability should be assigned to the outcome of zero flaws in the sheet selected. As we repeated this experiment over and over we would expect the relative frequency of the outcome *zero flaws* to approach more and more closely to $10/50 = 0.2$, the relative frequency of zero flaws in the fifty sheets.

The relative frequency concept of probability tells us how we might approximate a probability. Toss a coin n times and record the relative frequency of "heads" to approximate P(heads). The idea of relative frequency, as an experiment is repeated indefinitely, means that we can never know probabilities in an objective empirical sense—but we might be able to approximate them closely by collecting data on relative frequencies from repetitions of the experiment under constant conditions. The idea of probability as long-run relative frequency is fundamental to classical statistical methods.

Degree of Belief

Another way of assigning probability is according to the betting odds one would give on the outcomes of an experiment—that is, on one's *degree of belief* in a given outcome. If an individual is willing to bet even money on "heads" on the toss of a particular coin, he regards the chances of heads equal to the chances of tails, and for him $P(H) = 1/(1 + 1) = 1/2$. Another person, who perhaps knows that a given coin has two heads, would assess $P(H)$ as 1.0. A third person, who knows the coin has both a head and a tail side but believes the tosser is able to partially control the outcome, might assess the probability of heads as 0.3. These possibilities point out that probability, as a degree of belief, is personalistic. Individuals with different beliefs assign different probabilities to the same event, and the probabilities may be determined by quite subjective factors. The events to which probabilities are assigned can be unique—that is, beliefs are not restricted to repetitive events. The degree-of-belief concept of probability has a long history and it has been adopted by people particularly interested in individual decision-making. It is a foundation of so-called *Bayesian* decision methods, which we will encounter later in this text.

4.1

Sample Spaces and Events

   Often an experiment has a possible set of outcomes that cannot be further sub-divided. For example, while the result of a die toss may be an odd or an even number, each of these outcomes (or subsets) can be further divided. Similarly, while a roulette wheel may stop on black, red, zero, or double zero, the nondivisible set of outcomes is a set of 38 positions. The basic set of nondivisible outcomes of an experiment is called a *finite sample space*. The outcomes are called the *elements* of the sample space.

   Often the elements of a sample space can be regarded as equally likely. We may agree to regard each of the six die faces or each of the 38 roulette positions as equally likely outcomes. In terms of either the relative-frequency or degree-of-belief concepts of probability, this amounts to assuming an "honest" die or roulette wheel.

   We will use lower case *e*'s to denote the elements in a sample space. In symbols a sample space with N elements would be identified as

$$S = \{e_1, e_2, e_3 \ldots e_N\},$$

where the e_i are mutually exclusive, exhaustive, and indivisible.

   An *event* is a subset of a sample space. Consider a deck of 16 cards consisting of the jack, queen, king, and ace of each of the four suits—spades, hearts, diamonds, clubs. Our experiment consists of drawing a single card from this deck. The sample space can be represented as:

S	a	b	c	d
H	e	f	g	h
D	i	j	k	l
C	m	n	o	p
	J	Q	K	A

   Various events may be of interest to us. First are the *simple events*. These correspond to the individual elements in the sample space. For example, event a is the outcome jack of spades on the draw; event b is the outcome queen of spades. *Complex events* are subsets in the sample space made up of more than one simple event. The event *draw a spade* consists of the sample points $\{a, b, c, d\}$. The event *draw a black king* (spades and clubs are black) consists of the sample points $\{c, o\}$, and so on.

   When the elements in a sample space are regarded as equally likely, the probability of an event, E, can be obtained by taking the ratio of the number of elements contained in the event to the number of elements in the sample space.

$$\blacksquare \qquad P(E) = \frac{N(E)}{N(S)}, \qquad \blacksquare \qquad (4.1)$$

where

$$N(E) = \text{number of equally likely elements in the event } (E)$$

and

$$N(S) = \text{number of equally likely elements in the sample space } (S).$$

If we think of the occurrence of the event as a success, then $P(E)$ is the ratio of the number of outcomes qualifying as a success to the total number of possible outcomes of an experiment. If we accept a probability assignment of 1/16 to each of the sample points in our card draw, we then state

$$P(\text{Spade}) = 4/16 \quad \text{and} \quad P(\text{Black King}) = 2/16.$$

The expression for $P(E)$ above is justified by the additivity of probabilities for mutually exclusive and exhaustive subsets. We could as well say, for example,

$P(\text{Spade}) = P(\text{Jack of Spades} \cup \text{Queen of Spades} \cup \text{King of Spades} \cup \text{Ace of Spades}),$

$P(\text{Spade}) = P(\text{Jack of Spades}) + P(\text{Queen of Spades}) + P(\text{King of Spades})$
$\qquad\qquad + P(\text{Ace of Spades}),$

$P(\text{Spade}) = 1/16 + 1/16 + 1/16 + 1/16 = 4/16.$

The probability of an event can always be determined by summing the probabilities of the elements in the sample space contained in the event. Some examples from our card draw are:

(1) $A = \{\text{a jack}\} = \{a, e, i, m\},$
  $B = \{\text{a queen}\} = \{b, f, j, n\},$
  $P(A \cup B) = P(a) + P(e) + P(i) + P(m) + P(b) + P(f) + P(j) + P(n) = 8/16.$

(2) $A = \{\text{a jack}\} = \{a, e, i, m\},$
  $B = \{\text{a spade}\} = \{a, b, c, d\},$
  $P(A \cap B) = P(a) = 1/16.$

(3) $A = \{\text{a jack}\} = \{a, e, i, m\},$
  $B = \{\text{not a jack}\} = A',$
  $P(B) = P(A') = 1 - [P(a) + P(e) + P(i) + P(m)] = 1 - 4/16 = 12/16.$

4.2
_____

Summary

Sets and subsets of interest in probability and statistics are definable collections of units of observation in a statistical study. Various sets of interest can be manipulated symbolically by using the set relations of union, intersection, and complement. Venn diagrams are useful for visualizing set relations and for showing whether different statements about sets are equivalent.

From a mathematical standpoint, probability is nothing more than an assignment of numbers (following certain rules) to the subsets of a universal set. Some users restrict the application of probability to the outcome of "experiments" that can conceivably be repeated over and over again. Here the probability of a particular outcome is regarded as the long-run relative frequency of that outcome over many repetitions of the experiment. Other users interpret probability as a measure of degree of belief. Under this interpretation probabilities can be assigned to nonrepetitive events, and different individuals may assign different probabilities to the same event.

A sample space is a set of mutually exclusive, exhaustive, and indivisible outcomes of an experiment. If the elements in a sample space are assigned equal probabilities, then the probability of any event (subset) in the sample space is the ratio of the number of elements in the subset (event) to the number of elements in the sample space.

See Self-Correcting Exercises 4.A

Exercises

1. Twelve undergraduate students in a university got together to discuss the year's activities for the local Simulation Society. The students are identified by $\{i : i = 1, 2, \ldots, 12\}$. Upon inquiring about the students, the meeting's organizer found that the set of business majors was $\{1, 2, 3, 4, 5, 6\}$, the set of mathematics majors was $\{7, 8, 9\}$, the set of liberal arts majors was $\{7, 8, 9, 10, 11, 12\}$, the set of upper-division students was $\{1, 4, 7, 9, 10\}$, and the set of students who were members of the society was $\{1, 3, 7, 8\}$. Use the symbols B for business, M for mathematics, L for liberal arts, U for upper-division, and S for members of the society, and list the following subsets:

$$S \cup L; \quad B \cap U; \quad U \cup S; \quad U \cap S; \quad B \cap M; \quad L \cap M'; \quad U' \cap S'.$$

2. A gathering of twenty businessmen included 12 members of the local Chamber of Commerce. All twelve were engaged in selling at either the wholesale or retail level. Three sold at both wholesale and retail, and five at retail only. For the set of members of the Chamber of Commerce, find

$$N(R), \quad N(R'), \quad N(W), \quad \text{and} \quad N(W'),$$

where R stands for selling at retail and W stands for selling at wholesale.

3. Among nonmembers of the Chamber of Commerce at the gathering in Exercise 2,

$$N(R) = 6, \quad N(W) = 4, \quad N(W') = 4, \quad N(R') = 2, \quad \text{and} \quad N(R \cap W) = 2.$$

Find $N(R \cup W)$ and $N(R' \cup W')$.

4. From the information in Exercises 2 and 3, construct a three-way table presenting the enumeration of the twenty businessmen by chamber membership and mode of sale.

5. Draw a Venn diagram for the situation of Exercises 2–4, using C (for chamber members), R, and W for your three circles. Label the four mutually exclusive overlaps, or intersections, in the diagram with the appropriate set designations and find the number of businessmen in each intersection from your table in Exercise 4.

6. Draw a Venn diagram for A, B, and C in which $P(A \cap B) = 0$. Then express first $P(C)$ and then $P(B)$ as sums of mutually exclusive intersections. Given, further, that

$$P(A' \cap B' \cap C) = 0, \quad P(A' \cap B' \cap C') = 0.2, \quad P(A \cap C) = 0.1,$$
$$P(C') = 0.6, \quad P(A) = .4, \quad \text{and} \quad P(B) = .2,$$

find $P(B \cap C)$.

7. Three students were asked to assign subjective probabilities to a set of events. The set is exhaustive. Which of the following probability assignments are not valid and why?

   a. $P(E_1) = 1.0$, $P(E_2) = 0.5$, $P(E_1 \cap E_2) = 0.5$;

   b. $P(E_1) = 0.3$, $P(E_2) = 0.5$, $P(E_1 \cap E_2) = 0.2$;

   c. $P(E_1) = 0.2$, $P(E_2) = 0.7$, $P(E_1 \cap E_2) = 0.3$.

8. Draw a Venn diagram for subsets A, B, and C showing both $A \cap C$ and $B \cap C$ as the null set. Shade in the regions for the subset $C' \cap A \cap B'$ and the subset $(B \cup C) \cap A'$. Are these sets mutually exclusive?

9. Subsets A, B, and C comprise a universal set. If it is known only that $P(A \cap B \cap C) = 0$, are the events mutually exclusive? Suppose in addition it is known that $P(A \cap B) = 0$ and $P(B \cap C) = 0$. Are the events then mutually exclusive? Draw Venn diagrams to support your answers and then reread the statement on page 91.

10. Draw a representation of a sample space of five black balls numbered 1 through 5, five red balls numbered 1 through 5, and five white balls numbered 1 through 5. Assuming equally likely elements in the sample space, circle the following subsets and find their probabilities:

    {odd $\cup$ white},   {even $\cap$ black},   {(black $\cup$ white) $\cap$ (not an integer)}.

4.3   Conditional, Joint, and Marginal Probability

The elements in a sample space correspond to the *units of observation* relevant to a statistical study. *Events* correspond to characteristics of these units that are of interest to us. In the example of flaws on 50 galvanized steel sheets, the 50 sheets can be regarded for some purposes as the elements of the sample space. We are interested in the variable characteristic *number of flaws* in any sheet observed. If our *experiment* is to select one sheet from the 50 and we are able to do that in a way so that each element (sheet) in the sample space is equally likely, then we would assign probability 1/50 to each element (sheet). The probability of the event zero flaws is then 10/50, the sum of the probabilities for the equally likely elements that belong to the subset with zero flaws.

Often, as in Chapter 3, we are interested in the association between two characteristics of the observational units. Here we shall be concerned with whether or not the characteristics tend to occur together.

An example in Chapter 3 concerned recall among persons viewing advertising in a mail order catalog. There, recall tended to be associated with position of advertisements on the upper half of a page, as opposed to the lower half.

To examine association in a probability context, we select an example involving an experiment with a simple sample space. Consider the twelve face cards in a standard playing deck. There are two one-eyed jacks and one one-eyed king, indicated by the X's below:

$$
\begin{array}{c|ccc}
S & X & 0 & 0 \\
H & X & 0 & 0 \\
D & 0 & 0 & X \\
C & 0 & 0 & 0 \\
\hline
  & J & Q & K \\
\end{array}
$$

$H_0: \dfrac{\sigma_1^2}{\sigma_2^2} = 1$

$F_{.95, 10, 12} = 2.\cancel{20}$ >6

$n_1 = 10 \qquad S_1^2 = 20$

$n_2 = 12 \qquad S_2^2 = 80$

$F_{.95, 9, 11} = 2.90$

$\alpha = .10$

$F_{.95, 12, 10} = 2.91$

$F_{.95, 11, 9} = 3.10$

H_0: ~~of~~ Cars bought are evenly split between luxury, ~~vector~~ mid-range & economy.

Test: 150 cars → 25 lux
75 mid.
50 economy.

$\frac{1250}{50} = 25$

$x = \cancel{.00}\ 105$

$\chi^2_{.95,2} = 5.99$

H_0:

Second: 150 cars → ~~40~~ 40 lux
60 mid
50 economy

$\frac{200}{50} = 4$

$\alpha = \cancel{.10}\ .05$

$\chi^2_{.95,149} = 146.57$ $\chi^2_{.45,2} = 5.99$ $\chi^2_{.90,3} = 7.81$

$\frac{1450}{50} = 29$

$\chi^2_{.95,4} = 9.49$

Tests of Association

Actual ~~Expected~~

Expected

Hair
Lih dark
13x BDd

	Light	Dark	
high	15 8	5	20
IQ med	20 8	8	10
Low	8	12	20
	25	25	50

Expected:

10	10
5	5
10	10

In the experiment *draw a card* let us regard the sample points as equally likely. The probability of a one-eyed card, E_1, is clearly 3/12. But suppose we know that the card selected was a jack. We can then restrict our probability statement to that subset of the sample space. The probability of a one-eyed card, given that the card is a jack, is 2/4. In symbols,

$$P(E_1 | J) = \frac{P(E_1 \cap J)}{P(J)},$$

$$P(E_1 | J) = \frac{2/12}{4/12} = 2/4.$$

This probability means that in a long series of draws the relative frequency of a one-eyed card *when a jack is drawn* will approach two-fourths. Alternatively we could say if a jack is drawn, our degree of belief that the card is one-eyed is two-fourths.

We can now define a conditional probability, $P(E_j | E_i)$, in general terms.

Definition

The probability of the event E_j conditional on the event E_i is the probability of the intersection $E_i \cap E_j$ divided by the probability of the event E_i.

$$\blacksquare \quad P(E_j | E_i) = \frac{P(E_i \cap E_j)}{P(E_i)}. \quad \blacksquare \qquad (4.2)$$

Let us now tabulate the sample points in our example as follows.

	Card face			
	J	*Q*	*K*	*Total*
E_1 (one-eyed)	2	0	1	3
E_2 (other)	2	4	3	9
Total	4	4	4	12

This is essentially the same as the cross-classification table in Chapter 3. Our characteristic of interest, number of eyes, is the dependent variable. The independent variable is the card face, and we are concerned with how one-eyedness differs among the faces. In Chapter 3 we used the convention of finding the percentage distributions *within each class* of the independent variable. These correspond to the conditional probabilities of E_1 and E_2 given each face (in turn). The table follows.

	J	*Q*	*K*		*J*	*Q*	*K*
E_1	2/4	0/4	1/4	:	$P(E_1|J)$	$P(E_1|Q)$	$P(E_1|K)$
E_2	2/4	4/4	3/4	:	$P(E_2|J)$	$P(E_2|Q)$	$P(E_2|K)$
	4/4	4/4	4/4				

If we construct a table of relative frequencies, using the grand total rather than the column totals, the result is a table of probabilities for the *intersections* indicated by the column and row headings, shown below.

	J	Q	K		J	Q	K
E_1	2/12	0/12	1/12	:	$P(E_1 \cap J)$	$P(E_1 \cap Q)$	$P(E_1 \cap K)$
E_2	2/12	4/12	3/12	:	$P(E_2 \cap J)$	$P(E_2 \cap Q)$	$P(E_2 \cap K)$

These are often called *joint probabilities*, because they are associated with elements in a sample space defined by the intersections of two sets of events. The subset $E_1 \cap J$, for example, means that the outcome belongs at the same time (or jointly) to the class of one-eyed cards and to the class of jacks.

If the table of joint probabilities is added across and down, the resulting totals correspond to two more sets of probabilities. These are called *marginal probabilities*, from their location on the margin of the table. The boxed numbers in the following table show the marginal probabilities for our example.

	J	Q	K			J	Q	K	
E_1	2/12	0/12	1/12	3/12	:	$P(E_1 \cap J)$	$P(E_1 \cap Q)$	$P(E_1 \cap K)$	$P(E_1)$
E_2	2/12	4/12	3/12	9/12	:	$P(E_2 \cap J)$	$P(E_2 \cap Q)$	$P(E_2 \cap K)$	$P(E_2)$
	4/12	4/12	4/12	12/12	:	$P(J)$	$P(Q)$	$P(K)$	$P(S)$

The marginal probability, $P(E_1)$, is the probability of a one-eyed card over the entire sample space (that is, with no condition attached). It can be determined by the sum of probabilities for the mutually exclusive joint ways of getting a one-eyed card. These are a one-eyed jack, a one-eyed queen (an empty set), and a one-eyed king. Similarly, the marginal probability $P(J)$ can be obtained by summing $P(E_1 \cap J)$ and $P(E_2 \cap J)$.

In our example we were concerned with one-eyedness, so we looked at the conditional probabilities of E_1 given card face and E_2 given card face. We could have the situation that we knew whether or not a card drawn was one-eyed but wanted probabilities for the faces, given this knowledge. We could obtain these directly from the entries and row totals of the tabulation of sample points or from the table of joint probabilities along with the marginal probabilities of E_1 and of E_2.

	J	Q	K		J	Q	K
E_1	2/3	0/3	1/3	:	$P(J|E_1)$	$P(Q|E_1)$	$P(K|E_1)$
E_2	2/9	4/9	3/9	:	$P(J|E_2)$	$P(Q|E_2)$	$P(K|E_2)$

To consider a further aspect of conditional, marginal, and joint probabilities let us take as our sample space the nine face cards excluding the clubs. Again we will be interested in one-eyed cards, indicated by X's in the following table.

$$
\begin{array}{c|ccc}
S & X & 0 & 0 \\
H & X & 0 & 0 \\
D & 0 & 0 & X \\
\hline
 & J & Q & K
\end{array}
$$

Consider the events tabulated in the two tables below.

	(a)			
	J	Q	K	Total
E_1	2	0	1	3
E_2	1	3	2	6
Total	3	3	3	9

	(b)			
	S	H	D	Total
E_1	1	1	1	3
E_2	2	2	2	6
Total	3	3	3	9

In Table (a) we find

$$P(E_1|J) = 2/3, \quad P(E_1|Q) = 0/3, \quad \text{and} \quad P(E_1|K) = 1/3.$$

Knowing which face was drawn makes a difference in our belief that the card is one-eyed. However, in Table (b) we find

$$P(E_1|S) = P(E_1|H) = P(E_1|D) = 1/3.$$

Knowing which suit was drawn does not affect our belief that the card was one-eyed. In Chapter 3 we would have said that one-eyedness is not associated with card suit. In probability terms we say that the two sets of events are independent. This means that our belief about one-eyedness is not affected by (is independent of) any condition of card suit. Also our belief about card suits is not affected by knowledge of one-eyedness, because

$$P(S|E_1) = P(S|E_2), \quad P(H|E_1) = P(H|E_2), \quad \text{and} \quad P(D|E_1) = P(D|E_2).$$

Notice that in Table (b) the joint probability for any cell of the table can be obtained by multiplying the marginal probability for the row in which the cell is located by the marginal probability of the column in which the cell is located. For example,

$$P(E_1 \cap S) = P(E_1) \cdot P(S) = 3/9 \cdot 3/9 = 1/9.$$

This feature does not hold in a situation like Table (a). Indeed, this feature leads to a definition of independent sets of events. Consider a set of events that is a collection of mutually exclusive and exhaustive subsets of S. Let two such subsets be

$$A_i = \{A_1, A_2, \ldots, A_m\} \quad \text{and} \quad B_j = \{B_1, B_2, \ldots, B_n\}.$$

| Definition | Two sets of events are **independent** if the joint probabilities $P(A_i \cap B_j)$ are equal to the products of the marginal probabilities $P(A_i) \cdot P(B_j)$. | 4.3 |

4.4   Probability Laws and Their Use

We can now state some fundamental laws concerning the relations among probabilities illustrated in our simple examples. These laws are useful in computing probabilities of complex events from simpler events.

The first probability law is the *general addition law*. It was given as the third mathematical property in the definition of probability.

Statement │ The probability of the union of two events is the sum of the probabilities of the two events minus the probability of their joint occurrence.

$$\blacksquare \quad P(E_i \cup E_j) = P(E_i) + P(E_j) - P(E_i \cap E_j). \quad \blacksquare \quad (4.3)$$

The second law simply restates the additive property of probability for mutually exclusive events. It is often called the *special addition law*.

Statement │ The probability of the union of two mutually exclusive events is the sum of the probabilities for the individual events.

$$\blacksquare \quad P(E_i \cup E_j) = P(E_i) + P(E_j). \quad \blacksquare \quad (4.4)$$

The third law is the *general multiplication law*. For two events it is a restatement of the definition of conditional probability.

Statement │ The probability of the joint occurrence of two events is the product of the marginal probability of the first event and the conditional probability of the second event given the first:

$$\blacksquare \quad P(E_i \cap E_j) = P(E_i) \cdot P(E_j | E_i). \quad \blacksquare \quad (4.5)$$

The fourth law is a special case of the general multiplication law. If $P(E_j | E_i) = P(E_j)$, then we have a relation for two events that we previously described in connection with independent sets of events. This relationship is called the *special multiplication law*.

Statement │ The probability of the joint occurrence of two independent events is the product of the marginal probabilities of the two events.

$$\blacksquare \quad P(E_i \cap E_j) = P(E_i) \cdot P(E_j). \quad \blacksquare \quad (4.6)$$

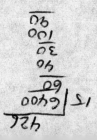

$$C = \frac{(80)^2}{15} = \frac{6400}{15} = 426\frac{2}{3}$$

$$\alpha = \cancel{.05} \ .01$$

$$SSA = \frac{1600}{5} + \frac{400}{5} + \frac{400}{5} - 427$$

$$= 320 + 80 + 80 - 427$$

$$= 480 - 427 = 53$$

$$SST = 556 - 427 = 129$$

A	B	C			
8	1	0		9	81
7	4	2		13	169
5	5	5		15	225
10	7	8		25	625
10	3	5		@18	324
40	20	20			
1600	400	400			

$$SSB = \frac{81}{3} + \frac{169}{8} + \frac{225}{3}$$

$$+ \frac{625}{3} + \frac{324}{3} - 427$$

$$= 27 + 56\frac{1}{3} + 75$$

$$+ 208\frac{1}{3} + 108 - 427$$

$$= 474\frac{2}{3} - 427$$

$$= 475 - 427 \circledast$$

$$= 48$$

6.93

	SS	df	MS	F	CV
A	53	2	26.5	4.19	$F_{.99, 2, 12}$
(R) W	76	12	6⅓		
T	129	14			

8.65

	SS	df	MS	F	CV
A	53	2	26.5	7.57	$F_{.99, 2, 8}$
B	48	4	12	3.43	$F_{.99, 4, 8}$
(R) W	28	8	3.5		
T	129	14			6.63

A useful device for visualizing many probability problems is the tree diagram, or probability tree. Consider our earlier example of the one-eyed card situation for the nine face cards, excluding clubs. We will view the relation between card faces and one-eyedness according to the sequence implied by the general multiplication law.

$$P(A_i \cap B_j) = P(A_i) \cdot P(B_j | A_i).$$

Here, A_i represents the card faces, J, Q, and K; and B_j represents one-eyed cards (E_1) and two-eyed cards (E_2). The probability tree starts with three branches leading to J, Q, and K. On these branches we place the marginal probabilities of the appropriate events, as in Figure 4.5.

The tree is completed by drawing, from each terminal of the first stage, two branches to represent the subsets of the second classification of events, E_1 and E_2. On these branches we place the conditional probabilities of E_1 and E_2 given the card face represented by the terminal of the first stage. The tree with these extensions is shown in Figure 4.6.

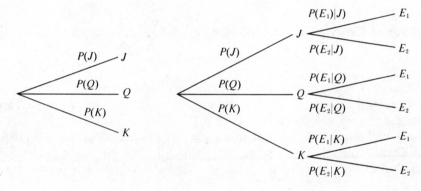

Figure 4.5   Single-Stage   **Figure 4.6**   Two-Stage Tree Diagram
Tree Diagram

We know from the general multiplication law that the products of probabilities lying on any path through the tree will give us the joint probability of the events lying on that path. The result of inserting the probabilities called for in the tree of Figure 4.6 and making the multiplications is shown in Figure 4.7.

Our special multiplication law applies to independent-event sets, such as we have in the example for one-eyedness and card suit. In this case we can place the marginal probabilities of E_1 and E_2 on the branches at the second stage, because

$$P(A_i \cap B_j) = P(A_i) \cdot P(B_j).$$

The tree appears in Figure 4.8.

The tree diagrams of Figures 4.7 and 4.8 depict the probability relationships of the general and special multiplication laws. The joint probabilities that the trees generated

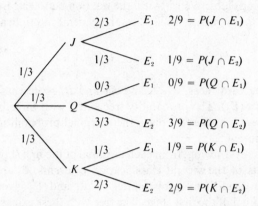

Figure 4.7   Tree Diagram Illustrating General Multiplication Law

could have been determined directly from the two tables on page 99. A glance back to the tables will confirm this for you. Our purpose was to introduce the tree diagram. We did not find any new probabilities.

The multiplication laws are more frequently used to calculate probabilities that we do not already know. We must know the probabilities that belong on the individual branches of the tree, and often this is the case. Consider a three-game playoff between

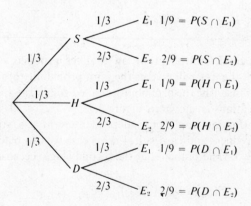

Figure 4.8   Tree Diagram Illustrating Special Multiplication Law

two teams and let A represent the event *team A wins a game* and B the event *team B wins a game*. The logical possibilities for the series can be diagrammed in the sequential tree of Figure 4.9.

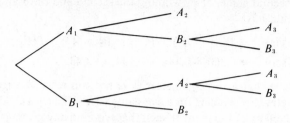

Figure 4.9   Tree Diagram for a Playoff Series

Each path along the tree represents a possible outcome for the series. For example, $A_1 B_2 A_3$ means A wins the first game, B wins the second, and A wins the third game (and the playoff!). If we have probability values for the individual events represented by the branches of this tree, we can then work out the probability of any desired path, or sequence of outcomes. An extension of the general multiplication law to K events serves us here. It is

$$P(E_1 \cap E_2 \cap E_3 \cdots \cap E_K)$$
$$= P(E_1) \cdot P(E_2|E_1) \cdot P(E_3|E_1 \cap E_2) \cdots P(E_K|E_1 \cap E_2 \cap E_3 \cdots \cap E_{K-1}).$$

If the events are independent,

$$P(E_2|E_1) = P(E_2), \qquad P(E_3|E_1 \cap E_2) = P(E_3),$$

and so forth, and we have the following multiplication theorem for K independent events.

$$P(E_1 \cap E_2 \cap E_3 \cdots \cap E_K) = P(E_1) \cdot P(E_2) \cdot P(E_3) \cdots P(E_K).$$

For example, if the probability of A winning any game is 0.6, what is the probability that A will win the series? Putting the probabilities on the relevant branches and multiplying the probabilities along each path gives the results shown in Figure 4.10.

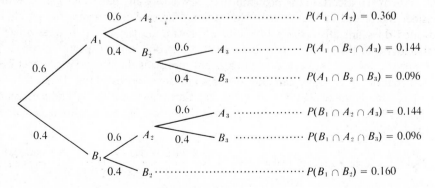

Figure 4.10   Probabilities for a Playoff Series with Independent Events

We can now apply the addition law for mutually exclusive events to find the probability that team A wins the series. Team A can win in three mutually exclusive ways. It can win the series by winning the first two games, by winning the first and third while losing the second game, or by winning the second and third after losing the first game. So we would have

$$P(A \text{ wins series}) = P(A_1 \cap A_2) + P(A_1 \cap B_2 \cap A_3) + P(B_1 \cap A_2 \cap A_3),$$
$$P(A \text{ wins series}) = 0.360 + 0.144 + 0.144 = 0.648.$$

We might conclude that a three-game series is not too reliable a test for the better team. (You might try this problem for a five-game series.) To try a conditional problem, suppose the psychology of the game is such that winning any given game increases the probability of winning the next game by 0.10. We would then have the situation shown in Figure 4.11. Now the probabilities on the branches are explicitly conditional—that is,

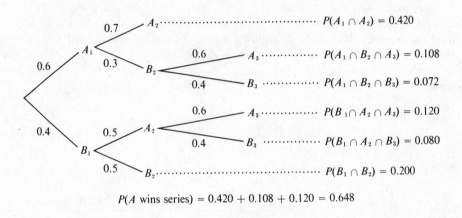

$$P(A \text{ wins series}) = 0.420 + 0.108 + 0.120 = 0.648$$

Figure 4.11   Probabilities for a Playoff Series with Dependent Events

$$P(A_2|A_1) = 0.7, \quad P(A_2|B_1) = 0.5, \quad P(A_3|A_1 \cap B_2) = 0.6, \quad P(A_3|B_1 \cap A_2) = 0.6,$$

and so forth. To ascertain the probability of A winning any particular game, we must consider the record of the series to that point. Interestingly enough, the psychology indicated does not affect the probability of A winning the series. It does affect the probability that the series lasts only two games, as one might expect.

By restricting our interest to terminals that represent the points in a subset of the sample space of interest, we can find certain conditional probabilities. In the foregoing example, what is the probability of A winning the series *given that the series lasts only two games*? In accordance with the general definition of a conditional probability, this probability is

$$P(A \text{ wins series}|\text{series is two games}) = \frac{P(A \text{ wins series} \cap \text{series is two games})}{P(\text{series is two games})}.$$

Looking at the joint probabilities for the different possible sequences for the series we see that (A wins series $\cap$ series is two games) is the event $A_1 \cap A_2$ with probability

0.420. The marginal probability for the condition (series is two games) can be found by adding the probabilities for the two mutually exclusive events that lead to this result, namely $A_1 \cap A_2$ and $B_1 \cap B_2$. So then we have

$$P(A \text{ wins series} \mid \text{series is two games}) = \frac{0.420}{0.420 + 0.200} = 0.677.$$

By contrast, we might want the probability that the series lasts two games *given that A wins the series.*

$$P(\text{series is two games} \mid A \text{ wins series}) = \frac{P(\text{series is two games} \cap A \text{ wins series})}{P(A \text{ wins series})}.$$

Here, the numerator event is again $A_1 \cap A_2$ with probability 0.420. The denominator event is satisfied by any of the mutually exclusive events

$$A_1 \cap A_2, \quad A_1 \cap B_2 \cap A_3, \quad \text{or} \quad B_1 \cap A_2 \cap A_3.$$

Finding the probabilities for these sequences from our extended tree diagram, we then have

$$P(\text{series is two games} \mid A \text{ wins series}) = \frac{0.420}{0.420 + 0.108 + 0.120} = 0.648. \qquad 4.4$$

Summary

When each element in a sample space is identified according to two classifications of events, the tabulation of elements takes the form of the cross-classification table of Chapter 3. We may then identify joint probabilities, marginal probabilities, and conditional probabilities. A joint probability is a cell count divided by the total elements in the table. A marginal probability is a total row or total column count divided by the total elements in the table. A conditional probability is a cell count divided by the total for the row (or column) in which the cell is located (depending on the condition). The two sets of events are independent if the probabilities for their intersections can be obtained by the products of appropriate marginal probabilities.

Four probability laws deal with relationships between joint, marginal, and conditional probabilities. The general multiplication law permits us to establish joint probabilities from the product of appropriate marginal and conditional probabilities. The special multiplication law applies when event sets are independent. The general addition law permits us to find the probability of a union of sets from appropriate marginal and joint probabilities. A special form of the addition law applies when the subsets in the desired union are mutually exclusive. Two subsets are mutually exclusive when their intersection has zero probability.

The relationships among joint, marginal, and conditional probabilities in a cross-classification table of elements in a sample space can be illustrated in a tree diagram. On the first set of branches are the marginal probabilities of one of the sets of events. On the second set of branches are the conditional probabilities of the second set of events given the first. The product of probabilities encountered in any route through a tree is the joint probability of the intersection of subsets represented by the route. The tree diagram is thus a visual representation of the general multiplication law for probabilities. In many problems joint probabilities are not known directly, and the appropriate multiplication law allows us to calculate them.

See Self-Correcting Exercises 4.B

Exercises

1. One hundred stocks in a portfolio were classed as either cyclical or growth stocks. Of 60 cyclical stocks, 45 had increased in price in the past month, while 15 of the 40 growth stocks had increased in price. If a stock is selected at random from the portfolio, what is the probability that it will be

   a. a cyclical stock that has not increased in price,

   b. a stock that has increased in price.

   c. Given that the stock selected has not increased in price, what is the probability that it is a cyclical stock?

2. Construct a two-way table enumerating the stocks in Exercises 1. Then find the two sets of marginal probabilities and check whether the direction of price movement is independent of type of stock.

3. A furniture salesman found that his performance (good versus bad) was independent of the weather (sunny or cloudy). The probability that he has a good day is 0.40, and the probability that the weather is sunny is 0.60. Find

   a. the probability that the weather is sunny and the salesman has a bad day,

   b. the probability that the weather is sunny when the salesman had a bad day.

4. A TV manufacturer trims his sets in either black or grey, and he can use either a walnut or a maple stain. A consumer survey showed that 20 percent of consumers desired a black trim with a walnut stain and that half of all consumers desired a black trim or a walnut stain. If it has already been decided to trim 30 percent of the sets in black, what proportion of the sets should be given a walnut stain?

5. The probability that a particular football team will run its " green " offense is 0.60 when a play begins inside its own 30-yard line and 0.25 otherwise. The probability that the fullback carries the ball given the " green " offense is 0.50, and 0.30 otherwise. If twenty percent of the team's offensive plays begin inside its own 30-yard line, what is the marginal probability that the fullback carries the ball?

6. In a game of dice the roller wins if the sum of the spots on two dice thrown total 7 or 11 on the first roll, and the nonroller wins if the sum of the spots is 2, 3, or 12 on the first roll. Find the probability that

   a. the roller wins on the first roll,

   b. the nonroller wins on the first roll,

   c. the game is undecided on the first roll.

7. John Eager is told by a prospective employer, " You are on trial for a year. One mistake is all right, but two mistakes and you're out!" John figures that there are four critical decisions to make during the year, and the probability that he will make any one of them correctly is 4/5, but he doesn't want to run more than a one-in-five chance of not lasting out the year. Assume independent events and draw a tree diagram for the process. Should John take the position?

8. A salesman figures that the probability is 0.6 that he will sell a prospect on the first call, and that on each subsequent call on a prospect who has not been sold, the probability of selling drops by 0.1. If the salesman is willing to make up to three calls on a prospect, what is the probability of selling to him? If the salesman persists beyond this, how much better can he do?

9. A production foreman is checking quality control for two processes, A and B, where process B follows process A. He knows that 11 percent of the production units going through the two processes are rejected after the completion of process B. The rejection rate for process A is 10 percent, but 80 percent of those units are corrected in process B. With the aid of a tree diagram, determine the rejection rate for process B for those production units that are acceptable after process A.

Glossary of Equations

$$P(E) = \frac{N(E)}{N(S)}$$

When the elements in a sample space are assigned equal probability, the probability of an event E can be found from the ratio of the number of elements in E to the number of elements in the sample space.

$$P(E_j \mid E_i) = \frac{P(E_i \cap E_j)}{P(E_i)}$$

The probability of the event E_j conditional on the event E_i is the probability of the intersection $E_i \cap E_j$ divided by the probability of the event E_i.

$$P(E_i \cup E_j) = P(E_i) + P(E_j) - P(E_i \cap E_j)$$

The probability of the union of two events is found by adding the probabilities of the two events and subtracting the probability of their joint occurrence (intersection).

$$P(E_i \cup E_j) = P(E_i) + P(E_j)$$

If E_i and E_j are mutually exclusive events, the probability of the union of E_i and E_j is the sum of the probabilities for the two events.

$$P(E_i \cap E_j) = P(E_i) \cdot P(E_j \mid E_i)$$

The probability of the joint occurrence (intersection) of two events is the product of the marginal probability of the first event and the conditional probability of the second given the first.

$$P(E_i \cap E_j) = P(E_i) \cdot P(E_j)$$

The probability of the joint occurrence (intersection) of two independent events is the product of the marginal probabilities of the two events.

5

Probability Distributions for Discrete Random Variables

In Chapter 4, which introduced probability and its basic uses, we saw how the concept of a set can be used to define subsets of a universal set. We then applied probabilities to the subsets of outcomes of an experiment. These subsets, called events, can be defined as conditional, joint, or marginal. The fundamental probability laws are used to find the probabilities of complex events from the probabilities of simple events. They also show the relations among marginal, conditional, and joint probabilities.

In this chapter, which extends the development of probability as a foundation for statistical inference, the idea of a random variable and its close connection with a sample space is the key element. The mean and variance are discussed as parameters of the probability distribution of a random variable, and some rules and formulas for counting elements in sample spaces are discussed. In all cases this chapter deals with discrete random variables only. Continuous random variables are reserved for Chapter 8.

5.1   Sample Space and Random Variable

A discount house handles a line of table radios. In the line are both AM and AM-FM models of radios, clock radios, and digital clock radios. Shown below are the relative frequencies of unit sales and the dollar gross margin obtained from a unit sale of each model.

	Model	Relative frequency of unit sales	Gross margin per unit
A1	AM radio	4/12	$ 8.00
A2	AM clock radio	2/12	16.00
A3	AM digital clock radio	1/12	16.00
F1	AM-FM radio	3/12	8.00
F2	AM-FM clock radio	1/12	16.00
F3	AM-FM digital clock radio	1/12	24.00

If we consider the experiment *select a unit sale*, the sample space is the six-element set that contains the different models. At the left above we see one assignment of numbers (probabilities) to the elements in this sample space. At the right we see another; these gross margin figures are numbers of concern to the merchandiser. The event *sale of a model A1 radio* means $8 to him, the event A2 means $16, and so forth. In this manner there corresponds to each element in the sample space a numerical value. This transformation of elements to numbers is the equivalent (from Chapter 1) of applying a measurement scale to units of observation.

Definition	A **random variable** is a numerical-valued function (rule) defined on the elements of a sample space.

Since our concern is with the numerical values created by our random variable, we would be interested in the probabilities of receiving $8 margin, $16 margin, and $24 margin. As a probability problem this offers no difficulty. We need only add the probabilities of the elements in the sample space that lead to the numerical outcomes of $8, $16, and $24. The answers are $P(\$8) = 7/12$, $P(\$16) = 4/12$, and $P(\$24) = 1/12$. If we let X equal the numerical outcome of the random variable, we can tabulate the results as follows:

X	$P(X)$
$ 8	7/12
16	4/12
24	1/12

This table is a probability distribution of values of a random variable. For our example it is the probability distribution of gross margin for a unit sale. To emphasize the relation of a probability distribution to the elements of a sample space, we recapitulate the process graphically in Figure 5.1.

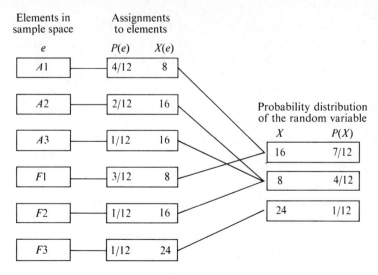

Figure 5.1   Probability Distribution of a Random Variable

Each element in the sample space has a pair of numbers assigned to it. The first is the probability assignment and the second is assigned by the random variable. We then collect the values of the random variable and their associated total probabilities. The relationship that pairs the values of the random variable with their probabilities is called the *probability distribution of the random variable*. The variable (X in this case) is often called *the random variable* (though strictly speaking that term refers to the rule of assignment rather than the numerical outcomes of the assignment).

A probability distribution of a random variable is a special kind of frequency distribution. One can say that it presents the long-run relative frequencies of values of a random variable for repeated trials of an experiment. This follows from the relative frequency concept of probability. Alternatively, we ·could say that it presents the degree of belief associated with the possible numerical outcomes of an experiment.

A random variable has a mean and a variance that we can calculate. The mean and variance of a random variable X are

$$\blacksquare \quad \mu_X = \sum [X \cdot P(X)], \quad \blacksquare \tag{5.1}$$

$$\blacksquare \quad \sigma_X{}^2 = \sum [(X - \mu_X)^2 \cdot P(X)]. \quad \blacksquare \tag{5.2}$$

For the gross margin associated with our radio example, Table 5.1 shows the mean

Table 5.1   Mean and Variance of a Random Variable

X	$P(X)$	$X \cdot P(X)$	$(X - \mu_X)^2$	$(X - \mu_X)^2 \cdot P(X)$
8	7/12	56/12	16	112/12
16	4/12	64/12	16	64/12
24	1/12	24/12	144	144/12
		144/12		320/12

$$\mu_X = 144/12 = 12 \qquad\qquad \sigma_X^2 = 320/12 = 26 \ 2/3$$

to be $12. The mean is also called the *expected value* of the random variable. Since probabilities are long-run relative frequencies, the mean can be interpreted as the average numerical outcome for a large number of repetitions of the experiment. If we made repeated selections of unit sales of the table-radio line, our average margin (per sale) would be $12. If we take a degree-of-belief view of probability, our interpretation of the mean would not have to contemplate repeated selections. We may only intend to make one selection. Nevertheless we can say that our *expectation* is a margin of $12. We know that the margin will be either $8, $16, or $24, but our belief is so distributed among these values as to produce an expectation (mean) of $12.

The experiment here can be viewed as a gamble for money outcomes. In cases like this, the mean is also called the fair (money) value of the gamble. Suppose there are twelve balls in an urn—seven with $8 written on them, four with $16, and one with $24 inscribed. Blindfolded, you have an opportunity to select one ball and receive the amount of money written on it. The mean of the probability distribution of money gain, that is, $12, is called the fair value of the gamble. Note that if you paid exactly $12 to play the game, your expected return *net* of the price paid would be zero.

Sums of Random Variables

Suppose we now consider an experiment (or sample) of two unit sales. We will be concerned with the total dollar margin for the two sales observed. How can we find the probability distribution of this random variable? One approach is to enumerate the sample space for a draw of two unit sales. The points in this sample space are the 36 different sequences of two models that can occur. These are identified in Table 5.2. This 6×6 table contains in each cell the total dollar margin that would result from a particular sequence of first and second sales.

Table 5.2   Total Dollar Margin for Two Sales

First sale \ Second sale		A-1 8	A-2 16	A-3 16	F-1 8	F-2 16	F-3 24
A-1	8	16	24	24	16	24	32
A-2	16	24	32	32	24	32	40
A-3	16	24	32	32	24	32	40
F-1	8	16	24	24	16	24	32
F-2	16	24	32	32	24	32	40
F-3	24	32	40	40	32	40	48

Now, we need to determine the probabilities associated with each point in the sample space. For independent draws we can take products of marginal probabilities. The format of Table 5.3 is convenient. We first enter the marginal probabilities and then multiply row times column marginal probabilities to obtain the joint probabilities for the sample sequences represented by each cell.

Table 5.3   Probabilities for Two Sales

First sale \ Second sale		A-1 4/12	A-2 2/12	A-3 1/12	F-1 3/12	F-2 1/12	F-3 1/12
A-1	4/12	16/144	8/144	4/144	12/144	4/144	4/144
A-2	2/12	8/144	4/144	2/144	6/144	2/144	2/144
A-3	1/12	4/144	2/144	1/144	3/144	1/144	1/144
F-1	3/12	12/144	6/144	3/144	9/144	3/144	3/144
F-2	1/12	4/144	2/144	1/144	3/144	1/144	1/144
F-3	1/12	4/144	2/144	1/144	3/144	1/144	1/144

To form the probability distribution of total margin for two sales, we need only write down (in order) the different sums that occur in Table 5.2 and find the total probability associated with each from Table 5.3. The result is shown below.

Total margin ($)	Probability
16	49/144
24	56/144
32	30/144
40	8/144
48	1/144

While the foregoing process emphasizes the elements in the sample space for selecting two unit sales, it is often unnecessarily tedious. We can find the probability distribution of total margin from two independently selected sales very easily from the probability distribution for a single sale. The process is shown in the tree diagram of Figure 5.2. By multiplying the probabilities on the branches for different routes through the tree, we obtain the probabilities for different numerical sequences of dollar margin.

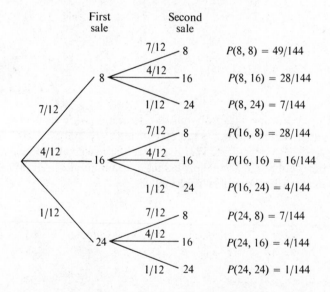

Figure 5.2   Tree Diagram for Generating Probability Distribution of a Sum

Some of these sequences yield the same total margin, and, in Table 5.4, we add their probabilities to obtain the probability distribution of total margin.

The process can also be shown in a tabular format (see Table 5.5). Place the marginal probabilities for first and second sales in the margins of a two-way table. Then multiply the appropriate row and column marginal probabilities to obtain the probabilities for the event intersections. This is appropriate because the selections are independent.

Table 5.4   Probability Distribution of Total Margin for
Two Sales

Total margin ($) $\sum X$		Probability $P(\sum X)$
16		49/144
24	28/144 + 28/144	= 56/144
32	7/144 + 16/144 + 7/144	= 30/144
40	4/144 + 4/144	= 8/144
48		1/144

Then sum the probabilities for the event intersections that lead to each different sum. For example, the sequences (24, 8), (16, 16), and (8, 24) all result in a total margin of $32. Thus,

$$P(\sum X = 32) = 7/144 + 16/144 + 7/144 = 30/144.$$

What we have found here is the probability distribution of the sum of two identically distributed and independent random variables. We say identically distributed because the probability distribution of dollar margin for the first sale is the same as that for the

Table 5.5   Tabular Method for Generating Probability
Distribution of a Sum

First sale \ Second sale		8 7/12	16 4/12	24 1/12
8	7/12	49/144	28/144	7/144
16	4/12	28/144	16/144	4/144
24	1/12	7/144	4/144	1/144

second sale. We are now ready to calculate the mean and variance of the probability distribution of our sum—the total margin. We could calculate these values from the frequency distribution of Table 5.4. Two important statements will give us the same results with less trouble.

Statement | The mean of the sum of n independent random variables is equal to the sum of their means.

$$\blacksquare \qquad \mu_{(X_1 + X_2 + \cdots + X_n)} = \mu_{X_1} + \mu_{X_2} + \cdots + \mu_{X_n}. \qquad \blacksquare \quad (5.3)$$

Here, subscripts indicate what variable we are talking about. They are necessary to clearly distinguish among the different means or expectations. For variances there is a similar statement.

Statement | The variance of the sum of n independent random variables is equal to the sum of their variances.

$$\blacksquare \qquad \sigma^2_{(X_1 + X_2 + \cdots + X_n)} = \sigma^2_{X_1} + \sigma^2_{X_2} + \cdots + \sigma^2_{X_n}. \qquad \blacksquare \quad (5.4)$$

In our case the random variable for the second sale was the same as for the first, so we would have

$$\mu_{X_1+X_2} = \mu_{X_1} + \mu_{X_2} = 12 + 12 = 24,$$

$$\sigma^2_{X_1+X_2} = \sigma^2_{X_1} + \sigma^2_{X_2} = 26\tfrac{2}{3} + 26\tfrac{2}{3} = 53\tfrac{1}{3}.$$

These are fairly potent statements because they provide two important parameters of the probability distribution of a random variable associated with a large sample space. For example, suppose there are 25 working days in a month. If there is no trend in sales over the month or systematic variation in sales by days of the week, the probability distribution of $\sum X$ for $n = 25$ sales will give the probability distribution of total dollar margin on the table-radio line for a month. We would have

$$\mu_X = 12, \qquad \sigma^2_X = 26\tfrac{2}{3}, \qquad n = 25;$$

$$\mu_{X_1+X_2+\cdots+X_{25}} = 25\mu_X = 25(12) = \$300;$$

$$\sigma^2_{X_1+X_2+\cdots+X_{25}} = 25\sigma^2_X = 25(26\tfrac{2}{3}) = 667.$$

The standard deviation of the probability distribution of total monthly margin for the table radio line would be

5.2

$$\sigma_{\Sigma X} = \sqrt{\sigma^2_{\Sigma X}} = \sqrt{667} = \$25.82.$$

Summary

A random variable is a rule of assignment of numerical values to the elements in a sample space. Since these numerical outcomes are what we often seek, we are interested in their probability distribution, called the probability distribution of the random variable. Since both the numerical value and the probabilities are originally associated with the elements in the sample space, they can be paired with each other in the form of the probability distribution of the random variable. The mean and variance of the random variable can then be found from formulas that use the probabilities.

The probability distribution of the summation of numerical outcomes for repeated trials of an experiment can be obtained in two ways. One is to enumerate the elements in the sample space for the repeated trials and then to associate the possible values of the summation with their probabilities. This can be a lengthy procedure, and a tree diagram showing the possible numerical values and their probabilities for each trial will produce the same final result. The mean and the variance of a random variable defined as the sum of n independent random variables can be found conveniently from appropriate formulas.

See Self-Correcting Exercises 5.A

Exercises

1. Consider a deck of sixteen cards consisting of the jack, queen, king, and ace of each of the four suits. Construct a representation of the sample space for the experiment *draw a card.*

2. Suppose in connection with the card draw in Exercise 1 that jack, queen, king, and ace count 1, 2, 3, and 4 points, respectively, and that an extra point is scored if a spade or heart is drawn. Assign point scores and probabilities to the sample points and find the probability distribution of the point score for a single draw.

3. Calculate the mean and variance of the probability distribution in Exercise 2.

4. Weekly demand for replacement part #062 at a supply facility has been found to be equiprobable for the number of demands 0, 1, and 2. Tabulate the sample space for total demands over a two-week period and (assuming independence) find the probability distribution of total demands for the two-week period.

5. Extend your results in Exercise 4 to find the probability distribution of total demand for a three-week period (again assuming independence).

6. Find the mean and the standard deviation of the probability distribution of total demand for a six-week period in the situation of Exercise 4.

7. In a particular contracting business the probability distribution of days lost in a year from work stoppages has a mean of 15 and a variance of 9, while the probability distribution of days lost from bad weather has a mean of 25 days and a variance of 16. Find the mean and standard deviation of the random variable *total time lost.* Discuss any assumption you may have made.

8. A strategy sometimes suggested in gambling is to bet half of your cumulated winnings on the next bet if you are ahead and to bet double the amount of your cumulated losses on the next bet if you are behind. Assume a $10 initial bet on a fair coin toss and follow this strategy through two more tosses (three tosses in all). Is it a paying strategy?

9. A real estate agency found that the probability of closing no deals on a working day was 0.6, and that the probability of closing 1, 2, 3, or 4 deals was 0.1 each. For 200 independent occurrences of this random variable, find the expected value and standard deviation of total closings.

10. A man has four bills in his wallet in denominations of $1, $5, $10, and $20. He selects one of the bills randomly and then selects a second bill without returning the first. Find the probability distribution of the total amount of money drawn.

Counting Finite Sample Spaces

5.3

It is useful to be able to count the elements in sample spaces corresponding to samples of different size. If any of N elements in an original set can be generated (or drawn) on each of the draws, the number of elements in the sample space for n draws will be N^n. This was the case in the radio-models example. In that example, $N = 6$, and there were $6^2 = 36$ different sequences for a sample of size 2. A sample of size $n = 3$ sales would generate a sample space of $6^3 = 216$ elements.

Suppose we have 50 student names on a class enrollment list. The individual names are cut out and placed in an urn. We can consider sampling from the urn *with replacement*. This means that after each draw, the name drawn is replaced prior to the next draw. Thus, any element (name) can be selected on any draw and the number of elements in the sample space is $N^n = 50^n$.

We might also sample the urn *without replacement*. This means that the element selected on any draw is not replaced before the next draw. Elements selected on a given draw cannot then be selected on succeeding draws. A little thought will verify that when $N = 50$ and sampling is without replacement, there will be 50(49) elements in the sample space for $n = 2$; 50(49)(48) elements for $n = 3$; 50(49)(48)(47) elements for $n = 4$, and so on. The reasoning here is that any one of N elements can be selected on the first draw; then only $N - 1$ are left to select from on the second draw, $N - 2$ remain for the third draw, and so forth. When sampling is without replacement, the number of elements in the sample space for a sample size of n is

$$N(N - 1)(N - 2) \cdots (N - n + 1).$$

Both of the examples above are applications of the so-called product rule for sets.

Statement	If event A can occur in $N(A)$ ways and event B can occur in $N(B)$ ways, then the number of ways in which $A \cap B$ can occur is $$N(A \cap B) = N(A) \cdot N(B).$$

We have seen how the product rule allows us to count the number of points in the sample space generated by a draw of n elements from a set of N elements. Suppose you have recordings of five musical selections that you wish to arrange on one tape. In how many ways can you program the selections? The answer is

$$N(N - 1)(N - 2) \cdots (N - n + 1), \quad \text{where } N = n = 5,$$
$$5(4)(3)(2)(1) = 120.$$

There are 120 different ways that you can program the five selections. This is called the *permutation* of five elements five at a time. If you wish to tape only three of the five recordings, the number of ways of selecting and ordering them would be the permutation of five things three at a time. Any of the five selections could be first, any of the remaining four could be second, and any of the remaining three could be placed third. The product rule would then give 5(4)(3) = 60 ways of selecting and ordering three from the five recordings.

A general formula for the number of permutations (possible ordered subsets) of n from among N elements is

$$\blacksquare \qquad {}_N P_n = \frac{N!}{(N - n)!}. \qquad \blacksquare \tag{5.5}$$

The operator "!", read "factorial," means the product of the integer indicated and all lesser integers; that is,

$$N! = N(N - 1)(N - 2) \cdots (N - N + 1)\dagger.$$

† Note, also, as a definition that $1! = 1$ and $0! = 1$.

In the example above,

$$_5P_3 = \frac{5!}{(5-3)!} = \frac{5(4)(3)(2)(1)}{2(1)} = 5(4)(3) = 60.$$

In answering some kinds of questions, order is of no concern. You might want to know how many different sets of three recordings could be selected from the five for inclusion on your tape. Here, you are concerned only with the different *combinations* of three that can be made from the five recordings. In general, the number of distinct combinations (possible subsets) of n from among N elements can be determined from

$$\blacksquare \qquad _NC_n = \frac{N!}{(N-n)!\,n!}. \qquad \blacksquare \qquad (5.6)$$

In our example,

$$_5C_3 = \frac{5!}{(5-3)!\,3!} = \frac{5(4)(3)(2)(1)}{2(1) \cdot 3(2)(1)} = 10.$$

Note in this example that there are 60 different programs when different orderings of the same set of musical selections are counted as different programs, but only 10 different programs when we do not count different orders of the same set. The relationship that prevails is

number of orders = number of sets × number of orders per set,

$$_NP_n = {_NC_n} \times {_nP_n}.$$

There are 10 different sets of selections, $_5C_3$, in our example. Each set of three selections can be arranged (or ordered) in $_3P_3 = 6$ ways. The product of 10 sets and 6 orders per set yields the 60 different orders, or permutations, of three from a set of five elements.

This can be clarified by setting out the list of combinations. Let the five musical selections be A, B, C, D, and E. The ten different programs, when order is not considered, that is, combinations, of three selections are:

$$\begin{array}{ll}
ABC, & ADE, \\
ABD, & BCD, \\
ABE, & BCE, \\
ACD, & BDE, \\
ACE, & CDE.
\end{array}$$

Each of these $_5C_3 = 10$ programs can be arranged, or ordered, in six ways. For example, the subset of selections ABC can be ordered as follows:

$$\begin{array}{ll}
ABC, & BCA, \\
ACB, & CAB, \\
BAC, & CBA.
\end{array}$$

Certain kinds of probability and counting problems that we have worked by using direct counting methods and/or the fundamental probability laws can also be worked using permutations and combinations. For example, consider the sample space for selecting two out of four recordings as a tree diagram in the manner of Chapter 4. Labeling the recordings A, B, C, and D, we would have the tree shown in Figure 5.3. Each distinct route through the tree diagram represents one of the $_4P_2 = 12$ different arrangements of two from the set of four recordings. Some of these arrangements, for example, A_1B_2 and B_1A_2, represent the same combination of selections. The number of *distinct* combinations is $_4C_2 = 6$. There are $_2P_2 = 2$ different ways of ordering each combination.

Given equal probabilities of selection at any stage, what is the probability that A and B are the two recordings selected? Using multiplication and addition rules, we obtain

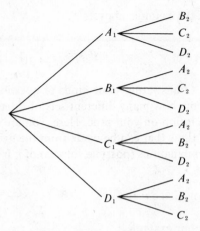

Figure 5.3   Tree Diagram Showing Selection of Two from Four Recordings

$$P[(A_1 \cap B_2) \cup (B_1 \cap A_2)] = \left(\frac{1}{4}\right) \cdot \left(\frac{1}{3}\right) + \left(\frac{1}{4}\right) \cdot \left(\frac{1}{3}\right) = \frac{2}{12}.$$

Notice that the denominator 12 counts the total number of permutations of 2 from among 4 elements. The numerator counts the number of these permutations that contain both A and B, which is $_2P_2$. Thus, the probability of selecting any specified two of four elements (where selection is without replacement and with equal probability at each stage) could be stated

$$P(2 \text{ of } 4) = \frac{_2P_2}{_4P_2} = \frac{2}{12}.$$

One could elect in a problem of this kind to count combinations rather than permutations. The total number of combinations possible is $_4C_2 = 6$, and the number of combinations of two elements that contain both A and B is $_2C_2$. The probability that A and B are the selections included is

$$P(2 \text{ of } 4) = \frac{_2C_2}{_4C_2} = \frac{1}{6}.$$

Summary

If n draws (with replacement) are made from a set of N elements, the number of elements in the sample space is N^n. If n draws are conducted without replacement from N elements, the number of elements in the sample space for n draws is

$$N(N - 1)(N - 2) \cdots (N - n + 1).$$

In general, if event A can occur in $N(A)$ ways and event B can occur in $N(B)$ ways, then the number of ways in which $A \cap B$ can occur is the product of $N(A)$ and $N(B)$.

5.3

Often we wish to count the number of different ordered subsets possible when n objects or elements are selected from N objects or elements. This is called the permutation of N things n at a time. When order is not relevant, the term combination is employed. The combination of n elements from among N elements is the number of different subsets (unordered) that can be formed from N things taken n at a time.

See Self-Correcting Exercises 5.B

Exercises

1. A research and development manager must create a three-man team that possesses the skills of a geologist, an engineer, and a mathematician. Available are two geologists, eight engineers, and three mathematicians. How many different combinations can be named for the team?

2. Three successive draws are to be made from an urn containing chips numbered 1 through 10. How many different sample sequences are there if

   a. sampling is with replacement?

   b. sampling is without replacement?

3. Express the result in Exercise 2(b) as the product of the number of different sample combinations and the number of permutations per sample combination.

4. In order to establish the potential success of a product, it has been decided to sell it in three cities on a test basis. The marketing research department has selected seven cities as possible test cities. How many different groups of test cities can be finally selected?

5. If the test marketing in Exercise 4 can be carried out in only one city at a time, how many different schedules can be set up for the three cities finally selected?

6. What is the meaning of the product of your answers in Exercises 4 and 5?

7. A union committee of six people is to be chosen from the supervisory employees in a factory. Although there are ten male and three female supervisors, the union leadership wants an equal split of male and female committee persons. How many different committees can be formed?

8. A box contains three red marbles and three blue marbles. Use tree diagrams to show the probability that two successive draws from the box will yield different colored marbles if

   a. the drawing is with replacement, and

   b. the drawing is without replacement.

9. Express the probability in Exercise 8(b) as a ratio of the product of two combinations to yet a third combination.

10. There are three positions open and ten candidates—six men and four women. Assume equal selection probabilities and use combinations to find

    a. the probability that three men will be selected, and

    b. the probability that one man and two women will be selected.

Glossary of Equations

$\mu_X = \sum [X \cdot P(X)]$

The mean (or expected value) of a random variable is the summation of products of values of the random variable and their probabilities.

$\sigma_X^2 = \sum [(X - \mu)^2 \cdot P(X)]$

The variance of a random variable is the summation of squared deviations of values of the random variable from their mean weighted by their corresponding probabilities.

$\mu_{(X_1 + X_2 + \cdots + X_n)} = \mu_{X_1} + \mu_{X_2} + \cdots + \mu_{X_n}$

The mean (or expected value) of the sum of n independent random variables is equal to the sum of their means.

$\sigma_{(X_1 + X_2 + \cdots + X_n)}^2 = \sigma_{X_1}^2 + \sigma_{X_2}^2 + \cdots + \sigma_{X_n}^2$

The variance of the sum of n independent random variables is equal to the sum of their variances.

$N(A \cap B) = N(A) \cdot N(B)$

The number of ways in which an event defined as the intersection of events A and B can occur is the product of the number of ways in which A can occur and the number of ways in which B can occur.

$_N P_n = \dfrac{N!}{(N - n)!}$

The permutation (number of distinct ordered subsets) of n elements selected from N elements $(n \leq N)$ is N factorial divided by $(N - n)$ factorial.

$_N C_n = \dfrac{N!}{(N - n)! \, n!}$

The combination (number of different subsets) of n elements selected from N elements $(n \leq N)$ is N factorial divided by the product of $(N - n)$ factorial and n factorial.

6

Some Discrete Sampling Distributions

The previous two chapters introduced basic probability concepts and laws, the relation between sample spaces and random variables, and the idea of sums of random variables for repeated experiments. We next extend these fundamentals to examine the behavior of repeated samples from specified populations. The population characteristics, or parameters, that we will deal with are means and proportions. The key concept in the chapter is the probability distribution of a sample statistic. This is the central concept on which all the methods of statistical inference depend.

The Binomial Sampling Distribution

You will recall from Chapter 1 that a binary population is one that can be represented by a set of zeros and ones. Given a universal set of units of observation, a zero is recorded for each unit not having a specified characteristic and a one for each unit having the characteristic. The prevalence of the characteristic among units of the set is measured as the proportion of units having the characteristic. Examples are the proportion of defective items in a production run, the proportion of buyers purchasing a given brand, the proportion of registered voters favoring a particular candidate, the proportion of autos equipped with antipollution devices, and so on.

6.1

If π (Greek lowercase pi) is defined as the probability of a success (observing a one) in a single draw from a binary population, a binary population can be visualized as follows:

X	$P(X)$
0	$1 - \pi$
1	π

Examples of binary populations with differing values of π are shown graphically in Figure 6.1.

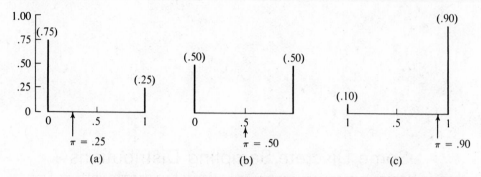

Figure 6.1   Binary Populations with Differing π

Given a binary population with a specified value of π, it could be shown on the initial branches of a tree diagram as

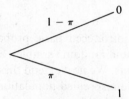

If repeated draws† are made from the binary population, the tree diagram for possible outcomes of the draws would appear as in Figure 6.2.

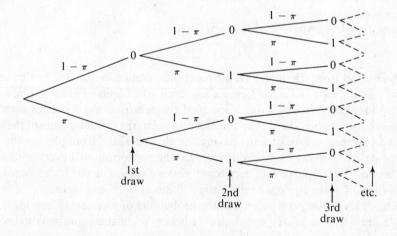

Figure 6.2   Repeated Draws from a Binary Population

† If the population is finite, the draws would be made with replacement. If the population is large in relation to the number of draws, what follows is correct to an approximation for draws without replacement.

We let the number of draws—that is, the size of the sample—be indicated by n. It is clear that, given a value for π, the probability of any sequence of zeros and ones through the tree can be calculated. For example, suppose it is known that 25 percent of the registered voters in a municipality voted in a recent mayorality election. Three registered voters, selected at random, are asked if they voted in the election. A "yes" answer is coded "1" and a "no" answer is coded "0." Our value of π (the probability of a "1" on any selection) is 1/4; the tree is shown in Figure 6.3.

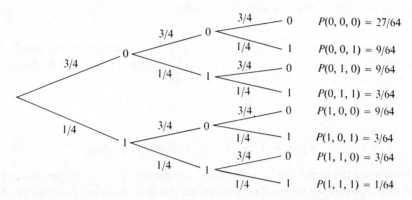

Figure 6.3   Outcomes for a Sample of $n = 3$ and $\pi = 1/4$

We now sum the probabilities for the different sample sequences leading to zero, one, two, and three persons who voted. The result, shown in Table 6.1, is the prob-

Table 6.1   Binomial Probability Distribution for $n = 3$, $\pi = 1/4$

Number of successes (persons voting) r		Probability $P(r)$
0		27/64
1	$9/64 + 9/64 + 9/64 =$	27/64
2	$3/64 + 3/64 + 3/64 =$	9/64
3		1/64

ability distribution of number of successes (r) in three independent draws from a binary population with $\pi = 1/4$. This would be called the binomial sampling distribution of r for $\pi = 1/4$ and $n = 3$.

Statistics and Parameters

In a random sample of n elements, the number of successes (r) is a *statistic*.

Definition	A **statistic** is a one-number transformation of the data in a sample. Similarly, a **parameter** is a one-number transformation of the data in a universe.

One possible sample of $n = 3$ from the binary population discussed above is $\{0, 1, 1\}$. For this sample, $r = 2$, the number of registered voters who voted.

The probability distribution of any statistic is also called its *sampling distribution*. Note that the sampling distribution of r in Table 6.1 tells us about the long-run relative frequency (or degree of belief) associated with each possible value of this statistic.

Formula and Tables for Binomial Probabilities

A formula is desired for $P(r)$ for $r = 0, 1, 2, \ldots, n$ for any given sample size. The multiplication rule for independent events can be applied to find the probability of any given sequence of zeros and ones. For example,

$$P(0, 1, 1, 1, 0) = (1 - \pi)(\pi)(\pi)(\pi)(1 - \pi)$$

and

$$P(1, 0, 1, 0, 1) = (\pi)(1 - \pi)(\pi)(1 - \pi)(\pi).$$

In both cases the terms can be collected to obtain $(\pi)^3(1 - \pi)^2$. In general, any particular sequence leading to specified r successes out of n trials will have the probability

$$\pi^r(1 - \pi)^{n-r}.$$

Since all sequences leading to a given r out of n have the same probability, we now need a rule for counting the number of (equiprobable) sequences. For example, how many different sequences of zeros and ones lead to three ones out of five draws? Here we visualize the five independent draws (in order of draw) as

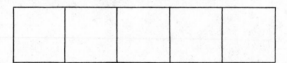

and ask in how many ways can three ones occupy three of the five boxes. This is the combination of 5 elements taken 3 at a time. In general we are interested in n elements (or sample positions) taken r at a time. The general rule for the combination of n elements taken r at a time is

$$_nC_r = \frac{n!}{(n - r)!\,r!}.$$

We are now able to write the *general binomial probability formula*,

$$\blacksquare \qquad P(r) = {_nC_r}\,\pi^r(1 - \pi)^{n-r}. \qquad \blacksquare \qquad (6.1)$$

The parameters that must be known before one can apply the binomial probability equation, are n, the number of trials, and π, the probability of a success on a single trial. This probability must remain constant over all trials. The trials must be independent of one another.

Because the binomial is such a commonly used sampling distribution, tables for binomial probabilities have been developed to do the work of the binomial formula. Appendix E gives such a table. Given specific values of π and n, we read from the table

6.1

6.2

any $P(r)$ desired. Let us check the values we calculated earlier for $P(r)$ for the binomial with $\pi = 1/4$ and $n = 3$. The headings of the table give π in decimal proportions, and we are concerned with $\pi = 0.25$. Follow that column down to the row where $n = 3$. There we read

n	r	π
		.25
$\vdots$	$\vdots$	$\vdots$
3	0	.4219
	1	.4219
	2	.1406
	3	.0156

These values of π are the decimal equivalents of the fractional probabilities calculated earlier. A standard notation for the probability 0.1406 above is

$$P(r = 2 \mid \pi = .25, n = 3) = .1406.$$

This reads, "The probability of $r = 2$ given $\pi = .25$ and $n = 3$ is .1406."

6.2

Let us look at some binomial sampling distributions for $n = 4$. We might, for example, be considering the possible outcomes of some samples of four parts from several large bins of parts in which different percentages of the parts are defective. We will consider the sampling situation when 10 percent, 30 percent, 50 percent, 70 percent, and finally 90 percent of the parts are defective. Table 6.2, which gives the distribu-

Table 6.2 Binomial Probabilities for $n = 4$ and Selected Values of π

Number of defectives r	True proportion defective in population (π)				
	.10	.30	.50	.70	.90
0	.6561	.2401	.0625	.0081	.0001
1	.2916	.4116	.2500	.0756	.0036
2	.0486	.2646	.3750	.2646	.0486
3	.0036	.0756	.2500	.4116	.2916
4	.0001	.0081	.0625	.2401	.6561
	1.0000	1.0000	1.0000	1.0000	1.0000

tions (you might check them with the table of binomial probabilities), illustrates some general features of binomial probability distributions:

(1) when $\pi = .50$, the binomial distribution is symmetrical;

(2) for $\pi < .50$, binomial distributions are positively skewed; and for $\pi > .50$ they are negatively skewed;

(3) the distributions for complementary values of π contain the same probabilities in reverse order.

Applications of Binomial Sampling Distributions

In a mathematical and logical sense we must have a parameter value before we can talk about the probability distribution of a related statistic. We must have the true proportion defective (π) before we can look up the probability distribution of number defective (r) in a sample of a given size (n). It may seem that we are beginning with the answer. However, in many situations we can profitably ask a series of "what if" questions about the parameter. The parameter value specified in any such question will be one that might exist and therefore is of interest to us.

Consider the production process turning out the bins of parts referred to earlier. Suppose that 15 percent defective ($\pi = .15$) is an acceptable quality level for the parts and a sample of $n = 4$ is selected from each bin. If no defectives are found in the four parts, the bin is declared acceptable. If any parts are found to be defective, the bin is declared unacceptable and the production line is shut down to look for a source of "trouble."

The binomial probabilities in Table 6.2 tell us something about this decision rule. We know that if the proportion of defective parts is .10, the probability of observing zero defectives and so declaring the bin acceptable is .6561. This seems quite low because it means the chances are better than 1 in 3 of stopping the production line and looking for a source of trouble when $\pi = .10$ (a better than minimum acceptable quality level). Also, when $\pi = .30$ (less than acceptable quality) the probability is .2401 of declaring the bin acceptable. Here we would fail to intervene in the process when it was less than satisfactory.

By increasing the sample size, one increases the ability to distinguish acceptable from unacceptable bins. To reduce both of the kinds of errors mentioned above, one will also have to change the maximum number of defectives that will lead to declaring the bin acceptable. Suppose we consider a sample size of $n = 15$. We are still concerned with the decision errors we might make with bins of $\pi = .10$ and $\pi = .30$. The relevant binomial probabilities are shown in Table 6.3. If we use $r = 1$ as our maximum acceptance number, we have

$$P(\text{accept bin} \mid \pi = .10) = .2059 + .3432 = .5491,$$

$$P(\text{reject bin} \mid \pi = .10) = 1 - .5491 = \underline{.4509},$$

Table 6.3   Binomial Probabilities for Acceptance
Sampling Example

		π	
n	r	.10	.30
15	0	.2059	.0047
	1	.3452	.0305
	2	.2669	.0916
	3	.1285	.1700
	4	.0428	.2186
		⋮	⋮

and

$$P(\text{accept bin}\,|\,\pi = .30) = .0047 + .0305 = \underline{.0352}.$$

If we use $r = 2$ as our maximum acceptance number, we will have

$$P(\text{accept bin}\,|\,\pi = .10) = .2059 + .3432 + .2669 = .8160,$$

$$P(\text{reject bin}\,|\,\pi = .10) = 1 - .8160 = \underline{.1840},$$

and

$$P(\text{accept bin}\,|\,\pi = .30) = .0047 + .0305 + .0916 = \underline{.1268}.$$

Is either of these rules better than the earlier one with a sample size of four? The first one ($r = 1, n = 15$) reduces only the probability of accepting a poor-quality bin. The second one ($r = 2, n = 15$) cuts down on both of the errors.

A rule with $r = 3, n = 15$ would reduce only the probability of rejecting a bin of 10 percent defectives.

Another application of the binomial distribution considers extreme values. Suppose you wish to check the "honesty" of a coin, that is, whether or not $P(\text{heads}) = 0.50$. You decide to toss it as a check. You regard extreme results in favor of either heads or tails as tending to cast doubt on the coin. You decide that should extremes occur whose probability, *given an honest coin*, is less than 0.01, you will reject the coin. In 20 tosses of the coin, what results should lead you to reject it?

We are concerned with the extreme, or "tail," probabilities for the binomial with $\pi = .50$ and $n = 20$. This is a symmetrical binomial, and the "tail" probabilities are

r	$P(r)$	r	$P(r)$
0	.0000	⋮	
1	.0000	16	.0046
2	.0002	17	.0011
3	.0011	18	.0002
4	.0046	19	.0000
⋮		20	.0000

The extreme values 0, 1, 2, 3, 4, 16, 17, 18, 19, 20 have a total probability of 0.0118, somewhat greater than your criterion of 0.01. Your rule should be to reject the coin if heads occurs either three or less times or 17 or more times. Events this extreme have a probability of 0.0026, which is less than 0.01.

It would be simpler to select a value of n whose tail probabilities for $\pi = .50$ came closer to 0.01. The binomial for $n = 8$ does this better than any other in the table. The test would be to toss your coin eight times, rejecting it only if zero heads or eight heads occurred. The probability of zero or eight heads, with an honest coin, is .0039 + .0039 = .0078.

Summary

The binomial sampling distribution is the probability distribution of the number of successes (r) in a sample of n independent draws from a binary population with probability of success on a single draw equal to π. With π and n given, the binomial formula or tables of binomial probabilities can be used to find the probability for any desired number of successes in the sample. While π is often not really known, it is frequently useful to check the probability of a certain sample result given a specified value of π. This procedure is particularly useful for evaluating the consequences of decision rules based on samples.

See Self-Correcting Exercises 6.A

Exercises

1. Use a tree diagram to find the binomial probabilities for $n = 3$, $\pi = 0.20$. Carry your work out in decimals and then check your results with Appendix E.

2. Use the binomial probability formula (6.1) to find the following probabilities. Check your answers by finding the same probabilities in Appendix E.
   a. $P(r = 2 \,|\, n = 4, \pi = 0.60)$,
   b. $P(r = 0 \,|\, n = 6, \pi = 0.20)$,
   c. $P(r = 4 \,|\, n = 5, \pi = 0.90)$.

3. Find the following binomial probabilities with the aid of Appendix E.
   a. $P(r \leq 3 \,|\, n = 8, \pi = 0.20)$,
   b. $P(r > 8 \,|\, n = 10, \pi = 0.70)$,
   c. $P(r > 2 \,|\, n = 12, \pi = 0.40)$.

4. A salesman for a computer manufacturer has a history of making successful sales calls one-third of the time. What is the probability that he will be successful on the first two and fail on the last two of the next four calls?

5. What is the probability that the salesman in Exercise 4 will be successful on two of his next four calls?

6. Parts coming off an assembly line have a probability of 0.15 of being defective. What is the probability that at least three of the next eleven parts are defective?

7. A student prepares for an exam by studying a set of ten problems. He can work seven of the ten. If the professor chooses randomly five of the ten problems for the exam, what is the probability that the student can work at least four of them? (The binomial is not applicable.)

8. In a random sample of fifteen students, ten are found to favor a particular candidate for public office. What is the probability of this sample outcome if
   a. 65 percent of all students favor the candidate?
   b. 45 percent of all students favor the candidate?

9. An acceptance sampling plan calls for lots to be accepted if a random sample of ten items yields two or fewer defectives. What is the probability of

a. rejecting a lot that has 5 percent defectives?

b. accepting a lot that has 40 percent defectives?

10. If the sample size in Exercise 9 is increased to fifteen, is there a maximum acceptance number that will result in decreasing the probabilities in (a) and (b)? If so, what is it?

Sampling Distributions of Means and Proportions

6.3

Up to now we have seen two kinds of probability distributions of the sum of a random variable for n independent draws from a population: the total-margin example of Chapter 5 and the binomial probability distribution discussed earlier in this chapter. The latter is a special case in that each draw can generate only one of two possible values, 0 or 1.

In many situations we are concerned about the sample mean generated by n independent draws from a population. This is another example of a statistic. The sample mean ($\bar{X}$) for n draws is

$$\blacksquare \quad \bar{X} = \frac{\sum\limits_{1}^{n} X}{n}. \quad \blacksquare \tag{6.2}$$

In the example of margin dollars from sales of table-model radios, the sample mean would be the average dollar margin per unit sold for n units. Keep in mind that if we know the probability distribution of dollar margin for a single unit sale, we know the population mean. Recalling the probability distribution, we calculate it again below.

$ margin X	Probability $P(X)$	$X \cdot P(X)$
8	7/12	56/12
16	4/12	64/12
24	1/12	24/12
		144/12

$$\mu_X = \sum [X \cdot P(X)] = 144/12 = 12 \text{ dollars.}$$

Given the underlying population (probability distribution of X), the population mean, μ_X, represents the average margin per sale over a long (infinite) series of sales. The average margin per sale for a finite number of n sales is a sample mean that will more or less closely approximate μ_X. The probability distribution of the sample mean enables us to be more specific about $\bar{X}$ as an approximation to μ_X. Ultimately, an understanding of the probability distribution of the sample mean (or sampling distribution of the mean) is the key to an understanding of statistical inference. *It is the single most important concept in statistics.*

In Chapter 5 we derived the following probability distribution of total dollar margin ($\sum X$) for a random sample of $n = 2$ sales.

$\sum X$	$P(\sum X)$
16	49/144
24	56/144
32	30/144
40	8/144
48	1/144

Since the sample mean is simply $\sum X/n$, we need only divide the values of the variable $\sum X$ by n to obtain values of the sample mean. This gives us the following probability distribution of the sample mean.

$\bar{X}$	$P(\bar{X})$
8	49/144
12	56/144
16	30/144
20	8/144
24	1/144

To emphasize what is involved in the sampling distribution of the mean, consider a random sample of $n = 3$ radio sales. To find the sampling distribution of the mean for $n = 3$ we would extend the tree diagram of Figure 5.2 (page 112) to include the third draw. Multiplying the probabilities on the appropriate branches would give us the probabilities for different margin sequences as follows:

$P(\ 8,\ \ 8,\ \ 8) = 343/1728$

$P(\ 8,\ \ 8,\ 16) = 196/1728$
$P(\ 8,\ 16,\ \ 8) = 196/1728$
$P(16,\ \ 8,\ \ 8) = 196/1728$

$P(\ 8,\ 16,\ 16) = 112/1728$
$P(16,\ \ 8,\ 16) = 112/1728$
$P(16,\ 16,\ \ 8) = 112/1728$
$P(\ 8,\ \ 8,\ 24) = \ 49/1728$
$P(\ 8,\ 24,\ \ 8) = \ 49/1728$
$P(24,\ \ 8,\ \ 8) = \ 49/1728$

$P(\ 8,\ 16,\ 24) = 28/1728$
$P(\ 8,\ 24,\ 16) = 28/1728$
$P(16,\ \ 8,\ 24) = 28/1728$
$P(16,\ 24,\ \ 8) = 28/1728$
$P(24,\ \ 8,\ 16) = 28/1728$
$P(24,\ 16,\ \ 8) = 28/1728$
$P(16,\ 16,\ 16) = 64/1728$

$P(16,\ 16,\ 24) = 16/1728$
$P(16,\ 24,\ 16) = 16/1728$
$P(24,\ 16,\ 16) = 16/1728$
$P(\ 8,\ 24,\ 24) = \ 7/1728$
$P(24,\ \ 8,\ 24) = \ 7/1728$
$P(24,\ 24,\ \ 8) = \ 7/1728$

$P(16,\ 24,\ 24) = 4/1728$
$P(24,\ 16,\ 24) = 4/1728$
$P(24,\ 24,\ 16) = 4/1728$

$P(24,\ 24,\ 24) = 1/1728$

From these results we can obtain the probability distribution of total margin, shown in (a) of Table 6.4. By dividing the sums by n we obtain the sampling distribution of the mean, shown in (b) of the table.

It is important to distinguish between the probability distribution of dollar margin for a single sale and the probability distribution of average margin per sale for a sample of n sales. One is the distribution of X for the underlying population, and the other is the distribution of $\bar{X}$ for repeated samples of size n taken from the X population. The

Table 6.4   (a) Sampling Distribution of a Sum and (b) Sampling
Distribution of a Mean

(a)		(b)	
Total margin $\sum X$	*Probability* $P(\sum X)$	*Average margin* $\overline{X}$	*Probability* $P(\overline{X})$
24	343/1728	24/3 = 8.00	343/1728
32	588/1728	32/3 = 10.67	588/1728
40	483/1728	40/3 = 13.33	483/1728
48	232/1728	48/3 = 16.00	232/1728
56	69/1728	56/3 = 18.67	69/1728
64	12/1728	64/3 = 21.33	12/1728
72	1/1728	72/3 = 24.00	1/1728

two distributions are compared in Figure 6.4. In (*a*) we see the distribution of individual X values in an underlying set of units of observation. The sampling distribution of the mean, shown in (*b*) of the figure, describes the variability that is possible among means of different samples of the same size taken from the population of X values.

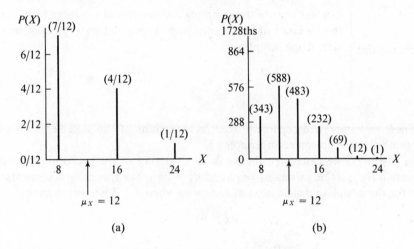

(a)   (b)

Figure 6.4   (a) Probability Distribution of X and (b) Probability
Distribution of $\overline{X}$ for a Sample of Size $n = 3$

The two distributions are related of course, because the distribution of $\overline{X}$ is derived (via sums of random variables) from the distribution of X. The most important relationships between them can be stated in the form of two statements for the parameters of the sampling distribution of the mean.

Statement | The mean (or expected value) of the sampling distribution of the mean ($\mu_{\overline{X}}$) is equal to the mean of the underlying X population (μ_X).

$$\blacksquare \qquad \mu_{\overline{X}} = \mu_X . \qquad \blacksquare \qquad\qquad (6.3)$$

In other words, the expected value of a sample mean is equal to the mean of the population sampled. In taking a sample of three randomly selected sales to determine the long-run average margin per sale, the *expectation* for the sample mean would be the population mean.

The second statement concerns the variance of the sampling distribution of the mean. While it may be nice to know that the expected value of the sample mean is equal to the population mean, we must recognize that the particular sample mean obtained for three sales may mislead us considerably about the true population, or long-run, mean. Suppose that without knowing the population mean, we tried to estimate it from three sales. Imagine, however, that the underlying population is in fact as shown in Figure 6.4(a). We will have to use $\overline{X} = \sum X/n$ from our sample of $n = 3$ to estimate the average margin per sale. From Figure 6.4(b) we see there is a considerable chance (343/1728) that we will estimate the average margin to be \$8.00, and a small chance (13/1728) that we will estimate it to be \$21.33 or more. Without knowing the underlying probability distribution of X, we would not be able to calculate these probabilities. Nevertheless, the example points out the importance of the variance of the sampling distribution of the mean.

Statement	The variance of the sampling distribution of the mean ($\sigma_{\overline{X}}^2$) is equal to the variance of the underlying X population (σ_X^2) divided by the size of the sample.

$$\blacksquare \qquad \sigma_{\overline{X}}^2 = \frac{\sigma_X^2}{n}. \qquad \blacksquare \tag{6.4}$$

These two statements can be proved as extensions of the statements on sums of random variables presented in Chapter 5.

In Chapter 5 we calculated the mean and variance of the probability distribution of margin for a single sale as $\mu_X = 12$ and $\sigma_X^2 = 26\frac{2}{3}$. The foregoing statements tell us that for the sampling distribution of the mean when $n = 3$ we would have

$$\mu_{\overline{X}} = \mu_X = 12,$$

$$\sigma_{\overline{X}}^2 = \frac{\sigma_X^2}{n} = \frac{26\frac{2}{3}}{3} = 8\frac{8}{9}.$$

The variance of the sampling distribution of the mean is often called more briefly the *variance of the sample mean*. Its positive square root is called the *standard error of the mean*. We can establish the formula

$$\blacksquare \qquad \sigma_{\overline{X}} = \frac{\sigma_X}{\sqrt{n}}. \qquad \blacksquare \tag{6.5}$$

For our example,

$$\sigma_{\overline{X}} = \frac{\sqrt{26\frac{2}{3}}}{\sqrt{3}} = \sqrt{8.89} = 2.98.$$

This is a measure of variability among the possible values of the sample mean that can be visualized as a distance along the horizontal axis in Figure 6.4(b).

The Chebyshev Inequality

In Chapter 1 we encountered a theorem of Chebyshev. The theorem states that the proportion of measurements in a set of data that lie within K standard deviations of the mean is not less than $1-(1/K^2)$, where K is at least 1. We can use the Chebyshev inequality to obtain a minimum probability for extreme values of a sample mean, whenever we know the mean and standard deviation of the underlying X population.

Statement	The probability that a sample mean will lie more than K standard errors from the population mean will be less than $1/K^2$, where $K \geq 1$.

$$\blacksquare \qquad P[|\overline{X} - \mu| > K\sigma_{\overline{x}}] < \frac{1}{K^2}. \qquad \blacksquare \qquad (6.6)$$

Applying the Chebyshev inequality to our sampling distribution of the mean for $n = 3$ we could make such statements as

$$P[|\overline{X} - \mu| > 2(2.98)] < \frac{1}{2^2}, \quad \text{or} \quad P[|\overline{X} - \mu| > 5.96] < \frac{1}{4};$$

and

$$P[|\overline{X} - \mu| > 3(2.98)] < \frac{1}{3^2}, \quad \text{or} \quad P[|\overline{X} - \mu| > 8.94] < \frac{1}{9}.$$

Sample means differing by more than $5.96 from the population mean of $12.00 are those less than $6.04 and those greater than $17.96. Our sampling distribution shows that such means ($18.67, $21.33, and $24.00) have a total probability of

$$\frac{69 + 12 + 1}{1728} = .0474.$$

This is less than 0.25 and so is in accord with the Chebyshev inequality. The only means more than $8.94 away from the population mean are $21.33 and $24.00 with probability $(12 + 1)/1728 = .0075$. The Chebyshev inequality says this probability will be less than one-ninth, as indeed it is.

The advantage of the Chebyshev inequality is that it gives us a limiting probability on any desired extreme sample means without our having to derive the entire sampling distribution. Suppose a sample of $n = 25$ sales were to be taken from our population with a mean of $12.00 and variance of $26\frac{2}{3}$. Calculating the entire sampling distribution of average margin per sale for $n = 25$ sales would be impractical. However we know that

$$\mu_{\overline{x}} = \mu_x = \$12.00,$$

$$\sigma_{\overline{x}} = \frac{\sigma_x}{\sqrt{n}} = \frac{\sqrt{26\frac{2}{3}}}{\sqrt{25}} = \frac{5.16}{5} = \$1.03.$$

Then the Chebyshev inequality tells us that the probability is less than 1/4 that a sample mean will lie outside the range

$$\mu_{\bar{X}} \pm 2\sigma_{\bar{X}}.$$

Thus

$$\$12.00 \pm 2(\$1.03) = \$9.94 \text{ to } \$14.06.$$

Similarly the probability is less than 1/9 for a sample mean outside the range

$$\mu_{\bar{X}} \pm 3\sigma_{\bar{X}}.$$
$$\$12.00 \pm 3(\$1.03) = \$8.91 \text{ to } \$15.09.$$

The Sampling Distribution of the Proportion

At the beginning of this chapter we considered a sample of three registered voters from a population in which 25 percent of the registered voters had cast ballots in a recent election. Here the population of X is the probability distribution of the binary population of zeros (nonvoters) and ones (voters) in the recent election.

X	$P(X)$
0	3/4
1	1/4

The mean of this population (a parameter) is

$$\mu_X = \sum [X \cdot P(X)] = 0(3/4) + 1(1/4) = 1/4 = 0.25.$$

This is also π, the proportion of "ones" in the population. The variance of our binary population is

$$\sigma_X^2 = \sum[(X - \mu_X)^2 \cdot P(X)] = \left(0 - \frac{1}{4}\right)^2\left(\frac{3}{4}\right) + \left(1 - \frac{1}{4}\right)^2\left(\frac{1}{4}\right),$$

$$\sigma_X^2 = \left(\frac{1}{16}\right)\left(\frac{3}{4}\right) + \left(\frac{9}{16}\right)\left(\frac{1}{4}\right) = \frac{12}{64},$$

$$\sigma_X^2 = \frac{3}{16} = .1875.$$

This variance happens to be $\pi(1 - \pi) = (1/4)(3/4) = 3/16$. What we have illustrated here *is always true for a binary population*, namely,

$$\blacksquare \qquad \mu_X = \pi, \qquad \blacksquare \tag{6.7}$$

$$\blacksquare \qquad \sigma_X^2 = \pi(1 - \pi). \qquad \blacksquare \tag{6.8}$$

With the aid of a tree diagram we then found (page 123) the probability distribution of $\sum X$ for three independent draws from this population. We called this sum r and found later that we could have simply read the probabilities for r from the binomial table for $\pi = 1/4 = .25$ and $n = 3$. The distribution (using the fractional probabilities) was

r	$P(r)$
0	27/64
1	27/64
2	9/64
3	1/64

The distribution of r, or *number* of persons voting out of three in this case, is a distribution of a sum of independent identical random variables. From Equations (5.3) and (5.4) in Chapter 5 the mean and variance of the sampling distribution of number voting out of three are

$$\mu_r = \mu_{(X_1 + X_2 + X_3)} = \mu_{X_1} + \mu_{X_2} + \mu_{X_3} = \frac{1}{4} + \frac{1}{4} + \frac{1}{4} = \frac{3}{4},$$

$$\sigma_r^2 = \sigma_{(X_1 + X_2 + X_3)}^2 = \sigma_{X_1}^2 + \sigma_{X_2}^2 + \sigma_{X_3}^2 = \frac{3}{16} + \frac{3}{16} + \frac{3}{16} = \frac{9}{16}.$$

We saw just above that for a binary population $\mu_X = \pi$ and $\sigma_X^2 = \pi(1 - \pi)$. Since each of the n draws is made from the same binary population, we can state general formulas for the mean and variance of the binomial sampling distribution of number of successes (r) out of n trials.

$$\blacksquare \qquad \mu_r = n\pi, \qquad \blacksquare \qquad \qquad (6.9)$$

$$\blacksquare \qquad \sigma_r^2 = n\pi(1 - \pi). \qquad \blacksquare \qquad (6.10)$$

Should we want to present our sampling distribution as the *proportion* of three persons who voted, we would divide the variable *number voting* (r) by three (n). It is customary to call this variable p (for proportion). We have

p	$P(p)$
0/3	27/64
1/3	27/64
2/3	9/64
3/3	1/64

Note that dividing $r/n = p$ is the same as dividing $\sum X/n = \overline{X}$. The sampling distribution of a proportion is a special case of the sampling distribution of a mean. It is special in that it relates to a binary population. We can then use Equations (6.3) and (6.4) to find the mean and variance of the sampling distribution of a proportion. Because the symbols for the binary sampling situation are very common, most statisticians use special formulas, which we shall now develop.

Equation (6.3) tells us

$$\mu_{\overline{X}} = \mu_X.$$

Then from

$$\overline{X} = \frac{r}{n} = p \quad \text{and} \quad \mu_X = \pi,$$

we have

$$\blacksquare \qquad \mu_p = \pi. \qquad \blacksquare \qquad (6.11)$$

Equation (6.11) says that the expected value of a sample proportion is equal to the underlying population proportion. From Equation (6.4),

$$\sigma_{\bar{X}}^2 = \frac{\sigma_X^2}{n},$$

but

$$\bar{X} = \frac{r}{n} = p \quad \text{and} \quad \sigma_X^2 = \pi(1 - \pi)$$

and

$$\blacksquare \qquad \sigma_p^2 = \frac{\pi(1 - \pi)}{n}. \qquad \blacksquare \qquad (6.12)$$

The positive square root of (6.12) is called the standard error of the sample proportion.

$$\blacksquare \qquad \sigma_p = \sqrt{\frac{\pi(1 - \pi)}{n}}. \qquad \blacksquare \qquad (6.13)$$

One advantage of transforming the sampling distribution of r to the sampling distribution of $p = r/n$ is that we can relate the sampling distribution of the proportion to the distribution of the binary population of X. In Figure 6.5 we do this for our

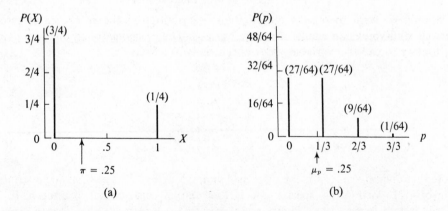

Figure 6.5   (a) Probability Distribution of a Binary Population and (b) Probability Distribution of Proportion (p) in a Sample of Size $n = 3$

sampling distribution of proportion voting in the same way that we related to the population distribution the sampling distribution of average margin for the random sample of three sales. The standard error of the sample proportion can be viewed as a distance on the p scale of Figure 6.5(b). In using binomial tables to look up the sampling distribution of p, values of r are converted to values of p by dividing them by n.

6.3

Summary

When the X's associated with a set of units of observation are known, the sampling distribution of mean X in repeated samples of a given size can be determined. This can be approached by listing the entire sample space for the draw of n units of observation, or by extending a tree diagram or equivalent table of the probability distribution of X through the required number of draws. When the units are drawn with replacement, the expected value of the sampling distribution of the mean equals the population mean and the variance of the sampling distribution equals the variance of the population divided by the sample size. The standard deviation of the sampling distribution of the mean is called the standard error of the sample mean. By the Chebyshev inequality, the probability that the sample mean will lie within K standard errors of the population mean is not less than $1 - (1/K^2)$.

In sampling from a binary population, the sampling distribution of the proportion is a special case of the sampling distribution of a mean. The expected value of the sample proportion is the population proportion (π), and the variance of the sampling distribution reduces to $\pi(1 - \pi)/n$. The standard deviation of the sampling distribution of a proportion is called the standard error of the sample proportion.

See Self-Correcting Exercises 6.B

Exercises

1. The law firms in a large city have the following probability distribution of number of partners. For a random sample of size $n = 2$, find the probability distribution of the average number of partners per firm.

Number of partners	2	3	4	5
Probability	0.4	0.3	0.2	0.1

2. Find the mean and variance of the population in Exercise 1. Calculate the mean and variance of the probability distribution of the sample mean from the distribution you obtained. Do the values agree with Formulas (6.3) and (6.4)?

3. Apply the Chebyshev inequality to the range of 2.0 standard errors from the population mean for the sampling distribution of the mean in Exercise 1. Compare the probability from the Chebyshev inequality to the actual probability for the sampling distribution.

4. Find the range that will include the sample mean with probability at least 0.75 if a sample of size $n = 9$ is taken from the population in Exercise 1; if a sample of size $n = 25$ is taken.

5. Twenty percent of the consumers in an urban market prefer the "Lo-Cal" variety of a nationally marketed soft drink, while 50 percent of the consumers have tried the "Lo-Cal" product. Find the mean and variance of each of the binary populations just mentioned.

6. Four consumers are selected at random in the situation of Exercise 5. Find the mean and variance of the sampling distribution of number of consumers who prefer the product, and the mean and variance of the sampling distribution of the number of consumers who have tried the product.

7. Find the mean and variance of the sampling distributions of the proportion of consumers who prefer and who have tried the product in the situation of Exercises 5 and 6.

8. A certain can-filling machine is known to produce 10 percent dented cans. If samples of 100 cans are chosen, what is the standard deviation of

   a. the proportion of dented cans?

   b. the number of dented cans in such samples?

9. A machine cuts cloth into 12-foot lengths with a standard deviation of 3 inches. If 100 such lengths are cut, what is the standard deviation of

   a. the total length of cloth cut?

   b. the mean length of cloth cut?

10. Find the probability distribution of the mean value per bill drawn in the example of Exercise 10, page 115, and then find the variance of the sample mean. Note that this value does not agree with Formula (6.4). The reason is that draws without replacement are not independent draws. The appropriate formula for sampling without replacement is

$$\sigma_{\bar{X}}^2 = \frac{\sigma_{\bar{X}}^2}{n}\left(\frac{N-n}{N-1}\right),$$

where N is the number of elements in the population.

Glossary of Equations

$P(r) = {}_nC_r\,\pi^r(1-\pi)^{n-r}$

The binomial probability formula for r successes in n independent draws from a binary population with probability of success equal to π on any single draw.

$\bar{X} = \dfrac{\sum\limits_1^n X}{n}$

The formula for the mean of a sample of n observations from a parent population.

$\mu_{\bar{X}} = \mu_X$

The mean (or expected value) of the sampling distribution of a mean is equal to the mean of the underlying X population.

$\sigma_{\bar{X}}^2 = \dfrac{\sigma_X^2}{n}$

The variance of the sampling distribution of a mean is equal to the variance of the underlying X population divided by the size of the sample.

$\sigma_{\bar{X}} = \dfrac{\sigma_X}{\sqrt{n}}$

The standard error of a sample mean is the positive square root of the variance of the sampling distribution of the mean.

$p[|\bar{X} - \mu| > K\sigma_{\bar{X}}] < \dfrac{1}{K^2}$

(The Chebyshev inequality.) The probability that a sample mean will lie more than K standard errors from the population mean will be less than $1/K^2$.

$\mu_X = \pi$

The mean of a binary population of X ($X = 1$ with probability π, and $X = 0$ with probability $1 - \pi$) is equal to π.

$\sigma_X^2 = \pi(1 - \pi)$

The variance of a binary population of X is the product of π and $1 - \pi$.

$\mu_r = n\pi$

The mean (or expected value) of the binomial sampling distribution of r successes in a sample of n draws from a binary population is n times the mean of the underlying binary population.

$\sigma_r^2 = n\pi(1 - \pi)$

The variance of a binomial sampling distribution of r is n times the variance of the underlying binary population.

$\mu_p = \pi$

The mean (expected value) of a sample proportion is equal to the underlying population proportion.

$\sigma_p^2 = \dfrac{\pi(1 - \pi)}{n}$

The variance of the sampling distribution of a proportion is equal to the variance of the underlying binary population divided by the size of the sample.

$\sigma_p = \sqrt{\dfrac{\pi(1 - \pi)}{n}}$

The standard error of a sample proportion is the positive square root of the variance of the sampling distribution of the proportion.

7

Elementary Decision Making

At this point, we are ready to apply probability to some problems of decision making. To make a decision is to adopt a course of action with which to face an uncertain state of nature. Uncertainty can be present because the state of nature that concerns us is a future state—as when a businessman decides to expand his plant in anticipation of a larger market five to ten years in the future. In other cases uncertainty may prevail because we have incomplete knowledge about a present state of nature— as when a sample is taken from a current production run to determine whether the product meets quality specifications. In gambling and games, which illustrate so well the elements of probability and decision making, uncertainty is deliberately created by the rules of play.

We begin this chapter with a gambling example that emphasizes the elements in a decision problem. Then we introduce an information strategy in the gambling problem to see how to determine the value of information in a decision problem. The second half of the chapter includes, among further aspects of decision theory, the important topics of subjective probability and subjective utility in which the personal beliefs and attitudes of the decision maker are incorporated.

A Game with Alternative Strategies

A gambler sets up a game using six urns whose outer appearances are the same. In one urn he places one gold coin and two copper coins; in another he places three copper coins; each of the remaining four urns contains two gold coins and one copper coin. For convenience we will call the three types of urns θ_1, θ_2, and θ_3.† They are depicted in Figure 7.1.

† Greek symbols are commonly used for states and parameters. This one is the Greek letter theta.

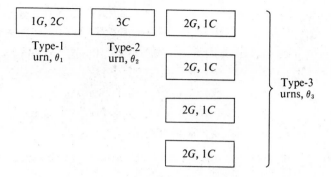

Figure 7.1   Representation of Urns in Gambler's Game

The player is to select an urn and he may elect to predict which type of urn he has selected. For a correct prediction of θ_1 or θ_2 he wins \$120, and for a correct prediction of θ_3 he wins \$30. As an alternative the player may simply choose (in advance) to keep either the gold coins or copper coins in the urn he has drawn. The gold coins are worth \$20 each and the copper coins, \$10 each. Our problem is to determine what is the best strategy to use in playing this game.

Expected Values of Alternative Strategies

7.1

We have then three possible true states and five available strategies. The player may (1) bet the urn is type 1, (2) bet the urn is type 2, (3) bet the urn is type 3, (4) keep the gold coins, or (5) keep the copper coins. To shorten our references to the different strategies we will call them S_1, S_2, S_3, S_4, and S_5. The payoffs from each strategy should each state occur can be arrayed in a "payoff table," as shown below.

States	Strategy payoffs				
	S_1	S_2	S_3	S_4	S_5
θ_1	120	0	0	20	20
θ_2	0	120	0	0	30
θ_3	0	0	30	40	10

We call these values V_{ij}, the return from the jth strategy should the ith state occur. For example, $V_{11} = 120$ and $V_{14} = 20$. This represents the fact that if urn θ_1 were drawn and we elected strategy 1, the payoff would be \$120. But if urn θ_1 were drawn and we elected strategy 4 (keep the gold coin(s)), the payoff would be only \$20. If one of the θ_3 urns is drawn and we elected the fourth strategy, we would receive a payoff worth \$40. This is V_{34}, the return from the third state given the fourth strategy.

The selection of any strategy (S_j) obviously is a gamble on the payoffs under that strategy. How do we decide among the strategies? The criterion known as Bayes' rule says to select the one with the greatest expected value. The payoffs can be designated by $V_{ij}|S_j$, and the expected value (or expected payoff) of the jth strategy can be written as

$$\blacksquare \quad EV(S_j) = \sum_i [(V_{ij}|S_j) \cdot P(\theta_i)]. \quad \blacksquare \quad (7.1)$$

If we accept probabilities for the states θ_1, θ_2, and θ_3 of 1/6, 1/6, and 4/6, respectively, we can find the expected value of each strategy. For the various strategies, we have

$$EV(S_1) = 120(1/6) + \quad 0(1/6) + \quad 0(4/6) = \$20,$$
$$EV(S_2) = \quad 0(1/6) + 120(1/6) + \quad 0(4/6) = \$20,$$
$$EV(S_3) = \quad 0(1/6) + \quad 0(1/6) + 30(4/6) = \$20,$$
$$EV(S_4) = \quad 20(1/6) + \quad 0(1/6) + 40(4/6) = \$30,$$
$$EV(S_5) = \quad 20(1/6) + 30(1/6) + 10(4/6) = \$15.$$

These calculations are often organized in tabular form by showing the payoff table along with probabilities of states.

States	$P(\theta_i)$	Payoffs S_1	S_2	S_3	S_4	S_5
θ_1	1/6	120	0	0	20	20
θ_2	1/6	0	120	0	0	30
θ_3	4/6	0	0	30	40	10
Expected values		20	20	20	30	15

$$\uparrow$$
$$EV(S_{\mathrm{opt}})$$

Strategy 4 (take the gold coins in the urn drawn) has the greatest expected value, and is said to be the optimal strategy for this game under Bayes' rule. In symbols we say $EV(S_{\mathrm{opt}}) = EV(S_4) = \$30$.

Since one interpretation of probability is long-run relative frequency, we can make a corresponding interpretation of expected value of a strategy. In weighting payoffs by probability in the expected-value calculations, we in effect calculate a mean, or average, return per play of the game. Because the weights are probabilities, the expected value is seen as the average return per play over a long run of plays.

Expected Value of Perfect Information

Let us consider now what our expected return from the game might be if we had a way of knowing which urn was drawn *before having to select a strategy*. Look again at the payoff table.

States	$P(\theta_i)$	Payoffs S_1	S_2	S_3	S_4	S_5
θ_1	1/6	120	0	0	20	20
θ_2	1/6	0	120	0	0	30
θ_3	4/6	0	0	30	40	10

Clearly, if we knew that θ_1 was drawn, we would select S_1 and gain \$120; if we knew θ_2 had been drawn, we would bet on it (S_2) and gain \$120; if we knew θ_3 had been drawn, we would take the gold coins (S_4) and gain \$40. In the table above, the maximum payoff, given each state, has been boxed in. We will call these $V_{max}|\theta_i$. Thus

$$V_{max}|\theta_1 = 120,$$
$$V_{max}|\theta_2 = 120,$$

and

$$V_{max}|\theta_3 = 40.$$

If we were to make many plays of the game, we know from $P(\theta_i)$ that θ_1 and θ_2 would each occur about 1/6 of the time, and θ_3 would occur 4/6 of the time. If we had a perfect predictor, we would know, prior to any play, which type urn would be drawn, and would elect our strategies as indicated earlier. Our expected gain, if we could so adjust strategies to states, would be

$$120(1/6) + 120(1/6) + 40(4/6) = 400/6, \quad \text{or } \$66.67.$$

This expectation, called the *expected value under certain prediction*, is obtained by finding the maximum payoff from any strategy given each state and then weighting these maximum returns by the probabilities of states.

Definition	The **expected value under certain prediction** is the expected return from perfect (or certain) adjustment of strategies to states. The states remain uncertain, but the decision maker is able to receive with certainty the maximum payoff given any state. $$\blacksquare \quad EV(CP) = \sum_i [(V_{max}|\theta_i) \cdot P(\theta_i)]. \quad \blacksquare \qquad (7.2)$$

Our best strategy when we had to choose in the face of uncertainty was S_4, with an expected value of \$30. Now we find that with a perfect predictor our expected return could be increased to \$66.67. The difference, \$36.67, is the expected value of perfect information $(EVPI)$. The expected value of the best strategy under certainty is abbreviated $EV(S_{opt})$, so we now have the following important definition.

Definition	The **expected value of perfect information** is the difference between the expected value under a certain, or perfect, prediction of states and the expected value of the optimal strategy under uncertain prediction. $$\blacksquare \quad EVPI = EV(CP) - EV(S_{opt}). \quad \blacksquare \qquad (7.3)$$

In our example,

$$EV(CP) = \$66.67, \qquad EV(S_{opt}) = \$30.00,$$

and

$$EVPI = \$66.67 - \$30.00 = \$36.67.$$

We should be willing to pay up to $36.67 per play of the game to obtain perfect information as to which urn was drawn. This would enable us to select the correct strategy each time. With perfect predictions we could expect to gain $66.67, while the best we can expect under uncertain prediction is $30. If an insider who knew how to identify the different types of urns offered to sell us this information for $20, it would pay us to buy it.

7.1

7.2 The Use of Information in Decision Making

Suppose now that we are given an opportunity to examine one coin from the urn drawn before having to select a strategy. Suppose also that the coin is copper. What is the best strategy in view of this information? Remember that the expected values of the several strategies depend on the payoffs under those strategies and on the probabilities of the states that produce the payoffs. If the coin drawn from the urn we selected were a gold one, we would know that the urn was not θ_2, for θ_2 contains no gold coin. The probabilities, $P(\theta_i)$, will change depending on what information we obtain, and so will the expected values of the different strategies.

Bayes' Theorem for Revising Probabilities of States

In Chapter 4 you encountered an application of conditional probabilities that allowed us to determine changed probabilities. This application is formally known as Bayes' theorem:

Statement

If an observed event can be produced by several causes, θ_1, θ_2, $\theta_3 \ldots, \theta_n$, the probability that the event was caused by θ_i is the ratio of the joint probability of θ_i and the event to the marginal probability of the event.

$$\blacksquare \qquad P(\theta_i | E) = \frac{P(\theta_i \cap E)}{P(E)}. \qquad \blacksquare \qquad (7.4)$$

Consider a probability tree showing the sequence of first choosing one urn from among the six urns and then picking a coin from the urn drawn in the first step. The tree is shown in Figure 7.2. We picked a copper coin. This is the observed event, and we want to know the probability that the event was "caused" by θ_1; by θ_2; by θ_3. Cause is used here in the sense of condition prior to an event. The copper coin could have been caused by drawing from θ_1, θ_2, or θ_3. The terminal branches that concern us are labeled C in Figure 7.2. Bayes' theorem tells us that

$$P(\theta_1 | C) = \frac{P(\theta_1 \cap C)}{P(C)}.$$

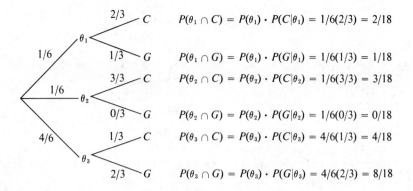

$$P(\theta_1 \cap C) = P(\theta_1) \cdot P(C|\theta_1) = 1/6(2/3) = 2/18$$

$$P(\theta_1 \cap G) = P(\theta_1) \cdot P(G|\theta_1) = 1/6(1/3) = 1/18$$

$$P(\theta_2 \cap C) = P(\theta_2) \cdot P(C|\theta_2) = 1/6(3/3) = 3/18$$

$$P(\theta_2 \cap G) = P(\theta_2) \cdot P(G|\theta_2) = 1/6(0/3) = 0/18$$

$$P(\theta_3 \cap C) = P(\theta_3) \cdot P(C|\theta_3) = 4/6(1/3) = 4/18$$

$$P(\theta_3 \cap G) = P(\theta_3) \cdot P(G|\theta_3) = 4/6(2/3) = 8/18$$

Figure 7.2   Probability Tree for Urn Problem

After drawing a copper coin, the only elements of the original sample space of concern to us are the points corresponding to that condition. These are the intersections that make up the marginal probability of a copper coin,

$$P(C) = P(\theta_1 \cap C) + P(\theta_2 \cap C) + P(\theta_3 \cap C).$$

Thus

$$P(C) = 2/18 + 3/18 + 4/18 = 9/18,$$

and we have

$$P(\theta_1 | C) = \frac{P(\theta_1 \cap C)}{P(C)} = \frac{2/18}{9/18} = 2/9.$$

Prior to observing the result of a coin draw, the probability of θ_1 was 1/6. After observing a copper coin our revised belief, or *posterior* (to the event) *probability*, is 2/9. Bayesians call the original probabilities *prior probabilities*. They are statements of degree of belief prior to the (current) evidence. Posterior probabilities are the statements of belief after the (current) evidence. Bayes' theorem accomplishes the transition whereby the earlier statement of belief is *revised* in accordance with the particular evidence observed—in this case a copper coin.

The revised, or posterior, probabilities for θ_2 and θ_3, given the observation of a copper coin, are

$$P(\theta_2 | C) = \frac{P(\theta_2 \cap C)}{P(C)} = \frac{3/18}{2/18 + 3/18 + 4/18} = 3/9,$$

$$P(\theta_3 | C) = \frac{P(\theta_3 \cap C)}{P(C)} = \frac{4/18}{2/18 + 3/18 + 4/18} = 4/9.$$

Revising the Strategies—Posterior Analysis

The calculation of the set of posterior probabilities through Bayes' theorem can be conveniently carried out in a tabular format. For the observation of a copper coin (let us use C) the calculations in this form are

θ_i	$P(\theta_i)$	$P(C\|\theta_i)$	$P(\theta_i \cap C)$	$P(\theta_i\|C)$
θ_1	1/6	2/3	2/18	2/9
θ_2	1/6	3/3	3/18	3/9
θ_3	4/6	1/3	4/18	4/9
			$\overline{9/18} = P(C)$	

Our belief that θ_1 was drawn has changed from 1/6 to 2/9 as a result of observing a copper coin. Our belief in θ_2 has increased from 1/6 to 3/9, and our belief in θ_3 is reduced from 4/6 to 4/9. Now we can reevaluate the strategies in view of our revised beliefs about the states.

States (θ_i)	$P(\theta_i\|C)$	S_1	S_2	S_3	S_4	S_5
				Strategies		
θ_1	2/9	120	0	0	20	20
θ_2	3/9	0	120	0	0	30
θ_3	4/9	0	0	30	40	10
Expected value given C		240/9	360/9	120/9	200/9	170/9

Strategy 2 with an expected value of

$$0(2/9) + 120(3/9) + 0(4/9) = 360/9 = \$40$$

is now optimal. We can abbreviate this by saying $EV(S_{\text{opt}}|C) = \$40$. Note that the expected value under certain prediction is now

$$EV(CP|C) = 120(2/9) + 120(3/9) + 40(4/9) = 760/9 = \$84.44,$$

and the expected value of perfect information is now

$$EVPI|C = EV(CP|C) - EV(S_{\text{opt}}|C) = \$84.44 - \$40.00 = \$44.44.$$

These compare with our earlier evaluations as follows:

Probabilities		$EV(S_{\text{opt}})$	$EV(CP)$	$EVPI$
Prior:	$P(\theta_i)$	\$30.00	\$66.67	\$36.67
Revised given C:	$P(\theta_i\|C)$	\$40.00	\$84.44	\$44.44

The expected value under certain prediction has increased, because our belief in θ_1 and θ_2 has been increased by the observation of a copper coin, and correctly predicting these states has large payoffs. Notice that five out of the nine copper coins in the six urns are in urns θ_1 and θ_2, and our revised probability for $\theta_1 \cup \theta_2$ is 5/9.

To emphasize the process of revising probabilities and strategies, let us see what our situation would be *if we were to observe a gold coin*. First, we carry out Bayes' theorem to obtain the probabilities of states (urns) posterior to observing a gold coin. Using the tabular format, these are

θ_i	$P(\theta_i)$	$P(G\mid\theta_i)$	$P(\theta_i \cap G)$	$P(\theta_i\mid G)$
θ_1	1/6	1/3	1/18	1/9
θ_2	1/6	0/3	0/18	0/9
θ_3	4/6	2/3	8/18	8/9
			$\overline{9/18} = P(G)$	

Then, the expected values of the different strategies would be

θ_i	$P(\theta_i\mid G)$	Strategies S_1	S_2	S_3	S_4	S_5
1	1/9	120	0	0	20	20
2	0/9	0	120	0	0	30
3	8/9	0	0	30	40	10
Expected value given G		120/9	0/9	240/9	340/9	100/9

The optimal strategy, should we observe a gold coin, is S_4, to keep the gold coin(s) in the urn drawn. This happens to be the same strategy that would be best when there is no opportunity to sample the urn drawn. The expected value is now 340/9 = \$37.78, rather than \$30.00 as before. This happens because the probabilities would change as a result of observing the gold coin. For the expected value under certain prediction and the expected value of perfect information, *should we observe a gold coin*, we have

$$EV(CP\mid G) = 120(1/9) + 120(0/9) + 40(8/9) = 440/9 = \$48.89,$$

$$EVPI\mid G = EV(CP\mid G) - EV(S_{\text{opt}}\mid G) = \$48.89 - \$37.78 = \$11.11.$$

The expected value of perfect information, posterior to observing a gold coin, is only \$11.11. This happens because there is now only a 1/9 probability of the urn actually being θ_1. If the urn is θ_1 rather than θ_3, we get a payoff under S_4 of 20 rather than the 120 that we could get with a perfect predictor of states. If we had a perfect predictor we would gain the 120. But it occurs with probability 1/9, so our expected *gain* from a perfect predictor is now (120 − 20)(1/9) = \$11.11.

We can now summarize the results of our revised strategies as follows.

Information event	Optimal strategy	$EV(S_{\text{opt}})$	$EV(CP)$	EVPI
Observe a copper coin	S_2	\$40.00	\$84.44	\$44.44
Observe a gold coin	S_4	37.78	48.89	11.11

Preposterior Evaluation of an Information Strategy

In the previous section we put ourselves in the position of having observed a copper (and then a gold) coin. We then reevaluated the expected values in accordance with the particular information observed. This is called posterior analysis—we evaluated our decision situation posterior to receiving specified information (however imperfect) about the states. We now want to examine a similar situation, but this time a decision must be made about whether or not to obtain the specified information before selecting a strategy.

Suppose that in this new situation, the five original strategies are available, but the gambler offers another alternative. For a $5.00 payment we can draw a coin from the urn selected and then decide on one of the five strategies. We will call this an information strategy because it provides some information about which urn was chosen. Our problem is to evaluate the information strategy to find whether the expected payoff under it justifies spending $5.00 per play to obtain the information. To do this we must find the optimal strategy for each possible information outcome. Then we must find the expected value of the set of contingent best strategies. The word coined to describe this analysis is *preposterior*.

We have already reevaluated the original strategies under the conditions of observing a copper coin and a gold coin. By selecting the best strategy given each outcome, we found that we could achieve an expected payoff given copper of $40.00 and an expected payoff given gold of $37.78.

The information option can now be seen as a gamble between an expected payoff of $40.00 given copper and of $37.78 given gold. One expected payoff prevails if a copper coin is drawn (and S_2 elected), and the other prevails if a gold coin is drawn (and S_4 elected). The expected value of the entire information strategy, abbreviated $EV(IS)$, is then

$$EV(IS) = EV(S_{\text{opt}}|C) \cdot P(C) + EV(S_{\text{opt}}|G) \cdot P(G).$$

We need the marginal probabilities of drawing a gold coin and of drawing a copper coin. These are in the denominator of the Bayes' theorem calculations of the revised probabilities. But to recall them we show below two tables of probabilities.

θ_i	$P(\theta_i)$	$P(Event\|\theta_i)$ G	C	θ_i	$P(\theta_i \cap Event)$ G	C
θ_1	1/6	1/3	2/3	θ_1	1/18	2/18
θ_2	1/6	0/3	3/3	θ_2	0/18	3/18
θ_3	4/6	2/3	1/3	θ_3	8/18	4/18
					9/18	9/18 ← Marginal probabilities of events

Note that

$$P(\theta_i \cap \text{Event}) = P(\theta_i) \cdot P(\text{Event}|\theta_i).$$

The expected value of the information strategy is then

$$EV(IS) = \$40.00(9/18) + \$37.78(9/18) = \$38.89.$$

The foregoing calculations can be illustrated with a tree diagram, Figure 7.3, that shows the information strategy in a time sequence.

If we use the option of selecting an urn and drawing a coin before electing a strategy, the first thing that happens is that we will draw either a gold or a copper coin. If the coin is gold, the optimal strategy is S_4, as we have already determined. Next we will

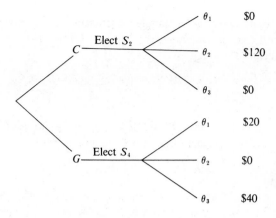

$$\begin{array}{c}
\\
\theta_1 \quad \$0 \\
\text{Elect } S_2 \\
C \quad\quad\quad\quad \theta_2 \quad \$120 \\
\\
\theta_3 \quad \$0 \\
\\
\theta_1 \quad \$20 \\
\text{Elect } S_4 \\
G \quad\quad\quad\quad \theta_2 \quad \$0 \\
\\
\theta_3 \quad \$40
\end{array}$$

Figure 7.3   Tree Diagram with Optimal Strategies in Urn Problem

find one of the states $(\theta_1, \theta_2, \theta_3)$ to be true. If the coin is copper, S_2 is optimal. And the tree diagram shows that one of the three states will be the true one in this case also. We find which state is true when the remaining coins in the urn selected are exposed. The tree diagram emphasizes that electing S_4 after observing a gold coin amounts to a gamble that we have already evaluated, namely, $EV(S_4|G)$. Similarly, electing S_2 after observing a copper coin amounts to another gamble, namely, $EV(S_2|C)$. Finally, we evaluate the gamble between observing G (gold) or C (copper) with their associated conditional expected values. The process can be summarized in a formula for the expected value of an information strategy, $EV(IS)$.

$$\blacksquare \quad EV(IS) = \sum_k [EV(S_{\text{opt}}|E_k) \cdot P(E_k)]. \quad \blacksquare \quad\quad (7.5)$$

Here, E_k stands for the possible information outcomes. In our example E_1 might stand for observing a gold coin and E_2 for observing a copper coin. One must determine S_{opt} for each E_k by reevaluating the basic strategies using posterior (to the event) probabilities determined from Bayes' theorem.

The Expected Value of Sample Information

Recall that when we could not examine a coin, the best strategy in the urn game was S_4, to take the gold coins in whatever urn was drawn. The expected value of this optimal strategy was \$30. Now, having determined the optimal strategies when we could sample the urn drawn, we find that our expected value for a play of the game is \$38.89. We can increase our expected gain by properly utilizing the information about states contained in a sample of one coin from the selected urn. The expected value of the information is this increase of \$8.89. Therefore if we could obtain this information for \$5.00, it would pay us to take it.

We can now state a formula for the expected value of sample information.

Definition	**The expected value of sample information** is the expected increase in payoff that accompanies the optimal use of that information in the selection of strategies.

$$\blacksquare \qquad EVSI = EV(IS) - EV(S_{opt}). \qquad \blacksquare \qquad (7.6)$$

In our example, $EV(IS) = \$38.89$ and $EV(S_{opt}) = \$30.00$ and $EVSI = \$8.89$. Recall the expected value of perfect information, where we had

$$EVPI = EV(CP) - EV(S_{opt}).$$

In both cases the expected value of information is the increase in expected payoff that can be achieved by using the information. Both the expected value under certain prediction and the expected value of an information strategy must be determined by looking forward to information outcomes. In calculating the expected value under certain prediction we have to ask ourselves what the optimal strategy would be, given certain information about each state. But because the states are uncertain, we must apply probabilities of states, $P(\theta_i)$, to these maximum payoffs. The calculation of the expected value of an information strategy is similar in concept. We must find the optimal strategy given each information outcome. Then, because the information outcomes are uncertain, we must weigh the expected values (of these optimal strategies) by probabilities for each information outcome.

7.2

Summary

When probabilities of states are known as well as payoffs from employing alternative strategies should different states occur, the expected payoffs from employing each strategy can be determined. Bayes' rule identifies as optimal the strategy with the largest expected payoff. After the maximum payoff given each state is found, the expected value under certain prediction can be found. The expected value of perfect information is the difference between the expected value under certain prediction and the expected value (payoff) of the optimal strategy.

When sample information about the states is received, our beliefs about states change. Bayes' theorem is used to calculate these posterior probabilities of states. Corresponding reevaluation of strategies may reveal that the optimal strategy has changed as a result of the information.

If possible sample outcomes are considered in combination with existing probabilities of states, a strategy to gather further information about states before making a decision can be evaluated. The expected value of the information strategy is found through preposterior analysis. The expected value of the sample information is the difference between the expected value of the information strategy and the expected value of the optimal strategy in the absence of the information.

See Self-Correcting Exercises 7.A

Exercises

1. A soft drink bottler presently packages his product in standard bottles. He is debating whether to change his packaging for next year. He can adopt a lift-top can (S_1), a screw-top bottle (S_2), or keep the old-style bottle (S_3). Profits from each strategy depend on what the bottler's major competitor does. The payoff table (in $10,000 units) and the probabilities of his competitor's actions are:

Competitor's action	Probability	Bottler's strategy		
		S_1	S_2	S_3
Uses old-style can	0.5	14	13	16
Uses screw-top bottle	0.3	9	8	6
Uses lift-top can	0.2	6	8	5

Determine the expected payoff from each of the strategies.

2. Find the expected value under certain prediction for the problem in Exercise 1. What is the expected value to the bottler of perfect advance information about which action his competitor will take?

3. The competitor in Exercise 1 currently uses old-style cans. Suppose it becomes known only that he is going to make a change. Revise the probabilities of states and the expected values of the bottler's strategies.

4. A friend has two dice. The first is a fair die and the second is a die so weighted that the probabilities of 1, 2, and 3 are 3/12 each, and the probabilities of 4, 5, and 6 are 1/12 each. You pick one of the dice at random, roll it twice, and obtain two "ones." What is the probability that you picked the fair die? What would the answer be if you had obtained two "sixes"?

5. Prior experience with shipments of four experimental components led an R & D department to establish the following probability distribution for number defective in a current shipment of four components.

Number defective	0	1	2	3	4
Probability	6/10	3/10	1/10	0	0

One component from the shipment is to be tested. Revise the probabilities for number of defectives in the shipment for the event that

a. the component inspected is good.

b. the component inspected is bad.

6. If the R & D department in Exercise 5 uses the components without any further checking, it will incur a cost of $500 for each defective component used. On the other hand, it can check all components at a testing cost of $50 per component and put any defectives in working order at a cost of $200 per defective. Evaluate these two strategies.

7. Reevaluate the strategies in Exercise 6 should one component be tested and found good. Remember that the test costs $50.

8. Reevaluate the strategies in Exercise 6 should one component be tested and found defective. Assume that the defective already found would be put in working order at a cost of $200.

9. Find the expected value of the information strategy of testing one component before deciding whether to accept or to test the remaining components. What is the expected value of the sample information?

Further Aspects of Decision Theory

The first half of this chapter presented the basic elements in Bayesian decision problems and the different expected values needed to deal with them. We go now to some related problems and issues that are likely to arise in applying these basic ideas.

A single example will serve as our vehicle for raising these additional aspects of decision theory. Our firm (A) has developed a product that gives it a unique capability to exploit a new market. The key to the product is a possibly patentable device. However, Firm B holds a patent on a somewhat similar device, and it is not clear whether Firm A's patent application would be granted. If we enter the market and our device proves not patentable, we will be liable to a suit for infringement of B's patent. An alternative strategy is to purchase rights from Firm B for use of this patented device in our product. The payoff table (in thousands of dollars) has been determined as follows:

	Strategy	
State	S_1 *Apply for patent*	S_2 *Purchase rights*
θ_1, Device patentable	60	50
θ_2, Device not patentable	30	50

If our firm applies for the patent and the device proves to be patentable, a $60,000 profit can be made in the market venture. Damages from a patent infringement suit, should the device not prove patentable, would reduce the profit to $30,000. On the other hand, should we purchase rights from Firm B, our net profit from the venture will be $50,000 whether or not our device could be patented.

Expected Value of a Decision Rule

Consider now an information seeking alternative. Before deciding whether to apply for a patent or purchase rights it might be useful to consult a patent attorney. The attorney is known to be 80 percent accurate in the sense that if the device is really patentable, the probability is 0.80 that he will say that it is and 0.20 that he will say it is not. Also if the device is not patentable, the probability is 0.80 that he will say it is not and 0.20 that he will say it is patentable. We want to consider the decision rule (DR) to apply for the patent if the attorney says the device is patentable and to purchase rights from Firm B if the attorney says the device is not patentable.

The decision rule amounts to two conditional gambles. If the device is patentable (θ_1), the probability is 0.80 that S_1 will be adopted and 0.20 that S_2 will be adopted. If the device is not patentable (θ_2), the probability is 0.20 for adopting S_1 and 0.80 for adopting S_2. The payoff table including the decision rule is as follows.

	Strategy		
State	S_1	S_2	DR
θ_1	60	50	$58 = 60(.80) + 50(.20)$
θ_2	30	50	$46 = 30(.20) + 50(.80)$

In the previous section preposterior revision of probabilities was carried out to determine optimal strategies conditional on the possible information outcomes. In the decision-rule situation the strategies to be adopted upon observing different information outcomes are dictated in advance. The decision rule may or may not represent the optimal use of the information (although it does here).

To determine the expected value of a decision rule it is easiest to regard the original states as primary and use the probabilities of information outcomes conditional on them. Suppose the probability that the device is patentable is 0.75 (this represents a subjective degree of belief held by the firm's officers). The expected values of S_1, S_2, and the decision rule (employing the lawyer's advice) are easily determined.

		Strategy		
State (θ_i)	$P(\theta_i)$	S_1	S_2	DR
θ_1	.75	60	50	58
θ_2	.25	30	50	46
Expected values		52.5	50.0	55.0

The expected value of the decision rule is $55 thousand, and this is $2.5 thousand better than applying for the patent now without using the decision rule. The lawyer's advice in conjunction with the rule for using it is worth $2.5 thousand.

A formula for finding the expected value of a decision rule can be written as

$$EV(DR) = \sum_i [E(V_{ik}|\theta_i) \cdot P(\theta_i)]. \tag{7.7}$$

Here, k indexes the information outcomes. For each information outcome the decision rule dictates adoption of one of the available strategies. This may be visualized by the "states first" tree diagram of Figure 7.4.

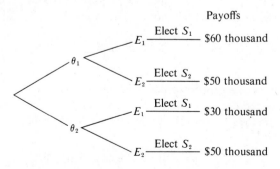

Figure 7.4   Tree Diagram for a Decision Rule

Here, E_1 is the information outcome that the lawyer says the device is patentable, and E_2 the outcome that he says it is not. The tree shows these outcomes leading to the strategies specified by the decision rule and the resultant payoffs depending on the true state. The statements about accuracy of the lawyer's advice are probabilities of the type $P(E_k|\theta_i)$, which belong on the second-stage branches. Following Formula 7.7 we would calculate

$$E(V_{ik}|\theta_1) = 60(.80) + 50(.20) = 58,$$
$$E(V_{ik}|\theta_2) = 30(.20) + 50(.80) = 46,$$
$$EV(DR) = 58(.75) + 46(.25) = 55.0.$$

Subjective Probability

You will notice a difference between the urn example and the patent problem in that there appears to be no objective way to establish probabilities of states in the patent problem. There is nothing in the patent problem that corresponds to the relative frequency of urn types. In Chapter 4 we saw that probabilities could be used as measures of degrees of belief as well as proportions of long-run frequencies. But how can one establish a measure of belief in a situation like the patentability of Firm A's device?

The answer is to find a gamble with objective probabilities that the decision maker regards as equivalent to the uncertain business prospect at hand. Equivalence here means indifference or lack of preference. To illustrate, the payoffs under S_1 are 60 if θ_1 and 30 if θ_2. We might ask the decision maker the following question.

There are 100 balls in an urn—50 red ones and 50 black. You have an opportunity to draw one ball (blindfolded). You will be paid $60 thousand if the ball drawn is red and $30 thousand if the ball drawn is black. Would you prefer this lottery to adopting S_1 in your decision problem?

If the answer is "no," we know (letting $W = P(\theta_1)$) that

$$60W + 30(1 - W) > 60(.50) + 30(.50),$$
$$30W > 15,$$
$$W > .50.$$

We would then ask the decision maker to consider a reference lottery with more red balls, say 80 red balls, and 20 black. Suppose he says he would prefer this lottery to S_1 in his decision problem. Then we know

$$60W + 30(1 - W) < 60(.80) + 30(.20),$$
$$W < .80.$$

Now we would reduce the proportion of red balls in the reference lottery. At some point in the process the decision maker would reach a point of indifference. If this occurred with a reference lottery of 75 red and 25 black balls, we would have the equation

$$60W + 30(1 - W) = 60(.75) + 30(.25),$$
$$W = .75.$$

At this point we have assessed the decision maker's degree of belief in the patentability of Firm A's device. The $\bar{0}.75$ probability for θ_1 and 0.25 for θ_2 can now be used in the same way that we used the probabilities of the different type urns in our earlier problem.

Recalling our payoff table and inserting the subjective probabilities of states, we have

		Strategies	
State (θ_i)	$P(\theta_i)$	S_1	S_2
θ_1: Device patentable	.75	60	50
θ_2: Device not patentable	.25	30	50

$$EV(S_1) = 60(.75) + 30(.25) = \$52.5 \text{ thousand},$$
$$EV(S_2) = 50(.75) + 50(.25) = \$50.0 \text{ thousand}.$$

The optimal strategy is S_1, to apply for the patent while going ahead with market development. The expected value under certain prediction is

$$EV(CP) = 60(.75) + 50(.25) = \$57.5 \text{ thousand}.$$

This means that if we had a way of getting certain (errorless) information about patentability before making a decision between applying for a patent (S_1) and purchasing rights (S_2), our expected payoff would be \$57.5 thousand—because our present belief is 0.75 that the information will reveal the device to be patentable and 0.25 that it will not. The expected value of this perfect information is the expected incremental gain from using it.

$$EVPI = EV(CP) - EV(S_{\text{opt}}),$$
$$EVPI = 57.5 - 52.5 = \$5.0 \text{ thousand}.$$

Sensitivity Analysis

Many businessmen are unwilling or unable to assess subjective probabilities in the manner of the preceding section. In such cases a decision-rule problem can be subjected to a sensitivity analysis. The object of a sensitivity analysis is to find out how the solution to a problem changes when various elements in the problem are varied. If we are unable to establish probabilities of states, we can ask how different values for these probabilities affect the optimum strategy solution.

Figure 7.5 portrays the elements of the patent problem. The horizontal axis is $P(\theta_1)$, the probability that the device is patentable. At $P(\theta_1) = 1.0$ we plot the payoffs under S_1 and under S_2. We also plot the expected payoff for the decision rule discussed earlier. At $P(\theta_1) = 0$ (when θ_2 is a certainty) we plot the payoffs under θ_2 for the three strategies. To recall the figures, see the payoff table on page 153.

Now we connect with straight lines the two payoffs under each strategy. Thus we have a line segment for S_1, another for S_2, and another for the decision rule (DR). Each line shows the relation between $P(\theta_1)$ and the expected payoff under a strategy.

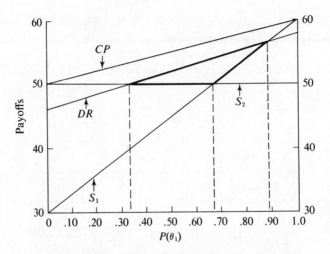

Figure 7.5   Elements of a Decision Problem

Note that the S_1 line lies below the S_2 line over the range of $P(\theta_1) < 2/3$ and that the S_1 line is higher when $P(\theta_1) > 2/3$. If $P(\theta_1) = 2/3$, we would be indifferent between S_1 and S_2. This probability level can be found algebraically by equating expressions for $EV(S_1)$ and $EV(S_2)$ and solving for the unknown $P(\theta_1)$. Letting $P(\theta_1) = W$, we would have

$$60W + 30(1 - W) = 50W + 50(1 - W),$$
$$60W + 30 - 30W = 50,$$
$$30W = 20,$$
$$W = 2/3.$$

When we examine the decision-rule line, we find that it lies above the other two lines over the range from $P(\theta_1) = 1/3$ to $P(\theta_1) = 8/9$. This is the range of values for $P(\theta_1)$ over which the decision rule is superior to either of the basic strategies. This superiority diminishes as $P(\theta_1)$ moves toward either limit—shown graphically by the shape of the triangle formed by the three intersecting lines. The range and shape of this triangle can be helpful to a decision maker unwilling or unable to assess subjective probabilities.

Also shown on the graph is the line corresponding to the expected value under certain prediction. This is the line of the equation

$$EV(CP) = 60 \cdot P(\theta_1) + 50 \cdot [1 - P(\theta_1)].$$

We can see that as long as there is any uncertainty about the true state, there is value to a decision rule based on perfect information. The $EV(CP)$ line lies above the line for the optimal strategy (at first S_2 and then S_1 as $P(\theta_1)$ increases) except at $P(\theta_1) = 0$ and $P(\theta_1) = 1.0$. At these values uncertainty disappears.

The graph can also give us a feeling for the effects of decision rules based on increased or decreased accuracy of information. Increased accuracy of information will move the $EV(DR)$ line up closer to the $EV(CP)$ line. The effect of this is to widen the range over which the decision rule is superior. Lower information accuracy will result in a more restricted range of $P(\theta_1)$ over which the decision rule has comparative value.

Subjective Utility

An assumption on which the criterion of maximizing expected monetary payoff rests is that the decision maker is equally desirous of each additional unit of money reward. Many decision makers do not always behave this way, and few students of decision theory believe that they should. Some individuals (and firms) strive for a modest reward and have less desire for additional money beyond this point than they have below it. Others have a low desire for early increments of reward and a greater desire for the increments as they begin to approach a level representing some goal of achievement or personal satisfaction. Indeed, both of these tendencies can exist in the same decision maker if a wide enough range of reward is considered. The result is an S-shaped total utility curve, such as shown in Figure 7.6. Here we see an early

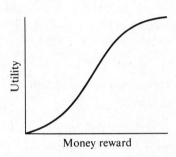

Figure 7.6   An S-Shaped Utility Curve

stage of increasing marginal utility of money reward, and a later stage of decreasing marginal utility for incremental money units.

If a decision maker does not have a straight-line utility curve over the range of payoffs in the payoff table for his decision problem, he should not use the previous methods of this chapter without modification. The modification is that he should employ utility values and expected utility criteria rather than expected money values. This raises the question of how to determine utility values, which we now turn to. We will use the example of the patent problem.

In determining utility we need to establish an interval scale of preference. Recall from Chapter 1 that on an interval scale equal differences anywhere on the scale represent equal amounts of the characteristic the scale is intended to represent. In a ten-pound scale of weight, the first pound (interval from $x = 0$ to $x = 1$) is the same in this sense as the last pound (interval from $x = 9$ to $x = 10$). A utility scale for monetary reward simply establishes the (possibly unequal) intervals of reward that are equal in terms of preference (desirability or utility).

The method of measuring subjective utility and the theory supporting it was first proposed by von Neumann and Morgenstern [1]. The problem reduces to properly positioning an intermediate payoff between the ends of a utility scale whose extremes are two other payoffs. This is the situation faced in our patent problem where there are only three payoff values, $30 thousand, $50 thousand, and $60 thousand. The units of the utility scale are arbitrary (as are units for weight, volume, temperature, and so forth), so we let the ends of the scale be zero and unity. Thus the utility of $30 thousand is established as zero and the utility of $60 thousand is established as unity. Symbolically, we write $U(30) = 0.0$ and $U(60) = 1.0$. The problem is to determine $U(50)$ for the decision maker.

The procedure for determining $U(50)$ is to present the decision maker with a reference lottery. We used these in connection with subjective probability. In the present case we ask the following question.

There are 100 balls in an urn, r of which are red and $100 - r$ of which are black. You receive a payoff of $60 thousand if you draw a red ball and $30 thousand if you draw a black ball. How many red balls must we place in the urn for you to be indifferent between receiving $50 thousand for certain and gambling on a draw from the urn with the payoffs as mentioned?

Suppose the answer is 90 red balls. We then know that the utility (degree of preference) of $50 thousand is equal to the expected utility of the gamble with a 0.90 probability of gaining $60 thousand and a 0.10 probability of gaining $30 thousand. Remember that we assigned $U(60) = 1.0$ and $U(30) = 0.0$. We have established that

$$U(50) = 0.90 \cdot U(60) + 0.10 \cdot U(30),$$

$$U(50) = 0.90(1.0) + 0.10(0.0),$$

$$U(50) = 0.90.$$

We have now positioned (or scaled) $U(50)$ between $U(30)$ and $U(60)$. It is important to notice that the reference lottery, in determining utility, does not associate the payoffs with the patent-decision situation. The choice in the reference lottery has nothing to do with belief in patentability of the device. Thus the utility determination is wholly separate from determination of subjective probability. In fact it is advisable that the two be disassociated as much as possible.

Once utilities are established, the decision problem is solved in terms of utility. Using $P(\theta_1) = 0.75$ as previously determined, we have

State (θ_i)	$P(\theta_i)$	Strategies		Decision rule (DR)
		S_1	S_2	
θ_1: Device patentable	0.75	1.00	0.90	$0.98 = 1.00(0.80) + 0.90(0.20)$
θ_2: Device not patentable	0.25	0.00	0.90	$0.72 = 0.00(0.20) + 0.90(0.80)$
Expected utility		0.75	0.90	0.915

If we ignore the decision rule for the moment, the optimal strategy is now S_2, to purchase rights, rather than S_1, which maximized expected money payoff. This happens, in part, because incremental dollar payoffs in the range from $50 thousand to $60 thousand have lower utility for the decision maker than the incremental dollar payoffs in the range between $30 thousand and $50 thousand.

Determination of the expected utility of the decision rule involving the lawyer's opinion would proceed as before once the utilities of the basic payoffs have been found. If θ_1 is true, following the lawyer's opinion will give us a 0.80 probability of gaining a utility of 1.00 and a 0.20 probability of gaining a utility of 0.90. Similarly, if θ_2 is true, the decision rule represents a gamble with 0.20 probability of gaining a utility of zero and 0.80 probability of gaining a utility of 0.90. Weighting the expected utilities of these gambles by the probabilities of states yields the expected utility of the decision rule.

7.3

Summary

A decision rule associates each possible information outcome with adoption of one from among the available strategies. The expected value of a decision rule is most easily calculated by first finding the expected values conditional on each state, and then combining these using the initial probabilities of states. In most business situations the probabilities of states are not objective, but are measures of the current belief of the decision maker about the states. These probabilities can be assessed by using reference lotteries that equate the subjective belief to an objective probability. If the decision maker hesitates to do this, an analysis of the sensitivity of optimal strategies to variations in probabilities of states can be helpful in selecting a strategy. In some situations decision makers may not accept monetary value as an interval measure of preference among payoffs. In these cases such an interval measure should be determined. The measure is called subjective utility because it reflects a scale of evaluation employed by the individual. Once a decision maker's utility scale is determined, utilities are used in analyzing strategies in the same way that monetary values would otherwise be used.

See Self-Correcting Exercises 7.B

Exercises

1. An executive is considering alternative advertising plans. If there is no marked change in economic climate, Plan A will have a payoff of $6000, while Plan B would have a pay off of $3000. If the economic climate changes markedly, Plans A and B would have payoffs of $4000 and $8000, respectively. The executive assesses the probability of no marked change in economic climate as one-half. Which plan is optimal?

2. If he wants to, the executive in Exercise 1 can follow a decision rule to select Plan A if his staff economist predicts no change and Plan B if change is predicted. In the past, when there has been no change the economist has so predicted 4/5 of the time. When there has been a marked change, he has predicted correctly only 3/5 of the time. Find the expected value of the decision rule.

3. Draw a graph like Figure 7.5 for the elements of the executive's problem in Exercises 1 and 2. How sure would the executive have to be about economic change before he could dispense with his economist's assistance?

4. A lumber yard can make 6 cents a board foot selling a certain variety of ungraded lumber. If it separates the lumber into grades 1 and 2, it can make 10 cents on the grade-1 lumber and 4 cents on grade 2. Over what range of quality in the ungraded lumber would it pay the yard to separate and sell graded lumber?

5. George is considering a business proposition whereby he could lose $200 or gain $400. He says he would not consider the proposition at all unless the probability of success exceeded 0.5, and that if he had another venture that yielded a sure $200, the probability of success in the uncertain venture would have to be 0.8 before he would prefer the gamble. Sketch George's utility curve for money gains from − $200 to $400, and describe George's behavior.

6. An insurance salesman handles two different insurance plans. His customers usually prefer one of the two types, and if he presents the wrong plan first, it tends to reduce the amount of coverage he will sell. To overcome this problem, he sends out a brief questionnaire before calling on a prospect. He has found that because of misunderstanding, twenty percent of those preferring each plan will misidentify their preference on the questionnaire. From the payoff table below, determine the expected value of presenting the plan indicated on the questionnaire.

| | | Presentation | |
Probability	Preference	Plan 1	Plan 2
Plan 1	0.6	40	20
Plan 2	0.4	30	40

7. Three individuals, A, B, and C, are offered for $3 a 50–50 chance of gaining $1 or $5. Their utility curves are

$$A: \quad U(X) = (X - 1)^2/16,$$

$$B: \quad U(X) = (X - 1)/4,$$

$$C: \quad U(X) = 5(X - 1)/4X.$$

Determine in each case whether or not the offer should be accepted.

8. Billing clerks each work on batches of 100 bills. For each bill that is incorrectly calculated, adjustment costs of $0.25 can be expected. The cost of a 100 percent check of the calculations would be $1.00 per batch of bills. Assume that past experience has shown that 70 percent of the batches have no errors and 30 percent of the batches have ten errors each. Find the expected values of the strategies "check all work" and "check no work."

9. A sampling plan for checking the billing operation in Exercise 8 is suggested. It consists of a sample of five bills (with replacement) from each batch. If no errors are found in a batch, the bills are sent out; if one or more errors are found, the entire batch is checked and all errors corrected. The cost of the sampling procedure is $0.20 per batch, and the cost of the checking operation (if required) is $1.00 per batch. Find the expected value of the decision rule.

Glossary of Equations

$$EV(S_j) = \sum_i [V_{ij} | S_j \cdot P(\theta_i)]$$

The expected value of a strategy is a weighted sum of payoffs under that strategy should different states of nature occur. The weights are the probabilities of the states of nature.

$$EV(CP) = \sum_i [(V_{max} | \theta_i) \cdot P(\theta_i)]$$

The expected value under certain prediction is a weighted sum of the maximum payoffs from any strategy given each state of nature. The weights are the probabilities of the states of nature.

$$EVPI = EV(CP) - EV(S_{opt})$$

The expected value of perfect information is the difference between the expected value under certain prediction and the expected value of the optimal strategy.

$$P(\theta_i | E) = \frac{P(\theta_i \cap E)}{P(E)}$$

The probability that an event was produced by one of several causes is the ratio of the joint probability of the cause and the event to the marginal probability of the event. (Bayes' theorem)

$$EV(IS) = \sum_k [EV(S_{opt} | E_k) \cdot P(E_k)]$$

The expected value of an information strategy is a weighted sum of the expected values of the optimal strategies given each information outcome. The weights are the marginal probabilities of the information outcomes.

$$EVSI = EV(IS) - EV(S_{opt})$$

The expected value of sample information is the difference between the expected value of an optimal information strategy based on the sample information and the expected value of the optimal strategy without the sample information.

$$EV(DR) = \sum_i [E(V_{ik} | \theta_i) \cdot P(\theta_i)]$$

The expected value of a decision rule is found by first calculating the expected payoff over the information outcomes given each state. These expected payoffs are then weighted by the initial probabilities of states.

8

Continuous Probability Distributions

In Chapters 4 to 6 we treated probability as a function of values of a random variable. However, we considered only discrete random variables. Their values were capable of being listed and their frequencies of occurrence counted. When all the elements in a sample space were listed, the probability distribution of a random variable associated with the space (summation of X, mean of X, and so on) could be determined by counting or by applying some elementary probability laws.

Discrete and Continuous Probability Distributions

For the sake of review, consider a draw from a card deck consisting of the thirteen spades. The sample space is $\{A, 2, 3, 4, \ldots, 9, 10, J, Q, K\}$. In many card games an ace has the value 1 and a jack, queen, or king count as a value of 10. If the elements in the sample space are equiprobable, the probability distribution of the random variable, X, where X is the customary point value associated with the cards, is

X	1	2	3	4	5	6	7	8	9	10
$P(X)$	1/13	1/13	1/13	1/13	1/13	1/13	1/13	1/13	1/13	4/13
$F'(X)$	1/13	2/13	3/13	4/13	5/13	6/13	7/13	8/13	9/13	13/13

Recall now the two methods explored in Chapter 2 for the graphic presentation of the relative frequency distribution of a discrete variable. First, we plotted the relative frequency of occurrence of values as point ordinates drawn at those values up to a height corresponding to their frequency. We called these values $f'(X)$. The second form of presentation was the cumulative relative frequency distribution, defined as $F'(X)$. Here we plotted, at each discrete value of X, the proportion of cases having a value of X equal to or less than this. If we have in mind an experiment *draw a single unit of observation*, then $f'(X)$ is the same as $P(X)$ and $F'(X)$ is the same as the cumulative probability of an observation equal to or less than the specified value of X. We will use $F'(X)$ to denote the cumulative probability distribution of a random variable X. In Figure 8.1 the graphs of $P(X)$ and $F'(X)$ for the card example are shown.

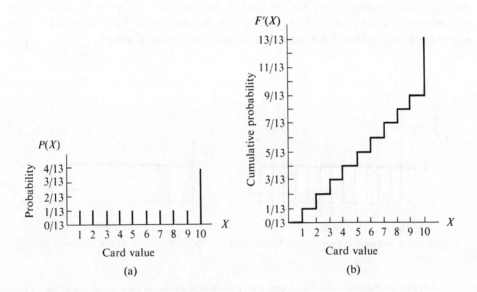

Figure 8.1   (a) Probability Distribution and (b) Cumulative Probability Distribution of a Discrete Random Variable

Consider now a slightly different kind of problem. A finely balanced spinner is mounted in the center of a board. Around the board is a scale of numbers from 1 to 12 arranged like the hours on a clock. Finer decimal divisions are indicated as well as the exact "hours." If the spinner is spun, what is the probability that it will come to rest between 2.0 and 3.0?

Spinning apparatus

Most students answer 1/12 with little hesitation. Without counting elements in the sample space, they assume that the spinner is as likely to come to rest at any one point as another and that the interval between 2.0 and 3.0 contains one-twelfth of the total points in the sample space. The points are, of course, not countable, because we are dealing with a continuous random variable. The spinner may come to rest anywhere along the continuum from 0.0 to 12.0. The possible values of the variable are restricted only by the practical limitations of our measurement apparatus.

A Continuous Probability Model

If you answered 1/12 to our question above, you intuitively applied a *continuous probability model*. A model is simply a representation of reality that is useful for certain purposes. In terms of outcomes of an actual measuring apparatus we cannot have an infinite sample space. Nevertheless the continuous model may be useful on one of two counts. First, the possible outcomes may be so numerous that the continuous model provides a reasonable approximation—and it is much more convenient than the discrete alternative. Second, rather than restrict our study to a particular apparatus of limited resolution, we might wish our answers to have relevance for any apparatus of this general kind regardless of how finely it can measure.

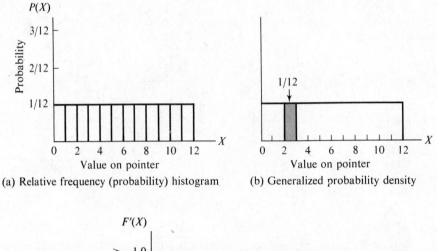

(a) Relative frequency (probability) histogram   (b) Generalized probability density

(c) Cumulative probability distribution

Figure 8.2   Graphs for a Continuous Rectangular Probability Model

In the probability model for a continuous random variable, the event is always *an interval* of the variable. Our problem is to find the probability of the event E where E is the subset of X in the interval from 2.0 to 3.0. This can be more briefly stated as find $P(2.0 \leq X \leq 3.0)$. The relative frequency histogram for the spinner problem is shown in Figure 8.2(a). To show $P(X)$ on the vertical scale we must first select equal intervals of X as a basis. We select an interval of length 1.0 for convenience. The scale of $P(X)$ then measures probability *per interval of* 1.0. To emphasize this we show the graph as it would appear in a relative frequency histogram with class intervals of 1.0. But the size of interval considered is arbitrary. If we took intervals of 0.5, the probabilities for the intervals would be 1/24; for intervals of 0.25 the probabilities would be 1/48; for intervals of 0.12 the probabilities would be 1/100, and so on. If one continues this process, it is clear that for intervals of zero width there is zero probability; in other words, the probability associated with a point, or exact value of X, on the scale is zero.

Because of the arbitrary nature of the size of the interval, it is customary in a graph of a continuous probability distribution to leave the vertical axis unlabeled. Since this axis represents the amount of probability per whatever unit interval of X we might specify, we call it the *generalized probability density*. The probability in any interval of X is measured by the proportion of area under the relative frequency curve within the interval of X. Thus panel (b) of Figure 8.2 shows no labeling on the vertical axis and the *area* under the distribution " curve " between $X = 2.0$ and $X = 3.0$ measures $P(2.0 \leq X \leq 3.0) = 1/12$.

The cumulative probability distribution of our continuous variable in the spinner problem is graphed in Figure 8.2(c). It is the equation

$$F'(X) = 0 + \left(\frac{1}{12}\right)X, \quad \text{where } 0 \leq X \leq 12.$$

The last clause indicates the range of X. The equation indicates that

$$F'(X = 0) = 0 + \left(\frac{1}{12}\right)(0) = 0,$$

which we know to be so. It says then that the additional probability cumulated per unit of X (or per interval of $X = 1.0$) is 1/12, which we accepted before. The total probability of 1.0 is cumulated at the upper limit of $X = 12$; that is,

$$F'(X = 12) = 0 + \left(\frac{1}{12}\right)(12) = 1.0.$$

The probability model for the spinner problem is a commonly encountered one. It can be described as a rectangular probability distribution. The cumulative relative frequency function of a rectangular probability distribution is a straight line.

Finding Probabilities from the Cumulative Probability Function

An advantage in expressing the cumulative probability function as an equation is that we can use the equation to find the probability for any desired interval of X. The general procedure is to find F' for the upper limit of the interval and to subtract from this F' for the lower limit of the interval. For example, the cumulative probability function for a rectangular distribution can be expressed as

$$F'(X) = \frac{X - X_{min}}{X_{max} - X_{min}},$$

where $X_{min} \leq X \leq X_{max}$. The values X_{min} and X_{max} indicate the range of the variable. In our spinner example $X_{min} = 0$ and $X_{max} = 12$. If we wished to find $P(2.5 \leq X \leq 3.7)$, we would calculate

$$F'(3.7) = \frac{3.7 - 0}{12 - 0} = 3.7/12,$$

$$F'(2.5) = \frac{2.5 - 0}{12 - 0} = 2.5/12,$$

and

$$P(2.5 \leq X \leq 3.7) = F'(3.7) - F'(2.5)$$
$$= 3.7/12 - 2.5/12 = 1.2/12 = 0.10.$$

Thus, the values X_{max} and X_{min} are the parameters of the particular rectangular distribution we are concerned with. Once they are specified we are able to use the equation for $F'(X)$ for a rectangular distribution to solve for a particular probability. In the chapter exercises some cumulative probability equations for distribution models other than the rectangular are provided.

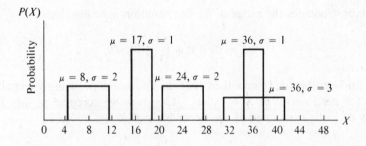

Figure 8.3   Five Rectangular Probability Distributions

Figure 8.3 emphasizes the idea of a family of continuous distributions all of the same form—in this case rectangular. The distributions vary in their central tendency and variability. Reading from the left the first distribution has a mean of 8 and a standard deviation of 2, the second has a mean of 17 and a standard deviation of 1, and so forth. The distributions are drawn with the same area, or total probability, so the distributions with smaller dispersion rise to higher probability densities. The third distribution has the same dispersion as the first but a greater mean. The last two distributions have the same mean, 36, but greatly different dispersions. One has a standard deviation equal to that of the second distribution ($\sigma = 1$), while the other has a standard deviation of 3, greater than the first and third distributions. The mean and standard deviation of a rectangular distribution are related in a fixed way to X_{max} and X_{min} that we used as parameters in the cumulative probability equation. The mean of a rectangular distribution is always equal to $(X_{max} + X_{min})/2$, and the standard deviation is always equal to $(X_{max} - X_{min})/\sqrt{12}$.

The Normal Distribution Model

A continuous probability model of crucial importance in statistical theory and practice is the *normal probability distribution*, or so-called *normal law of error*. The normal probability model has been found to provide a suitable description of the variability encountered upon making repeated measurements of many phenomena in the physical and biological sciences. In the nineteenth century some social scientists held the view that the normal law of error would be found to describe satisfactorily all manner of variable characteristics in their fields of inquiry—that is, it would be found to be a sort of immutable law of nature. While this view is no longer common, the normal distribution does have a general applicability in describing the variability in sampling distributions of continuous variables. The idea of a sampling distribution is very much like that of making repeated measures of a constant physical phenomenon. The true mean of a population is analogous to the true measure of a physical property, and the concept of means from repeated samples is analogous to repeated imperfect measures of the physical phenomenon.

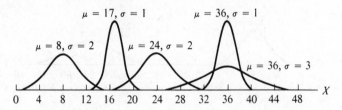

Figure 8.4   Five Normal Probability Distributions

Figure 8.4 shows the normal distributions with the same means and standard deviations as the rectangular probability distributions in Figure 8.3. All are symmetrical, bell-shaped distributions with the following properties:

1. The mean, median, and modal values of a normally distributed variable are identical.

2. The probability within an interval from μ to $(\mu + x)$ is the same as the probability within the interval from $(\mu - x)$ to μ. Remember that x is a deviation from the mean: $x = X - \mu$.

3. The curves are asymptotic to the X axis—the probability density approaches, but never reaches, zero. This means there is some probability, however small, that X exceeds $\mu + x$, however large, as well as an equal probability that X is less than $\mu - x$, however small.

In Figure 8.4 the several normal distributions should appear to be located like the rectangular distributions in Figure 8.3. In fact they are, and their standard deviations parallel those of Figure 8.3. Thus we have five normal distributions with the following parameters.

(1) $\mu = 8$, $\sigma = 2$,

(2) $\mu = 17$, $\sigma = 1$,

(3) $\mu = 24$, $\sigma = 2$,

(4) $\mu = 36$, $\sigma = 1$,

(5) $\mu = 36$, $\sigma = 3$.

The Standard Normal Distribution

The normal distribution is a fixed-form distribution. The mean and standard deviation serve to completely identify a normal distribution in the same way that specifying X_{max} and X_{min} serve to completely identify a rectangular distribution. If one properly adjusts for the scale factors of location and spread (μ and σ), the result is a standard form of the normal distribution. The adjustment is made by transforming the scale of the normal distribution to a z scale, where

$$\blacksquare \qquad z = \frac{X - \mu}{\sigma}. \qquad \blacksquare \qquad (8.1)$$

When transformed to the variable z, all normal distributions are alike. Figure 8.5 shows the z scale for the first and last distributions from Figure 8.4. When the units of z are spaced the same, the two distributions look exactly the same. Any normally distributed z variable has a mean of zero and a standard deviation of unity (1.0). This distribution is termed the *standard normal distribution*.

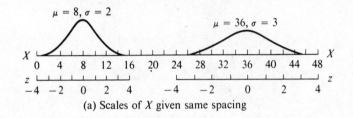

(a) Scales of X given same spacing

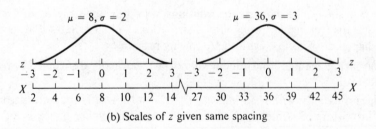

(b) Scales of z given same spacing

Figure 8.5   Two Normal Distributions from Figure 8.4 with z Scales

Appendix Table A contains cumulative probabilities for the standard normal distribution. For any value of z the table gives the cumulative probability up to that level of z. We call this a "lower tail" probability. For example, at $z = -1.50$ we read 0.0668. Thus

$$F'(z = -1.50) = 0.0668,$$

or we can write

$$P(z \le -1.50) = 0.0668.$$

Graphically, we have found the proportionate area shown in Figure 8.6. Note that if we wanted $P(z < -1.50)$ the answer would be the same, because there is no probability (area) at any exact value of a continuous variable; that is, $P(z = 1.50) = 0$.

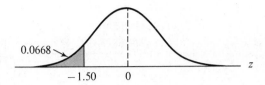

Figure 8.6   A Lower Tail Probability under the Standard Normal
Distribution

One might want the probability that z exceeds some value. We call this an "upper tail" probability. As an example consider $P(z > 0.26)$. The table gives

$$F'(z = 0.26) = P(z \leq 0.26) = 0.6026.$$

Since the total probability under the curve is 1.0, the probability we want is the complement of this:

$$P(z > 0.26) = 1 - P(z \leq 0.26) = 1 - 0.6026 = 0.3974.$$

This is shown graphically in Figure 8.7.

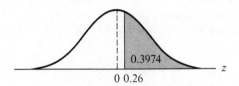

Figure 8.7   An Upper Tail Probability under the Standard Normal
Distribution

A third situation is a bounded interval. An example is to find the probability that z lies in the interval from -0.50 to 2.00. Here we find the cumulative probability associated with the upper end of the interval and subtract out the cumulative probability associated with the lower end. In symbols

$$P(-0.50 \leq z \leq 2.00) = F'(z = 2.00) - F'(z = -0.50),$$
$$P(-0.50 \leq z \leq 2.00) = 0.9772 - 0.3085 = 0.6687.$$

This is shown graphically in Figure 8.8.

Sometimes we need to find the value of z that has an associated cumulative probability. For example, find the value of z for which $F'(z) = 0.10$. Here one must find the cumulative probability 0.10 (or the closest value to 0.10) in the body of the table and read the associated value of z from the margin of the table. Looking through the table we find the cumulative probability .1003 at $z = -1.28$. The value -1.28 is the 10th percentile of z, or the value of z which has a probability of 0.10 of not being exceeded and a probability of 0.90 of being exceeded. A common way to refer to such a cutoff value is to write $z_{.10} = -1.28$.

Suppose we want the value of z that is exceeded with probability 0.05. First we recognize that this is $z_{.95}$. Looking for the cumulative probability .95 in the table we find it eventually at 1.64, that is, $z_{.95} = 1.64$.

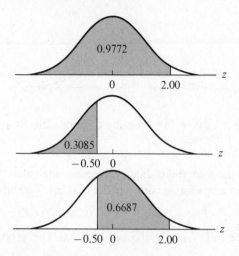

Figure 8.8   Finding the Probability for a Bounded Interval under the Standard Normal Distribution

Finally, suppose we want a value of z such that the total probability associated with values less than $-z$ or greater than z is, say 0.05. Since the normal curve is symmetrical, we recognize that these cutoff values of z are $z_{.025}$ and $z_{.975}$. Consulting the table we find $z_{.025} = -1.96$ and $z_{.975} = 1.96$. This is illustrated graphically in Figure 8.9.

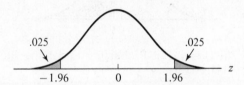

Figure 8.9   The 5 Percent Most Extreme Values of z under the Standard Normal Distribution

Probabilities under a Normal Distribution of X

To find probabilities under a normal distribution other than the standard normal, it is necessary to convert the particular normal distribution to the standard form. This is done by converting the relevant values (X) to the standard form through the expression $z = (X - \mu)/\sigma$. This amounts to transforming the scale of the X variable to standard deviation units. For example, consider a normal distribution of X with a mean of 20.0 and a standard deviation of 4.0. What is the probability that X is equal to or less than 17.0? The steps in obtaining the answer are:

1. Problem: $P(X \leq 17.0) = ?$
2. Convert X to z:

$$z(X) = \frac{X - \mu}{\sigma},$$

$$z(17.0) = \frac{17.0 - 20.0}{4.0} = -0.75.$$

3. Solve equivalent standard normal probability problem:

$$P(z \le -0.75) = F'(z = -0.75) = 0.2266.$$

4. State answer in terms of X:

$$P(X \le 17.0) = P(z \le -0.75) = 0.2266.$$

Graphically, these steps represent a substitution of the z scale for the X scale in order to use the table of probabilities for the standard normal distribution. Having done this we then convert back to the original X scale of the problem. (See Figure 8.10.)

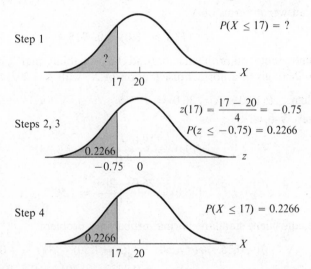

Step 1

$P(X \le 17) = ?$

17 20

Steps 2, 3

$z(17) = \dfrac{17 - 20}{4} = -0.75$

$P(z \le -0.75) = 0.2266$

0.2266

-0.75 0

Step 4

$P(X \le 17) = 0.2266$

0.2266

17 20

Figure 8.10   Graphic Representation of Normal Probability Problem and Its Solution

One can visualize the procedure in one graph with the z scale shown as well as the X scale, as in Figure 8.11.

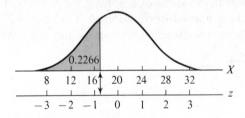

0.2266

8   12   16   20   24   28   32 X

-3 -2 -1 0   1   2   3 z

Figure 8.11   Representation of Normal Probability Problem Using X and z Scales

Since the solution is carried out in terms of z values, the mechanics of dealing with normally distributed X variables are the same as those already presented. For an "upper tail" problem let us find the probability that X exceeds 18.0 in a normal population of X with a mean of 20.0 and a standard deviation of 4.0.

1. Problem: $P(X > 18.0) = ?$

2. Convert X to z:

$$z(X) = \frac{X - \mu}{\sigma},$$

$$z(18.0) = \frac{18.0 - 20.0}{4.0} = -0.50.$$

3. Solve equivalent standard normal probability problem:

$$P(z > -0.50) = 1 - F'(z = -0.50) = 1 - 0.3085 = 0.6915.$$

4. State answer in terms of X:

$$P(X > 18.0) = 0.6915.$$

For a bounded-interval problem let us find the probability that X lies in the interval from 19.0 to 26.0, given a normal distribution of X, with $\mu = 20.0$ and $\sigma = 4.0$.

1. Problem: $P(19.0 \leq X \leq 26.0) = ?$
2. Convert X to z:

$$z(19.0) = \frac{19.0 - 20.0}{4.0} = -0.25,$$

$$z(26.0) = \frac{26.0 - 20.0}{4.0} = 1.50.$$

3. Solve equivalent standard normal probability problem:

$$P(-0.25 \leq z \leq 1.50) = F'(z = 1.50) - F'(z = -0.25)$$
$$= 0.9332 - 0.4013 = 0.5319.$$

4. State answer in terms of X:

$$P(19.0 \leq X \leq 26.0) = 0.5319.$$

To find a value of X having a specified cumulative probability, one must follow the same process of converting to terms of z, solving the problem in terms of z, and converting back again. In a normal distribution of X with $\mu = 20.0$ and $\sigma = 4.0$ let us find the value of X that is not exceeded with probability 0.80 (the 80th percentile of X).

1. Problem: $X_{.80} = ?$
2. Convert X to z:

$$z_{.80} = ?$$

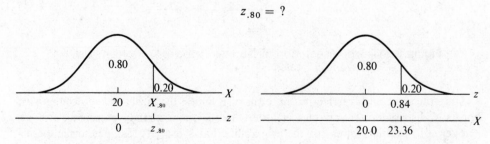

Figure 8.12   Graphic Representation of Procedure for Finding a Percentile of a Normal Distribution

3. Solve equivalent standard normal probability problem:

$$z_{.80} = 0.84.$$

4. State answer in terms of X:

$$X = \mu + z\sigma,$$

$$X_{.80} = 20.0 + z_{.80}(4.0),$$

$$X_{.80} = 20.0 + 0.84(4.0) = 23.36.$$

In the final step we deal with the z expression stated as an equation for X, because X is our unknown. .

$$z = \frac{X - \mu}{\sigma},$$

■ $X = \mu + z\sigma.$ ■   (8.2)

We know μ and σ because they are given as the parameters of the normal distribution of interest to us. The value of z that we want, $z_{.80}$ in the problem above, comes from the specification of the cutoff point. We find $z_{.80}$ by looking for the cumulative probability 0.80 in the standard normal probability table and noting at what z value this cumulative probability appears.

8.2

Summary

In a probability model of a continuous random variable, probability within any interval of the variable is measured by the proportion of total area under the distribution curve within the given interval. When the *cumulative* probability function for the variable is expressed as an equation, the probability within an interval can be obtained by working with the equation. Some probability models have a fixed, or standard, form that can be adapted to particular situations by substituting certain parameters of the given situation into a general expression related to the fixed-form model. Both the rectangular and normal probability models are of this type. The fixed form of the normal distribution model is incorporated in tables of cumulative probabilities under the standard normal distribution. The key to using this fixed form to find probabilities under a particular normal distribution is conversion of the X scale of the variable to the standard z scale. This conversion involves the mean and standard deviation of the X population through $z = (X - \mu)/\sigma$. One may want to find the probability associated with a particular interval of X, or he may need to find the value of X that has a specified cumulative probability associated with it. Examples of both types of problems were carried out.

See Self-Correcting Exercises 8.A

Exercises

1. A machine cuts wood blanks used in making pencils. The minimum length of the blanks is 8.00 inches. The mean blank length is 8.15 inches. If all blank lengths between the minimum and the maximum length are equally probable, what is the maximum length of wood blanks for this machine?

2. In the situation of Exercise 1, what is the probability that a blank will measure
   a. 8.10 inches to the nearest .005 inches?
   b. 8 inches exactly?

3. Based on a household survey in a particular community, a researcher announced that the cumulative probability distribution of the amount of money desired for a "guaranteed annual income" could be expressed by $F'(X) = 0 + X^2/16$ $(0 \leq X \leq 4)$, where X is income in $ thousands. Find the probability for the intervals

$$0 \leq X \leq 1, \quad 1 \leq X \leq 2, \quad 2 \leq X \leq 3, \quad 3 \leq X \leq 4.$$

   How would you describe the distribution?

4. Find the following probabilities under the standard normal distribution:
   a. $P(0 \leq z \leq 1.25)$ b. $P(-0.15 \leq z \leq 0.36)$ c. $P(z \geq 2.50)$ d. $P(z < -0.52)$.

5. For a normal distribution with $\mu = 25.0$ and $\sigma = 4.0$, find:
   a. $P(X \leq 27.0)$ b. $P(22.0 \leq X \leq 28.0)$ c. $P(X > 35.0)$.

6. Find the following "percentiles" for the standard normal distribution:
   a. $z_{.10}$ b. $z_{.60}$ c. $z_{.90}$ d. $z_{.99}$ e. $z_{.995}$.

7. A machine fills bottles of wine with an average of 0.200 gallons. The standard deviation of fills is 0.008 gallons. If the fills are normally distributed, the shortest 10 percent of the fills will lie below what value?

8. If the wine bottler wants to guarantee that no more than 1 percent of the fills will be short of 0.200 gallons and he cannot change the standard deviation, what mean fill will he have to provide?

9. An improved bottle filling machine has a standard deviation for fills of 0.005 gallons. What percentage saving will this secure for the bottler in meeting the guarantee in Exercise 8?

10. Completion times for an assembly operation among a group of workers had a mean of 80 seconds and a standard deviation of 10 seconds. Can you count on a completion time of 100 seconds being in the extreme ten percent of all completion times
    a. if the distribution of times is normal?
    b. if nothing is known about the shape of the distribution?

Sampling Distributions for Continuous Variables

In the last section we found that the sample space represented by a truly continuous variable is infinite. The values of X comprising the population cannot be listed. Nevertheless, the probability distribution for the population can be portrayed by an equation or by the graph of the probability function.

In examining the behavior of statistics in samples drawn from such populations, the method used in Chapter 6 for studying sampling distributions for discrete variables is not available. We cannot list the elements in a sample space in order to count up the associated occurrences of specified values of a statistic. Given this difficulty two avenues are open. However, one would be a purely mathematical development that would have to draw on methods not assumed as a prerequisite of this text. The second approach, through simulated draws from a population, works with a discrete approximation to the continuous distribution. We will first talk about sampling from normally distributed populations, presenting results as theorems without mathematical proof. Then we will introduce sampling from nonnormal populations where we will show results of simulated draws to reinforce the credibility of stated general results.

Sampling a Normal Population of X

8.3

Let us return to our example normal population with a mean of 20.0 and standard deviation of 4.0. Suppose this X population represents the lengths of life (in hours) of a specified type of flashlight bulb. Suppose now that four bulbs are selected at random. What is the probability distribution of the total length of life of the four bulbs?

We are concerned with an experiment that calls for the result of $\sum X$ upon four independent draws from our underlying X population. Our statistic is $\sum X$ and we want to describe the sampling distribution of this statistic.

Statement	If n independent draws are made from a normally distributed X population with mean μ_X and standard deviation σ_X, the probability distribution of $\sum X$ for the n draws is a normal distribution with the following parameters.

$$\mu_{\Sigma X} = n\mu_X, \tag{8.3}$$

$$\sigma_{\Sigma X}^2 = n\sigma_X^2. \tag{8.4}$$

Formulas (8.3) and (8.4) were presented in Chapter 6 for sampling from discrete populations. Additionally, the statement says that the sampling distribution of $\sum X$ for n independent draws from a normal distribution of X is also a normal distribution. The formulas provide us with the mean and variance of this sampling distribution. For our example of four draws we have

$$\mu_{\Sigma X} = n\mu_X = 4(20.0) = 80.0,$$
$$\sigma_{\Sigma X}^2 = n\sigma_X^2 = 4(4.0)^2 = 64.0.$$

The probability distribution of the total length of life of four randomly selected bulbs is then completely identified by saying:

1.   The probability distribution is normal.
2.   Its mean is 80.0   ($\mu_{\Sigma X} = 80$).
3.   Its standard deviation is 8.0   ($\sigma_{\Sigma X} = \sqrt{64.0}$).

With this information we can answer a question such as "what is the probability that four successive bulbs (independently selected) last less than a total of 70.0 hours?" Following our previous four-step method we have:

1. Problem: $P(\sum X < 70.0) = ?$
2. Convert $\sum X$ to z:

$$z(\sum X) = \frac{\sum X - \mu_{\Sigma X}}{\sigma_{\Sigma X}},$$

$$z(70.0) = \frac{70.0 - 80.0}{8.0} = -1.25.$$

3. Solve equivalent standard normal probability problem:

$$P(z < -1.25) = F'(z = -1.25) = 0.1056.$$

4. State answer in terms of $\sum X$:

$$P(\sum X < 70.0) = 0.1056.$$

Sometimes, as above, the sum of X for n independent draws will be a statistic of interest. If we take four bulbs along on a camping trip, we are interested in how much light (in hours) we may have. More generally, in statistical work the sample mean, $\sum X/n$, will be of paramount interest.

Statement	If n independent draws are made from a normally distributed X population with mean μ_X and standard deviation σ_X, the probability distribution of the sample mean ($\overline{X}$) for the n draws is a normal distribution with the following parameters.

$$\blacksquare \qquad \mu_{\overline{X}} = \mu_X, \qquad \blacksquare \tag{8.5}$$

$$\blacksquare \qquad \sigma_{\overline{X}}^2 = \frac{\sigma_X^2}{n}. \qquad \blacksquare \tag{8.6}$$

For our experiment, where $\mu_X = 20.0$, $\sigma_X = 4.0$, and $n = 4$, we would have

$$\mu_{\overline{X}} = \mu_X = 20.0,$$

$$\sigma_{\overline{X}}^2 = \frac{\sigma_X^2}{n} = \frac{(4.0)^2}{4} = 4.0.$$

We can now identify the sampling distribution of mean length of life of four independently selected bulbs by saying:

1. The probability distribution of $\overline{X}$ is normal.
2. Its mean is 20.0 ($\mu_{\overline{X}} = 20.0$).
3. Its standard deviation is 2.0 ($\sigma_{\overline{X}} = \sqrt{4.0}$).

Figure 8.13 shows our original X population along with the two sampling distributions we have developed. Here we show all three against a scale of hours. The sampling distribution of $\sum X$ for four draws has a wider spread ($\sigma_{\Sigma X} = 8.0$) than the original

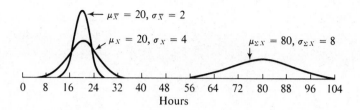

Figure 8.13   A Probability Distribution of X along with Sampling Distribution of ΣX and Sampling Distribution of $\overline{X}$ for $n = 4$.

population ($\sigma_X = 4.0$). The sampling distribution of $\overline{X}$ for four draws is centered about the mean of the X population ($\mu_{\overline{X}} = \mu_X$) but has a smaller spread ($\sigma_{\overline{X}} = 2.0$). The spread of the sampling distribution of $\overline{X}$ depends on the size of sample, n. More formally

$$\blacksquare \quad \sigma_{\overline{X}} = \frac{\sigma_X}{\sqrt{n}}. \quad \blacksquare \tag{8.7}$$

The probability distribution of the mean length of life in a random sample of 100 bulbs in our example is a normal distribution with

$$\mu_{\overline{X}} = \mu_X = 20.0,$$

$$\sigma_{\overline{X}} = \frac{\sigma_X}{\sqrt{n}} = \frac{4.0}{\sqrt{100}} = 0.40.$$

With this information we can answer such a question as "what is the probability that the mean life of a sample of 100 bulbs will vary from the population mean by more than 1.0 hours?" The key is to transform the variable $\overline{X}$ to the standard normal variable, z, through the equation

$$\blacksquare \quad z = \frac{\overline{X} - \mu_{\overline{X}}}{\sigma_{\overline{X}}}. \quad \blacksquare \tag{8.8}$$

The two-scaled representation of our problem is shown in Figure 8.14.

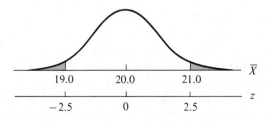

Figure 8.14   Representation of Normal Sampling Distribution Problem Using $\overline{X}$ and z Scales

1. Problem: $P[(\overline{X} < 19.0) \cup (\overline{X} > 21.0)] = ?$
2. Convert $\overline{X}$ to z:

$$z = \frac{\overline{X} - \mu_{\overline{X}}}{\sigma_{\overline{X}}},$$

$$z(19.0) = \frac{19.0 - 20.0}{.40} = -2.5,$$

$$z(21.0) = \frac{21.0 - 20.0}{.40} = 2.5.$$

3. Solve equivalent standard normal probability problem:

$$P[(z < -2.5) \cup (z > 2.5)] = F'(z = -2.5) + 1 - F'(z = 2.5)$$
$$= .0062 + (1 - .9938),$$

$$P[(z < -2.5) \cup (z > 2.5)] = .0062 + .0062 = 0.0124.$$

4. State answer in terms of $\overline{X}$:

$$P[(\overline{X} < 19.0) \cup (\overline{X} > 21.0)] = 0.0124.$$

A problem to find a particular "cutoff" value of $\overline{X}$ parallels this type of problem shown previously for a normal probability distribution of X. Let us find the 75th percentile of $\overline{X}$, or $\overline{X}_{.75}$. This is a value of $\overline{X}$ so located that the probability of its not being exceeded is 0.75. The key to solving for $\overline{X}$ is the expression

$$\overline{X} = \mu_{\overline{X}} + z\sigma_{\overline{X}}. \tag{8.9}$$

Applying our four-step procedure, we have:

1. Problem: $\overline{X}_{.75} = ?$
2. Convert $\overline{X}$ to z:

$$z_{.75} = ?$$

3. Solve equivalent standard normal probability problem:

$$z_{.75} = 0.67.$$

4. State answer in terms of $\overline{X}$:

$$\overline{X} = \mu_{\overline{X}} + z\sigma_{\overline{X}},$$
$$\overline{X}_{.75} = 20.0 + 0.67(.40),$$
$$\overline{X}_{.75} = 20.268.$$

We have dealt in this section with the normal probability distribution of $\sum X$ and the normal probability distribution of $\overline{X}$. In the preceding section we solved some probability problems relating to the distribution of an underlying X population. The key to solving any normal probability problem is the z expression for the distribution involved. All z expressions represent the deviation of a random variable from its expected value expressed in standard deviation units. The three expressions for the distributions we have studied are given below. Also given are the equivalent expressions for a value of the random variable. We used the second form in solving for percentiles of the random variable.

Random variable

X	ΣX	$\bar{X}$

$$z = \frac{X - \mu_X}{\sigma_X} \qquad z = \frac{\Sigma X - \mu_{\Sigma X}}{\sigma_{\Sigma X}} \qquad z = \frac{\bar{X} - \mu_{\bar{X}}}{\sigma_{\bar{X}}}$$

$$X = \mu_X + z\sigma_X \qquad \Sigma X = \mu_{\Sigma X} + z\sigma_{\Sigma X} \qquad \bar{X} = \mu_{\bar{X}} + z\sigma_{\bar{X}}$$

8.3

Sampling a Nonnormal Population of X

Many populations encountered in economic and business situations are not normally distributed. Skewness to the right is common for variables such as profit rates of firms, incomes of families, and consumption rates of individuals; and one may encounter distributions skewed to the left as well. Further, distributions that are symmetrical are not necessarily normal in shape. They may have a central clustering but be more (or less) peaked (or bunched up in the middle) than a normal distribution. They may even be U-shaped, with greater frequency density at the extremes than near the mean. To help you appreciate a very important theorem to be stated soon, we will now summarize a limited simulation of draws from a nonnormal X population.

Table 8.1 and Figure 8.15 show a relative frequency distribution of processing times

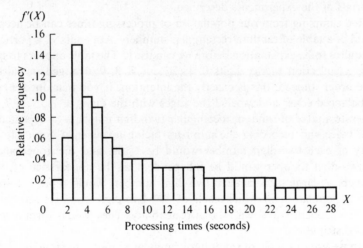

Figure 8.15   Relative Frequency Distribution of Job Processing Times

for a large number of jobs processed by a computer facility in the last three months. The variable, processing time, is inherently continuous, but the times have been recorded only to the nearest second. The distribution is highly skewed to the right, reflecting the fact that most jobs use a modest amount of processing time but a few jobs use comparatively large amounts of time. The distribution has a mean of 9.4 seconds and a standard deviation of 6.9 seconds. That is to say, the parameters of the X population of times are

$$\mu_X = 9.4, \qquad \sigma_X = 6.9.$$

Table 8.1 Relative Frequency Distribution of Job Processing Times (in Seconds) for a Computer Facility

X	f'(X)	X	f'(X)
1	.02	15	.03
2	.06	16	.02
3	.15	17	.02
4	.10	18	.02
5	.09	19	.02
6	.06	20	.02
7	.05	21	.02
8	.04	22	.02
9	.04	23	.01
10	.04	24	.01
11	.03	25	.01
12	.03	26	.01
13	.03	27	.01
14	.03	28	.01
		Total	1.00

To study the behavior of sample means from this skewed population, let us ask what might happen if we randomly picked three jobs during the period in question and calculated the mean processing time for the three jobs. Later we shall examine the same question for a random selection of six and of nine jobs. We do this by simulating repeated trials of the experiments described.

Simulated sampling from our population of processing times can be accomplished with the aid of a table of random rectangular numbers. Appendix F, a portion of such a table, requires some explanation before we can use it. The table appears to be nothing more than a collection of the digits 0, 1, 2, ..., 7, 8, 9 arranged in pairs with no discernible order. Indeed, this is exactly the intention. If you machined a cylinder to have ten balanced edges and labeled the edges with the digits 0, 1, 2, ... 7, 8, 9, you could generate a table of random rectangular two-digit numbers by repeatedly rolling the device twice and recording the numbers facing up when it comes to rest. The probability of each two-digit number would be 1/100, and the appearance of any specified two-digit number would be independent of the appearance of other two-digit numbers. A large table of such two-digit numbers would represent an approximation to a population in which each of 100 two-digit numbers (00 through 99) appeared independently and with equal probability. They would form a rectangular probability distribution.

How can we use this table of random rectangular numbers to simulate independent draws from our nonrectangular population of processing times? The procedure is to allow appropriate intervals of the random digits to stand for specified values of our processing time variable. For example, in our population the processing time of 1 second occurs with probability .02. If we let two two-digit random numbers, say 00 and 01, stand for a 1-second processing time, then about two out of one hundred "processing times" in the random number table will be 1 second. A processing time of 2 seconds has probability .06 in Table 8.1. If we let the sequence of two-digit random numbers from 02 to 07 inclusive stand for a 2-second processing time, then such times have a relative frequency in the random number table of about 6/100. By designating inclusive random number sequences in proportion to the probabilities for our X population, we effectively place our population in the random number table. The entire transformation table for our population is shown in Table 8.2.

The $f'(X)$ values from Table 8.1 are repeated so that the correspondence between proportions of assigned random numbers and probabilities can be seen.

Table 8.2 Table for Transforming Random Rectangular Numbers to Values of the Variable X

Random number	X	$f'(X)$	Random number	X	$f'(X)$
00–01	1	.02	77–79	15	.03
02–07	2	.06	80–81	16	.02
08–22	3	.15	82–83	17	.02
23–32	4	.10	84–85	18	.02
33–41	5	.09	86–87	19	.02
42–47	6	.06	88–89	20	.02
48–52	7	.05	90–91	21	.02
53–56	8	.04	92–93	22	.02
57–60	9	.04	94	23	.01
61–64	10	.04	95	24	.01
65–67	11	.03	96	25	.01
68–70	12	.03	97	26	.01
71–73	13	.03	98	27	.01
74–76	14	.03	99	28	.01

Now, to simulate independent draws from the population of processing times, all we do is read successive two-digit numbers from the random-number table. The table can be entered anywhere and read in any direction. In this simulation we enter at the upper left of the table and read the rows. The first three two-digit numbers are 10, 09, and 73. Consulting our transformation table, these random numbers correspond to the X values of 3, 3, and 13 seconds, respectively. These are the values in our first sample of three processing times, and the mean of the three processing times is $(3 + 3 + 13)/3 = 6\frac{1}{3}$ seconds. If we wish to study the sampling distribution of the mean for a sample of size $n = 3$, we must draw many such samples. That is, we must simulate repeated trials of the experiment *draw three processing times and calculate their mean*. We read 120 sets of three two-digit random numbers, thus repeating our experiment 120 times. Some results are shown in Table 8.3.

By the simulation method we generated 120 means from samples of three processing times each. These means were then grouped into a frequency distribution with a class interval of 1.0 seconds. The result is shown in Figure 8.16. While we might need more than 120 trials to remove some of the chance irregularities in this approximation to the sampling distribution of the mean for samples of size $n = 3$, it is clear that the basic shape of the distribution retains some of the positive skewness of the parent X population. When the simulation is done for samples of size $n = 6$ and $n = 9$, however,

Table 8.3 Results of Simulated Draws of 120 Random Samples of Three Observations Each

Trial No. 1		Trial No. 2		$\cdots$	Trial No. 120	
Random number	X	Random number	X		Random number	X
10	3	25	4		76	14
09	3	33	5		49	7
73	13	76	14		69	12
	$\bar{X} = 6\frac{1}{3}$		$\bar{X} = 7\frac{2}{3}$	$\cdots$		$\bar{X} = 11$

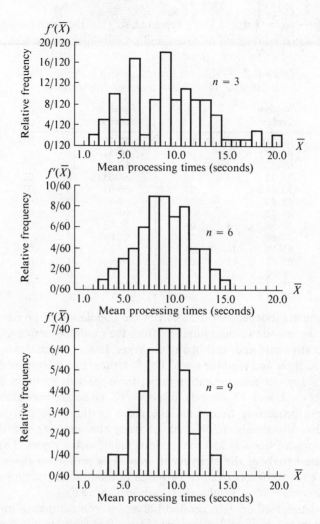

Figure 8.16   Simulated Sampling Distributions of the Mean for Samples of Different Size from Population of Processing Times

the skewness in the resulting sampling distributions diminishes markedly. (For samples of size $n = 6$ we used 60 successive sets of six random numbers each, and for samples of size $n = 9$ we used 40 successive sets of nine random numbers each.)

We see here the operation of a very remarkable and important theorem in statistics, the *central limit theorem*. Even though our simulation was limited to 40 trials, the distribution of means of samples of size $n = 9$ from our highly skewed population is highly suggestive of the normal curve shape.

Statement	As the sample size is increased, the sampling distribution of the mean of a continuous variable approaches the normal distribution form regardless of the distribution form of the underlying population.

In our simulated example the normal sampling distribution form appeared to emerge at a very modest sample size. For most populations it is safe to assume a normal sampling distribution of the mean for samples of $n = 20$ or more.

The existence of the central limit theorem greatly extends the applicability of normal sampling distribution theory. Consider our processing time population. If we had been asked to estimate the mean processing time from the mean of a sample of 100 jobs, how good might such an estimate have been? We know

$$\mu_{\bar{X}} = \mu_X = 9.4, \quad \sigma_{\bar{X}} = \frac{\sigma_X}{\sqrt{n}} = \frac{6.9}{\sqrt{100}} = 0.69.$$

From the central limit theorem we can be assured that the probability distribution of $\bar{X}$ for $n = 100$ is normal. What is the probability that the mean of a sample of 100 processing times will differ from the population mean by no more than 1.0 seconds?

1. Problem: $P(8.4 \leq \bar{X} \leq 10.4) = ?$

2. Convert $\bar{X}$ to z:

$$z(\bar{X}) = \frac{\bar{X} - \mu_{\bar{X}}}{\sigma_{\bar{X}}},$$

$$z(8.4) = \frac{8.4 - 9.4}{0.69} = -1.45,$$

$$z(10.4) = \frac{10.4 - 9.4}{0.69} = 1.45.$$

3. Solve equivalent standard normal probability problem:

$$P(-1.45 \leq z \leq 1.45) = F'(z = 1.45) - F'(z = -1.45) = 0.9265 - 0.0735,$$
$$P(-1.45 \leq z \leq 1.45) = 0.8530.$$

4. State answer in terms of $\bar{X}$:

$$P(8.4 \leq \bar{X} \leq 10.4) = 0.8530.$$

We should visualize the problem as the graph of Figure 8.17.

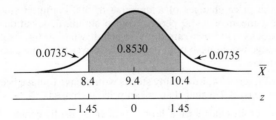

Figure 8.17   Normal Probability That a Sample Mean Is within a Given Range from the Population Mean

The chances are better than 85 out of 100 that a sample of 100 job times would have yielded a sample mean within 1.0 second of the true population mean. This may be sufficient accuracy for some purposes but not for others. That is another question. Knowledge of sampling distribution theory has at least answered the first necessary question.

Summary

The sampling distribution of the mean for samples of any given size drawn from a normally distributed X population is a normal distribution.

The central limit theorem assures us that when the sample size is moderate ($n \geq 20$), the sampling distribution of the mean will be a normal distribution (approximately) even when the underlying population is not normally distributed. This theorem greatly extends the applicability of normal sampling distribution theory, because many X populations encountered in economic and social data are not normally distributed.

The mean and standard deviation of the sampling distribution of a mean are known from formulas involving the mean and standard deviation of the underlying X population. If, in addition, the sampling distribution is normal, then probability questions about the sample mean can be answered by using the table of areas under the normal distribution.

See Self-Correcting Exercises 8.B

Exercises

1. Given a normally distributed population with $\mu = 80$ and $\sigma = 10$ and an experiment of four independent draws from the population, what is the probability that $\sum X$

   a. exceeds 350?

   b. lies between 300 and 350?

   c. fails to exceed 275?

2. A secretary has found that the time her boss takes for appointments is normally distributed with a mean of 13 minutes and a standard deviation of 3 minutes. What is the probability that four randomly selected appointments will take less than an hour?

3. If the secretary wants to be 99 percent sure that four appointments will not exceed one hour, how much reduction in average time taken for appointments is required?

4. Over a period, the price changes of a universe of 500 common stocks were normally distributed with a mean of $+4.0$ percent and a standard deviation of 8.0 percentage points. If four stocks had been randomly selected, what is the probability that their average price change would have been negative? Answer the same question for a random selection of 25 stocks.

5. In the situation of Exercise 4, how many stocks would have to have been selected to make the probability at least 0.975 that their mean price increased?

6. A financial manager in charge of a large number of profit centers has observed that monthly expenditures as a percent of budget are normally distributed with a mean of 100.0 and a standard deviation of 3.0. Yearly performance is viewed in terms of average percent of budget for the twelve months. If each center's twelve-month record were made up of independent draws from the monthly population, what average percent of budget would cut off the lowest 5 percent of the centers?

7. Discuss what the independence assumption means in the context of Exercise 6.

8. The amount of additives in a certain food product has a right triangular distribution over a range of 0 to 12 percent by weight. Such a distribution has a mean of 4 and a variance of 8. For samples of size $n = 5$ from this population, what is the minimum probability that the sample mean will lie within two percentage points of the population mean?

9. For a sample of size $n = 16$ in the situation of Exercise 8, what confidence do we have that the sample mean will lie within two percentage points of the population mean

   a. according to the Chebyshev inequality?

   b. if a normal sampling distribution can be applied?

10. Consider a sample of size $n = 32$ in the example of Exercise 8. What is the probability that the sample mean will differ from the population mean by more than ten percent.

Glossary of Equations

$$z = \frac{X - \mu}{\sigma}$$

The z transformation for a random variable, X, is the scale of deviation of X from the mean of X expressed in standard deviation units.

$$X = \mu + z\sigma$$

A rearrangement of the z transformation equation to solve for an unknown X. This is used in solving for a percentile of X in a given normal population of X.

$$\mu_{\Sigma X} = n\mu_X$$

The expected value of $\sum X$ in n independent draws from a population with mean μ_X is $n\mu_X$.

$$\sigma^2_{\Sigma X} = n\sigma^2_X$$

The variance of $\sum X$ in n independent draws from a population with variance σ^2_X is $n\sigma^2_X$.

$$\mu_{\bar{X}} = \mu_X$$

The expected value of $\bar{X}$ in n independent draws from a population with mean μ_X is μ_X.

$$\sigma^2_{\bar{X}} = \frac{\sigma^2_X}{n}$$

The variance of $\bar{X}$ in n independent draws from a population with variance σ^2_X is σ^2_X divided by n.

$$\sigma_{\bar{X}} = \frac{\sigma_X}{\sqrt{n}}$$

The same as the formula immediately above but stated in terms of standard deviation rather than variance. The standard deviation of the sample mean, $\sigma_{\bar{X}}$, is also called the standard error of the mean.

$$z = \frac{\bar{X} - \mu_{\bar{X}}}{\sigma_{\bar{X}}}$$

The z transformation for a random variable, $\bar{X}$, is the scale of deviation of $\bar{X}$ from the expected value of $\bar{X}$ expressed in standard deviation units.

$$\bar{X} = \mu_{\bar{X}} + z\sigma_{\bar{X}}$$

A rearrangement of the z transformation equation immediately above to solve for an unknown $\bar{X}$. This is used in solving for a percentile of $\bar{X}$ in situations where a normal sampling distribution of $\bar{X}$ applies.

9

Estimation of Population Means

In the first two chapters of this book we discussed useful properties of *statistical populations*—that is, complete sets of observations of a single variable. Some properties defined for populations were central location, dispersion, and skewness. We went on to develop *parameters*, or measures of these properties. The population mean (μ) was shown to be an important parameter for measuring central location, for example. Our concern was statistical description. We had complete populations at our disposal and we sought meaningful characteristics of these populations.

In the last five chapters one major objective has been to learn how samples behave when we know the statistical population from which they come. We discussed measures of useful properties of the observations in a sample and called such measures *statistics*. We learned that the sample mean ($\overline{X}$) is an important statistic for measuring the central location of a sample, for instance. Probability theory made it possible for us to find the sampling distribution for this statistic, given the statistical population from which it came and the size of the sample under consideration. Our pattern of reasoning was from facts about a population to conclusions about a sample statistic.

We now want to reverse the direction of our thinking: *Given observations that constitute a single sample, what can we infer about some characteristic of the parent population from which the sample came?* We want to know which calculations from a single set of sample observations will be most effective for estimating the unknown value of a parameter of interest. This is an important problem whenever one has only sample data available.

Point Estimation

Suppose a market researcher must estimate the current mean weekly income per household (μ) in a certain city. Having selected a random sample of 80 households, he has obtained the observations of weekly income that constitute his data. How can he best estimate the value of the parameter he seeks? Should he calculate the sample mean or the sample median or the sample mode, to mention only a few of the many possible statistics he might calculate? Suppose he arbitrarily elects to calculate the sample mean ($\overline{X}$), finds the value to be $104, and takes this value as his estimate of the value of the parameter he seeks (μ). In statistical language, he has chosen to use the sample mean ($\overline{X}$) as his *estimator* of the population mean (μ), and his *point estimate* of the population mean is $104.

In the example above, the selection of the sample mean as the estimator of the population mean was arbitrary. It would be preferable to have a sound theoretical basis for selecting an estimator, and such a basis is provided by estimation theory.

Estimation theory has been developed by studying the behavior of different sample estimators with respect to the parameter being estimated. The goal is to determine how well each estimator conforms to several desirable characteristics for estimators. One such characteristic is absence of bias. Under random sampling conditions, the sampling distribution of values of an unbiased estimator will be centered upon the parameter being estimated. For example, in Figure 8.13, we saw that the various values of the sample mean ($\overline{X}$) from random samples of four observations are centered upon 20, the mean (μ) of the population from which the samples came. If we use $\overline{X}$ as an estimator of μ, we know that it will be correct on the average, even though no sample may have a mean precisely equal to 20.

Definition	An **unbiased estimator** is a statistic having an expected value the same as that of the parameter being estimated. In other words, an unbiased estimator is correct *on the average*.

In the preceding example, we saw that the sample mean satisfies the definition of an unbiased estimator of the population mean. From the discussion in Chapter 8 we know that the same would be true if the sampling distribution were based on samples of any size other than four observations. In fact, for samples of any one size selected randomly from any statistical population, the expected value of the sample mean is equal to the population mean. Hence the sample mean is an unbiased estimator of the population mean for any statistical population whatever. This general absence of bias would not hold for the sample median or the sample mode, for instance, as estimators of the population mean.

Even though the sample mean is correct *on the average*, a random sample having a mean equal to the population mean may, in certain circumstances, occur very infrequently. In other words, even though it is true in general that $\mu_{\overline{X}} = \mu$, it may seldom or never be true that $\overline{X} = \mu$. For example, consider the probability (sampling) distribution of $\overline{X}$ in Figure 6.4 (page 131). Figure 6.4(a) shows the parent population to contain the values 8, 16, and 24 in such proportions that the mean (μ) is 12. For random samples of three observations each, (b) shows that the only possible values a sample mean can take on are 8, $10\frac{2}{3}$, $13\frac{1}{3}$, 16, $18\frac{2}{3}$, $21\frac{1}{3}$, and 24. Even though the distribution of these seven values is such that $\mu_{\overline{X}} = 12$, an individual sample that has a mean of 12 can never occur. The estimator is correct on the average but we can never get a sample for which the estimate has the correct value.

As we have indicated, absence of bias is an important property considered by statisticians in choosing among alternative point estimators. Reference 1 contains descriptions of most of the additional properties, which usually are left to more advanced courses.

Interval Estimation—Standard Deviation Known

9.1   Even though some point estimators have the desirable properties just mentioned, all still have one major shortcoming. In any given instance, there is nothing to tell us how much reliance to place in the estimate obtained. An interval estimate, on the other hand, carries with it a measure of the confidence we can place in it. Such an estimate consists of a range of values, rather than just one, for our estimate of the underlying parameter.

Three things determine how we must proceed when we want to make an interval estimate of the population mean. These are: our knowledge about the shape of the population distribution, our knowledge about the value of the population standard deviation (σ), and the number of observations in our random sample. We will discuss interval estimation under each combination of these characteristics. The entire discussion is summarized in a table on the last page of the chapter.

Normally Distributed Population

A food packing plant has a machine that fills No. 303 cans with cream style corn. A dial is set by the operator to control the amount of corn deposited in each can. Experience has shown that, regardless of the dial setting, machines of this type fill cans with a standard deviation (σ) of 0.80 ounces. In other words, the statistical population of fill weights made at any fixed dial setting has a standard deviation of 0.80 ounces. In addition, the fill weights at any one setting are assumed to be normally distributed, because experience with many large samples from this type of machine has shown this to be the appropriate shape.

Suppose the operator finds the dial on the filling machine to be loose. After he tightens it and sets the dial at a particular reading, he is not sure what fill weight this reading indicates. He must take a sample of fill weights and use the sample to estimate the mean of the population of fill weights at this setting. He will then loosen the dial, move it to this estimate without changing the amount of fill delivered by the machine, and retighten it.

The operator arbitrarily decides to fill 25 cans at the fixed setting and weigh the contents of each can. After doing this, he calculates the sample mean ($\overline{X}$) and finds it to be 17.60 ounces. If the 25 content weights are treated as independent random observations, 17.60 ounces can be used as an unbiased point estimate of the population mean (μ). On the other hand, the sampling theory covered in Chapter 8 can be used to form a different type of estimate of the population mean—an interval estimate. We will first describe how to make such an estimate and the rationale behind it. Then we will compare it with a point estimate.

We have the mean of a single random sample that we assume came from a normally distributed population of fill weights. The mean is 17.60 ounces and there are 25 observations in the sample. We also know that the standard deviation of the population is 0.80 ounces.

We can now refer to the statement on page 176 and the discussion following it. We conclude from this material that our sample mean of 17.60 belongs to a distribution of all possible means that could arise from random samples of 25 observations each. Furthermore, we can state three characteristics of this distribution of sample means:

1. It is normally distributed.
2. It has the same mean as does our objective, the mean of the parent population, even though we don't know the numerical value of either mean, that is, $\mu_{\bar{X}} = \mu$.
3. It has a standard deviation ($\sigma_{\bar{X}}$), called the *standard error of the mean*, the value of which can be found from the standard deviation of the parent population ($\sigma = 0.80$ ounces), and the sample size ($n = 25$). Specifically,

$$\sigma_{\bar{X}} = \frac{\sigma}{\sqrt{n}}, \quad \text{or} \quad \sigma_{\bar{X}} = \frac{0.80}{\sqrt{25}}, \quad \text{or } 0.16;$$

that is, the standard error is 0.16 ounces.

We know that we have one observation ($\bar{X}$) belonging to a distribution for which we know two of the three determining characteristics. We know the shape of the distribution (normal) and we know its dispersion ($\sigma_{\bar{X}} = 0.16$ ounces). But we do not know where our observation lies with respect to the distribution's mean ($\mu_{\bar{X}} = \mu$).

The single sample mean we have observed ($\bar{X}$) could be some place well below the distribution's mean ($\mu_{\bar{X}}$), very near the population mean, or well above it. From our work with normal probabilities, however, we can be confident that our observation lies within, say, three standard errors of the population mean. This circumstance is illustrated in Figure 9.1.

In Figure 9.1, the distribution to the left is built on the assumption that our sample mean ($\bar{X}$) of 17.60 ounces lies exactly three standard errors above the population mean ($\mu_{\bar{X}}$). To be in this position the standard normal deviate (z) must be $+3$ in the expression

$$z = \frac{\bar{X} - \mu_{\bar{X}}}{\sigma_{\bar{X}}}.$$

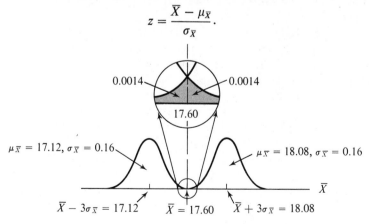

Figure 9.1   Two Possible Sampling Distributions of the Mean

We already know that $\overline{X}$ is 17.60 and $\sigma_{\overline{x}}$ is 0.16 in this expression. Consequently we can state that

$$3 = \frac{17.60 - \mu_{\overline{x}}}{0.16},$$

from which

$$\mu_{\overline{x}} = 17.12 \text{ ounces.}$$

If the mean of the population of fill weights is 17.12 ounces, the mean of the random sample we selected (17.60) is three standard errors greater. Similarly, for our sample mean to lie exactly three standard errors below the population mean, the standard normal deviate must be -3 and $\mu_{\overline{x}}$ must be 18.08 ounces. We can designate 18.08 and 17.12 as the limits of our interval estimate of the population mean. Furthermore, by using normal curve theory, we can get a measure of how much confidence we can place in this estimate.

In Appendix A, the probability of a randomly selected observation falling three standard deviations or more above the mean is seen to be 0.0014. Hence, only 14 times in 10,000 would we expect to get a sample mean in the tail of the distribution above 17.60 ounces in Figure 9.1 when the population mean is 17.12 ounces. Similarly, only 14 times in 10,000 would we expect to get a sample mean in the tail of the distribution below 17.60 ounces when the population mean is 18.08 ounces. Now 17.12 is 0.48 ounce (0.5 to one decimal) below 17.60, and 18.08 is half an ounce above it. As a result the operator can change the dial to 17.60 ounces and be virtually certain that it is reading within a half ounce of the true mean fill at the machine's current setting. He can be highly confident that the population mean lies in the interval which has 17.12 ounces as its lower limit and 18.08 ounces as its upper limit. This interval centered on 17.60 ounces ($\overline{X}$) and running from three standard errors below $\overline{X}$ to three standard errors above it constitutes an *interval estimate of the population mean*.

Next consider Figure 9.1 in the light of what would happen were we to set the lower limit of our interval estimate even farther to the left than 17.12 and the upper limit even farther to the right than 18.08. Then the sum of the two normal tail areas would come to less than 0.0028 and we could place even greater confidence in this wider interval estimate. Conversely, bringing the two limits in so that they are less than 0.48 ounces away from 17.60 will establish a narrower interval, in which we would place less confidence.

Using the reasoning just developed, we can construct a measure of our degree of confidence in a particular interval estimate. We shall call this measure the *confidence coefficient*. As with probabilities, a confidence coefficient can range from 1 to 0. When we are certain that an interval contains the population mean, we have complete confidence in that estimate and assign to it a confidence coefficient of 1. When we are certain an interval does *not* contain the population mean, we assign to it a confidence coefficient of 0. Between these extremes we find the confidence coefficient by subtracting the sum of the tail areas, such as those illustrated in Figure 9.1, from 1. Hence, for the situation shown in that figure, the confidence coefficient is 0.9972, or $1 - 0.0028$. Compare this coefficient with the value we would get by setting the limits for our interval estimate at 17.28 and 17.92 ounces. These limits are only two standard errors, rather than three, from our sample mean (17.60). The sum of the tail areas would increase to 0.0456, and the confidence coefficient would drop to 0.9544 for this narrower interval.

Before formally summarizing our discussion of interval estimates, we must consider another major influence—the effect of variations in the number of sample observations on an interval estimate of the population mean. Instead of 25 observations, suppose the operator in our example had chosen a random sample of 100 observations. Further suppose that the mean ($\overline{X}$) of this larger sample turned out to be 17.30 ounces, rather than 17.60. In this case, the standard error of the mean ($\sigma_{\overline{X}}$) would be only 0.08 ounces, because

$$\sigma_{\overline{X}} = \frac{\sigma}{\sqrt{n}} = \frac{0.80}{\sqrt{100}}, \quad \text{or } 0.08$$

An interval with limits three standard errors above and below the new sample mean would have a lower limit (X_L) of

$$X_L = \overline{X} - 3\sigma_{\overline{X}} = 17.30 - 3(.08), \quad \text{or } 17.06 \text{ ounces,}$$

and an upper limit (X_U) of

$$X_U = \overline{X} + 3\sigma_{\overline{X}} = 17.30 + 3(.08), \quad \text{or } 17.54 \text{ ounces.}$$

For this larger sample of 100 observations, the length of the interval estimate is only 0.48 ounces (17.54 − 17.06), as compared with an interval length of 0.96 ounces (18.08 − 17.12) based on the sample of 25 observations. Nonetheless, the confidence coefficients are 0.9972 for both intervals, because both sets of limits are three standard errors from their respective sample means. This example shows that by increasing the sample size, we can maintain a given degree of confidence while narrowing the length of an interval estimate. More information allows us to narrow the interval without loss of confidence. Alternatively, given a larger sample, we could elect to maintain the same length of interval and increase the value of the confidence coefficient. Changes in the opposite direction would accompany a decision to select a smaller, rather than a larger, sample.

In the situation just discussed, we are given the values of a sample mean ($\overline{X}$), the number of observations upon which it is based (n), and the standard deviation (σ) of the population from which it came. We also have reason to assume that the population is normally distributed. In this situation we can form an interval estimate by executing the steps in the following procedure:

Rules

1. Select a relatively large value (usually 0.90 or greater) for the confidence coefficient.† Set this value equal to $1 - \alpha$ and solve for α, where α is the sum of the areas in the tails of the sampling distributions corresponding to those shown in Figure 9.1.

2. From Appendix A find the magnitude of the standard normal deviate (z) that cuts off a tail area equal to $\alpha/2$. Call this magnitude $z_{\alpha/2}$ and disregard its algebraic sign.

3. Find the standard error of the mean ($\sigma_{\overline{X}}$) by dividing the population standard deviation by the square root of the sample size; that is,

$$\sigma_{\overline{X}} = \frac{\sigma}{\sqrt{n}}.$$

† The coefficient must be large enough to represent a high level of confidence in the estimate yet not be so high that the cost of sampling is excessive. Judgment and trial and error can be used to find a satisfactory value.

4. Find the lower (X_L) and upper (X_U) limits of the interval estimate by subtracting $(z_{\alpha/2} \cdot \sigma_{\bar{X}})$ from and adding $(z_{\alpha/2} \cdot \sigma_{\bar{X}})$ to the sample mean $(\bar{X})$; that is,

$$X_L = \bar{X} - z_{\alpha/2} \cdot \sigma_{\bar{X}} \quad \text{and} \quad X_U = \bar{X} + z_{\alpha/2} \cdot \sigma_{\bar{X}}.$$

5. Alternatively, we can say that the $(1 - \alpha)$ **confidence interval estimate** of the population mean (μ_X) is defined symbolically as

$$\blacksquare \quad \bar{X} \pm z_{\alpha/2} \cdot \sigma_{\bar{X}}. \quad \blacksquare \qquad (9.1)$$

Suppose in our can-filling example we are beginning with the sample mean of 17.60 ounces, based on 25 observations from a normal population with a standard deviation of 0.80 ounces. Furthermore, suppose we elect a confidence coefficient of 0.95. Then α is 0.05 and $\alpha/2$ is 0.025. In Appendix A, we see that a tail area of 0.025 is cut off by a standard normal deviate (z) of 1.96. Hence, $z_{\alpha/2}$ is 1.96. In addition, the standard error $(\sigma_{\bar{X}})$ is 0.16 ounces, as before. It follows from Equation (9.1) that the 0.95 confidence interval estimate of the population mean is $17.60 \pm 1.96(0.16)$, or 17.60 ± 0.31 ounces. From this we obtain a lower limit of 17.29 ounces and an upper limit of 17.91 ounces for the interval.

Some values commonly used in setting confidence interval estimates when the situation is as described in this section are shown in Table 9.1. The value of $z_{\alpha/2}$ from Appendix A associated with each value of $1 - \alpha$ is also listed.

Table 9.1 Values Commonly Used for Confidence Interval Estimates

$1-\alpha$	$z_{\alpha/2}$
0.90	1.645
0.95	1.96
0.9544	2.00
0.99	2.58
0.9972	3.00

Unspecified Population Distribution

Twenty or More Observations in the Sample   In the previous section the population was assumed to be normally distributed. In such case *the sampling distribution of the mean* is normally distributed *for any sample size whatsoever.* Consequently, Equation (9.1) and the rules which accompany it apply to any sample size.

We now turn to the case for which we are given the sample mean $(\bar{X})$, the sample size (n), and the population standard deviation (σ) as before. But we cannot assume the population to be normally distributed. In this case we can make use of the central limit theorem (page 182) *provided that the number of observations in the sample is at least* 20. For all but extremely skewed populations, this theorem and a minimum sample size of 20 assure us that the sampling distribution of the mean will be essentially normally distributed, even though the parent population is not.

A look at the rules stated in conjunction with Equation (9.1) shows they require only that the sampling distribution of the mean, and not the parent population, be normally distributed. If the sampling distribution of the mean is normally distributed, then we can use Appendix A to find the value of $z_{\alpha/2}$ in Equation (9.1). Since the central limit theorem provides the assurance that the distribution is essentially normal, we can also apply the steps of the rule in the previous section to the case being considered here.

Rule	If the value of the sample mean ($\overline{X}$) and the value of the population standard deviation (σ) are known and if the sample size (n) is at least 20, then the procedure leading to Equation (9.1) should be used to set a $(1 - \alpha)$ confidence interval estimate of the population mean (μ). No knowledge of the shape of the parent population distribution is presumed.

9.1

Fewer Than 20 *Observations in the Sample*   In this case the sample size (n) is less than 20 and the value of the population standard deviation (σ) is known, but the form of the population distribution either is unknown or is known to be heavily skewed. Here we *cannot* assume that the sampling distribution of the mean is normal in shape. We are forced to assume that the distribution of the sample means is still too close in shape to that of the parent population for Appendix A to apply. Therefore, we are not able to use the procedure of the previous two sections to form a confidence interval estimate of the population mean (μ). However, we do have enough information to use another procedure based on Chebyshev's inequality.

9.2

By way of review, Chebyshev's inequality states that, for any statistical population, the proportion of the distribution outside an interval running from K standard deviations below the mean to K standard deviations above the mean is less than $1/K^2$. In the case at hand, the sampling distribution of the mean qualifies as a statistical population because it consists of all possible values of $\overline{X}$ for the stipulated population and sample size. Furthermore, since we know the value of the standard deviation (σ) for the parent population, we can find the standard error ($\sigma_{\overline{x}}$) of the sampling distribution of the mean even though this distribution is not normally distributed. This is all we need know to apply the following procedure when the values of the sample mean ($\overline{X}$) and of the population standard deviation (σ) are known and the sample size (n) is less than 20.†

---

† A closely related discussion involving the use of Chebyshev's inequality and the sampling distribution of the mean appears in Chapter 6.

Rules

1. Set the degree of confidence desired equal to $1 - (1/K^2)$ and solve for K. In this expression $(1/K^2)$ is the maximum area in the tails of the sampling distributions, and K is half the width of the confidence interval expressed in standard deviation units.

2. Find the standard error of the mean $(\sigma_{\bar{X}})$ from the equation

$$\sigma_{\bar{X}} = \sigma/\sqrt{n}.$$

3. The $1 - (1/K^2)$ **confidence interval estimate of the population mean** is found by substituting the numerical values just found in the expression

$$\bar{X} \pm K \cdot \sigma_{\bar{X}}. \tag{9.2}$$

As an example of the application of these rules we will again consider the can-filling illustration, in which $\sigma = 0.8$ ounces, $n = 25$, and $\bar{X} = 17.60$ ounces. We will now assume that we do not know that the parent population is normally distributed. We want the degree of confidence to be 0.99. Then

$$0.99 = 1 - \frac{1}{K^2},$$

from which

$$K = 10 \text{ standard deviations}$$

and

$$\bar{X} \pm K \cdot \sigma_{\bar{X}} = 17.60 \pm 10\left(\frac{0.8}{\sqrt{25}}\right).$$

In this case the 0.99 confidence-interval estimate runs from 16.0 ounces to 19.2 ounces, a length of 3.2 ounces. When we could assume the parent population to be normal, even the 0.9972 confidence-interval estimate ran from 17.12 to 18.08 ounces, a length of only 0.96 ounces. We pay a heavy price when we cannot assume that the sampling distribution is normal.

An interval estimate has one very great advantage over a point estimate. We can associate a degree of confidence with an interval estimate, whereas we have no information at all concerning the extent to which we can trust a point estimate.

9.2

Summary

When the only data available regarding a statistical population are the observations in a random sample, techniques for estimating the population mean (μ) are needed.

The sample mean ($\bar{X}$) is a *point estimator* of the population mean. One advantage of this estimator is that it is unbiased—that is, it has an expected value equal to the population mean. This is true even though there are situations in which it is impossible to get a value of the *point estimate* ($\overline{X}$) that equals the mean of the population to be sampled.

When constructed as described, an interval estimate of the population mean is accompanied by a measure of the degree of confidence that can be placed in the estimate. All such estimates discussed so far assume that the value of the standard deviation (σ) of the population is available. When the sampling distribution of the mean is known to be normally distributed, interval estimates are constructed by using normal curve areas. When the normality of the sampling distribution of the mean cannot be assumed, a technique based on Chebyshev's inequality applies.

See Self-Correcting Exercises 9.A

Exercises

1. In a survey of household incomes for a certain neighborhood, a student selects a random sample of households, determines their individual incomes, and calculates the mean of the sample to be $9,152. He then states that the mean income for the entire neighborhood is also $9,152 because the sample mean is an unbiased estimator of the population mean. Do you agree? Discuss.

2. A certain procedure is used to measure the hardness of water in parts of calcium carbonate per million parts of water. If the procedure is applied repeatedly on the same water source, the resulting measurements will be normally distributed with a standard deviation of 6 ppm (parts per million). Nine independent readings are made of the hardness for a certain water source. The sample mean is 20.13 ppm. Find the 0.99 confidence interval estimate of the true hardness.

3. A statewide study has shown that weights of twelve-year-old boys are normally distributed with a standard deviation of 10 pounds for any major city in the state. A random sample of 225 such weights in one of these cities has a mean of 85 pounds. Find the 0.9972 confidence interval estimate of the mean weight of all twelve-year-old boys in the city.

4. Balances for commercial accounts in a branch bank are known to be distributed with moderate positive skewness. The standard deviation of the distribution has remained essentially constant at $210 for the past three years, and there is no reason to assume it has changed this year. A random sample of 49 accounts has a mean balance of $832 today. What is the 0.95 interval estimate of the mean balance for all accounts?

5. Suppose that in Exercise 4 the random sample had contained only 9 observations instead of 49. Find the 0.75 interval estimate of the population mean if the sample mean is $800.

6. A widely used achievement test is given to over a thousand freshmen entering a university. A random sample of 16 scores is selected from this group and the sample mean score is found to be 132. Nationwide, the test is known to have normally distributed scores with a standard deviation of 20 points. Use these national characteristics to find the 0.95 confidence interval estimate of the mean score of the entering freshmen at the university.

7. In Exercise 6, what would be the effect if the sample size were increased to 100? What would the length of the new 0.95 interval estimate be?

Interval Estimation—Standard Deviation Unknown

9.3   Our previous discussion of interval estimation has been based on the assumption that the value of the standard deviation (σ) for the sampled parent population is available. Now we will consider interval estimation when this assumption cannot be made.

Normally Distributed Population

Thirty or More Observations in the Sample   When we know that the population from which our random sample came is normally distributed, we also know that the sampling distribution of the mean ($\overline{X}$) is precisely normally distributed. The sampling distribution of the mean has this shape *regardless of the number of observations in the sample*.

Unfortunately, however, we often do not know the value of the population standard deviation (σ). Hence, we cannot find the value of the standard error of the sample mean ($\sigma_{\overline{X}}$). Instead, we must obtain a point estimate of the population standard deviation from our sample. We will call this the *sample standard deviation* and give it the symbol s.

Definition	The **sample standard deviation** (s) is found from the relationship $$\blacksquare \qquad s = +\sqrt{\frac{\sum [(X - \overline{X})^2]}{n - 1}}, \qquad \blacksquare \qquad (9.3)$$ where $\overline{X}$ is the sample mean and n is the number of observations in the sample.

In Equation (9.3) note that the sample mean ($\overline{X}$), rather than the population mean (μ), must be taken as the reference point for the deviations. Note also that the divisor is ($n - 1$) rather than n. Were we to divide by n, the result would be a biased point estimator of the population variance (σ^2), whereas dividing by ($n - 1$) makes s^2 an *unbiased estimator* of σ^2. It does not follow, however, that the sample standard deviation (s) is an unbiased estimator of the population standard deviation (σ).

We can now make an interval estimate of the population mean (μ).

Rule	Given a random sample of at least 30 observations (n) from a normally distributed population with unknown standard deviation (σ), find the sample mean ($\overline{X}$) and the sample standard deviation (s). Then estimate the value of the standard error of the mean ($\sigma_{\overline{X}}$) from the relationship $$s_{\overline{X}} = \frac{s}{\sqrt{n}}. \qquad (9.4)$$

Finally, form the $(1 - \alpha)$ confidence interval estimate of the population mean from the expression

$$\overline{X} \pm z_{\alpha/2} \cdot s_{\overline{X}}, \tag{9.5}$$

where $z_{\alpha/2}$ is as defined in conjunction with Equation (9.1).

In spite of the fact that the sampling distribution of the mean is normal for any sample size (n), we must apply the above procedure only when n is at least 30. The estimate s is a fluctuating sample value rather than the fixed population value σ. For large samples, this fluctuation is negligible and we are usually safe when we make a direct substitution of $s_{\overline{X}}$ for $\sigma_{\overline{X}}$ in Equation (9.1). This is not an acceptable procedure for small samples, however.

Fewer Than 30 *Observations in the Sample*   Except for Expression (9.2), all of our interval estimates up to this point have been based upon the fact that the deviate (z), where

$$z = \frac{\overline{X} - \mu_{\overline{X}}}{\sigma_{\overline{X}}},$$

follows the standard normal distribution for all possible samples of a fixed size (n). Even in Expression (9.5) this relationship was the basis, because we merely took the sample standard deviation (s) to be negligibly different from the population standard deviation (σ).

In the current situation we must take a new approach, based on a new statistic.

Definition

For a random sample of fewer than 30 observations from a normally distributed population with unknown standard deviation, a confidence interval estimate of the population mean is based on the statistic t,

$$t = \frac{\overline{X} - \mu_{\overline{X}}}{s_{\overline{X}}}, \tag{9.6}$$

where $s_{\overline{X}}$ is $s/\sqrt{n}$, and s is as defined in Equation (9.3).

Although the z statistic is normally distributed for any sample size, the t statistic can be treated as being normally distributed *only* when the t values come from all possible samples, each of the same fixed size above 30. Values of t from all possible samples of a size less than 30 will be distributed in a manner that resembles a standard normal distribution but has greater dispersion. This is plausible when we note the fluctuating sample value $(\overline{X})$ in the numerator of Equation (9.6) and another fluctuating sample value (s) in the denominator, while the expression for z contains such a value only in the numerator. Furthermore, the additional dispersion increases as the sample size decreases. Consequently, there is a different t distribution for every sample size as contrasted with the single standard normal z distribution for any sample size.

A t distribution is shown in Figure 9.2, along with the standard normal distribution (z) for comparison. As in the case of this z distribution, t distributions are symmetrical about a mean of 0 and are bell-shaped. The t distribution is labeled with the symbol d.f., which stands for *degrees of freedom*. The number of degrees of freedom which apply to a sample of n observations is $n - 1$. Hence, in Figure 9.2, the t distribution labeled "d.f. = 2" is appropriate for a sample of three observations.

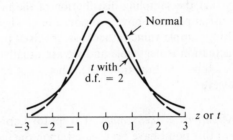

Figure 9.2   Normal Distribution and a t Distribution

Some insight into the degree-of-freedom concept can be had from examining Equation (9.3) and recalling that s is contained in the relationship for t [Equation (9.6)]. The denominator $s_{\overline{X}}$ of Equation (9.6) is a sample estimate of the dispersion in the population distribution of sample means ($\overline{X}$). We can reason that the reference point for deviations in the population standard deviation (σ) is the population mean (μ), while the reference point in Equation (9.3) is the sample mean ($\overline{X}$). The value of $\overline{X}$ tends to follow the location of the sample from which it comes, while μ is stationary. In a sample of n observations we would have n degrees of freedom on which to base an estimate of σ if we knew μ. If we do not know μ, we must use one degree of freedom in the sample to establish a reference point ($\overline{X}$) from which to measure deviations. This leaves only $n - 1$ degrees of freedom available for the estimate of population dispersion. Consequently, in the sample standard deviation (s) as defined in Equation (9.3) there are only ($n - 1$) degrees of freedom. As a result, a t distribution based on a sample of n observations has only ($n - 1$) degrees of freedom associated with it.

Since t distributions are continuous distributions similar to the standard normal distribution, we could construct a table of areas under a particular t distribution in the same manner as was done in Appendix A for the normal distribution. Because the shapes of t distributions are different for different degrees of freedom, however, we would need one such table for every degree of freedom. This is not practical, so we must take a different approach.

Appendix B lists the degrees of freedom for 34 different t distributions down the left side. At the top, the column headings are areas under the distributions, or probabilities, cumulated from the left. These correspond in concept to the entries in the *body* of Appendix A. In Appendix B, the entries in the body of the table are values of t that cut off the areas given at the top of the columns. These correspond in concept to the values of z shown at the left side and top of Appendix A.

An example of how to read Appendix B is shown graphically above the table. The values on this graph come from the second row of the table, which applies to a t distribution with two degrees of freedom. Moving to the right in that row until we come to the column headed $t_{.90}$, we read 1.886. For a t distribution with two degrees of freedom, there will be 0.90 of the total area to the left of a value of $+1.886$ on the t scale. Similarly, we see in the same row that, because of symmetry, there will be 0.10 of the total area to the left of a value of -1.886 on the t scale. For a t distribution with three degrees of freedom, however, the table shows that we would only have to go out to a t value of $+1.638$ to cut off 0.90 of the area under the curve. Finally, the values on the bottom row are standard normal curve values. This reflects the fact that the shape of a t distribution approaches that of the standard normal distribution as the degrees of freedom grow large without limit, or "approach infinity (∞)."

We are now ready to discuss how to make a confidence interval estimate of a population mean (μ) when we have a small sample from a normal population for which we do not know the standard deviation.

Rule	Given a random sample of fewer than 30 observations from a normally distributed population with unknown standard deviation (σ), find the sample mean ($\overline{X}$) and the sample standard deviation (s). Then estimate the value of the standard error of the mean ($\sigma_{\overline{X}}$) by finding $s_{\overline{X}}$ as defined in Equation (9.4). Finally, form the $(1 - \alpha)$ confidence interval estimate of the population mean (μ) from the expression

$$\blacksquare \quad \overline{X} \pm t_{\alpha/2} \cdot s_{\overline{X}}, \quad \blacksquare \tag{9.7}$$

where $t_{\alpha/2}$ is the magnitude of t from Appendix B that cuts off a tail area equal to $\alpha/2$.

A consumer research organization is interested in estimating the minimum impact force necessary to bend a certain type of automobile bumper a given amount. The bumper is supported at both ends and a weight is dropped on it from a specified height. Suppose that, in six independent experiments, the minimum weights in pounds needed to bend bumpers of this type the specified amount were 104, 110, 106, 107, 103, and 106.

The left side of Table 9.2 shows the calculations necessary to get the sum of squared deviations from the mean of our sample ($\overline{X}$) to use in Equations (9.3) and (9.4). Actually, on the right side of the table we find s^2 and $s_{\overline{X}}^2$, so that we only need take one square root at the end of our calculations.

Table 9.2 Sample Estimate of the Standard Error of the Mean

X	x	x^2
104	-2	4
110	4	16
106	0	0
107	1	1
103	-3	9
106	0	0
Total $\overline{636}$	$\overline{0}$	$\overline{30}$
$\overline{X}=106$		

$$s^2 = \frac{\Sigma x^2}{n-1} = \frac{30}{5} = 6$$

$$s_{\overline{X}}^2 = \frac{s^2}{n} = \frac{6}{6} = 1$$

$$s_{\overline{X}} = \sqrt{1} = 1 \text{ pound}$$

If we assume that the X population is normally distributed, we can use Expression (9.7) to form a confidence interval estimate of μ, the mean force necessary to bend such bumpers. Since there are six observations in our example, the t distribution with five degrees of freedom applies. Suppose we choose 0.99 as our degree of confidence. Then $\alpha/2$ is 0.005. In Appendix B for five degrees of freedom, we see that $t_{\alpha/2}$ is 4.032, or 4.0 for practical purposes. Then, from Expression (9.7), the 0.99 confidence interval estimate of μ is

$$\overline{X} \pm t_{\alpha/2} \cdot s_{\overline{X}} = 106 \pm 4.0(1).$$

This interval estimate extends from 102 pounds to 110 pounds.

For samples from normally distributed populations with unknown standard deviations, we form interval estimates of μ with Expression (9.7) for small samples and Expression (9.5) for large samples. Strictly speaking, t-distribution theory and Expression (9.7) apply when σ is unknown regardless of the sample size. As a practical matter, however, t-distributions for 29 degrees of freedom or more are considered to have approached the standard normal distribution sufficiently for us to ignore the slight difference in dispersion.

9.3

Unspecified Population Distribution

Fifty or More Observations in the Sample   When we know neither the shape nor the standard deviation of the population being sampled, we can use a technique already presented to form a confidence interval estimate of the population mean. We again use Expression (9.5):

$$\overline{X} \pm z_{\alpha/2} \cdot s_{\overline{X}}.$$

Since we can no longer assume a normally distributed population, we cannot assume that the sampling distribution of the mean is precisely normal for any sample size. We can only rely on the central limit theorem to establish an essentially normal distribution of sample means. Of even more importance is the fact that, since the sample is also being used to calculate $s_{\overline{X}}$ as an estimate of the population standard error of the mean ($\sigma_{\overline{X}}$), we must have at least 50 observations in the sample. Then we can be reasonably safe in assuming that $s_{\overline{X}}$ differs from $\sigma_{\overline{X}}$ only trivially and that the sampling distribution of $\overline{X}$ is essentially normal. On the other hand, if we are working with a distribution of incomes, for example, which can be presumed to have severe positive skewness, a minimum sample size of 100 usually is required for safety.

Fewer Than 50 Observations in the Sample   There is only one general course of action that is safe. Either make enough observations to use the technique just described, or make no confidence interval estimate of the population mean (μ). One might consider substituting the estimated standard error ($s_{\overline{X}}$) for the population standard error ($\sigma_{\overline{X}}$) in Chebyshev's inequality and using the technique presented in conjunction with Expression (9.2). But $s_{\overline{X}}$ fluctuates so much in small samples that such a procedure can be very risky.

Summary

When the population standard deviation (σ) is unknown and a confidence interval estimate of the population mean (μ) is required, one of two approaches is taken.

The first approach is taken when the population is normally distributed. The distribution of sample means can be transformed to a t distribution, which can be used to establish an interval for a specified degree of confidence. As a practical matter, the standard normal deviate (z) is substituted for t when the sample contains at least 30 observations.

The second approach is taken when the population cannot be presumed to be normally distributed. In this approach a point estimate is substituted for the standard error of the mean ($\sigma_{\bar{x}}$) and the central limit theorem is invoked to assure normality of the distribution of sample means. There must be at least 50 observations in a sample when this approach is used.

Summary Table

Interval Estimation of the Population Mean

Population	Sample size	σ Known	σ Unknown
Normal	Large ($n \geq 30$)	Confidence coefficient: $(1 - \alpha)$	$\bar{X} \pm z_{\alpha/2} \cdot \dfrac{s}{\sqrt{n}}$ †
Normal	Small ($n < 30$)	Interval estimate: $\bar{X} \pm z_{\alpha/2} \cdot \dfrac{\sigma}{\sqrt{n}}$	$\bar{X} \pm t_{\alpha/2} \cdot \dfrac{s}{\sqrt{n}}$
Not normal	Large	Observations: $n \geq 20$ Interval estimate: $\bar{X} \pm z_{\alpha/2} \cdot \dfrac{\sigma}{\sqrt{n}}$	Observations: $n \geq 50$ Interval estimate: $\bar{X} \pm z_{\alpha/2} \cdot \dfrac{s}{\sqrt{n}}$
Not normal	Small	Observations: $n < 20$ Coefficient: $1 - \dfrac{1}{K^2}$ Interval estimate: $\bar{X} \pm K \cdot \dfrac{\sigma}{\sqrt{n}}$	Observations: $n < 50$ Interval estimate: None available

† z is an approximation for t.

See Self-Correcting Exercises 9.B

Exercises

1. The mean daily wage of a random sample of 25 unskilled workers in a city is $7.50. The sum of squared deviations of the 25 observations from the sample mean is 240.
   a. What is the value of the sample variance?
   b. What is the value of the sample standard deviation?

2. Refer to Exercise 1.
   a. For what value is the answer to part a an estimate? Is it unbiased?
   b. For what value is the answer to part b an estimate, and is it unbiased?

3. Use the t table in Appendix B to answer the following questions.
   a. What are the values of t for a t distribution with 17 degrees of freedom such that five percent of the area is in each tail?
   b. In a t distribution with 21 degrees of freedom, what is the absolute value of t that will be exceeded with the probability 0.05?

4. How many degrees of freedom are there in a t distribution in which the probability of t being greater than $+2.718$ is 0.01?

5. The distribution of lifetimes for a type of transistor has moderate positive skewness. A random sample of 100 from a large shipment of these transistors has a mean lifetime of 1,190 hours and a standard deviation of 90 hours. What is the 0.95 confidence interval estimate of the mean lifetime for the shipment?

6. In a factory, a new work-station design is being tested for productivity. Observations are made on output during each of 16 hours. The sample mean is 236 units of output per hour, and the sample standard deviation is 20 units per hour. Assume that the population of hourly outputs is normally distributed and find the 0.99 confidence interval estimate of the population mean hourly output.

7. The net weight printed on boxes of one brand of butter is one pound. A random sample of 36 of these boxes has been selected from several supermarkets. The sample mean net weight is 15.95 ounces and the sample standard deviation is 0.12 ounces. What is the 0.99 confidence interval estimate of the mean net weight?

8. Nickel-cadmium battery packs for a certain brand of cassette recorder are very expensive to produce. It has been suggested that a single random sample of only 5 such packs be selected from a large production lot in order to obtain a confidence interval estimate of the mean lifetime for the lot. The sample standard deviation must be used in place of the unknown population value.
   a. What shape can the population distribution be expected to have?
   b. What is the correct course of action with respect to the recommended procedure?

10

Testing Hypotheses about Population Means

In the previous chapter we considered how to use the data from a sample to estimate the value of the mean of the population from which a sample came. We had no preconception as to the value of the population mean. We simply wanted the best estimate of its value that could be obtained from a random sample.

The type of hypothesis testing discussed in this chapter begins with a statement about the value of the population mean. We have some basis upon which to form a preconception, or hypothesis, about what the value is. We then test the hypothesis.

The form of hypothesis test to be described consists of combining the data from a random sample with relevant sampling theory to make a decision about the hypothesis. The alternatives are to accept the statement about the value of the population mean or to reject it because some other values are much more plausible.

In hypothesis testing, as in estimation, drawing a conclusion about the value of the population mean from a sample that comprises only part of the population always entails a risk. Estimation presents the risk that our sample estimate of the parameter is seriously in error. In hypothesis testing we run the risk that the sample evidence leads us to the wrong alternative concerning the statement being tested.

Two-Tailed Tests

In the type of test to be described now, we want to guard against a bad decision in either direction. If we hypothesize that the population mean is a certain value, we want a test which rejects that hypothesis when the true mean lies substantially above or below that value. The section that follows this, by way of contrast, will consider one-tailed tests, wherein our only concern is whether or not the true mean differs from the hypothesized value in a single direction.

10.1

Known Population Standard Deviation

Normally Distributed Population   A cement manufacturing plant has just installed a new machine to fill 100-pound sacks. Although the manufacturer of the filling machine sets machines for the correct fill, vibration in shipment sometimes changes the weights delivered at the 100-pound setting. Therefore, the cement manufacturer is advised to check.

The population standard deviation of fill weights for this type of machine is 0.90 pounds. Regardless of the setting, this standard deviation applies almost without exception to any given machine of this type. Similarly, any machine of this type can be assumed to deliver a normally distributed population of fill weights at any single setting. The cement manufacturer has reason to believe that his new filling machine belongs to the majority of such machines. If so, it will deliver a mean fill weight of 100 pounds. To play safe, he decides to state that the population mean (μ) is 100 pounds and to treat this as a hypothesis to be tested.

To test the hypothesis that the population mean (μ) is 100 pounds, the manufacturer will fill four sacks and treat their content weights as a random sample of four observations. If the mean ($\overline{X}$) of these four weights is far greater than 100 pounds, he will reject his original hypothesis in favor of a conclusion that the population mean (μ) is greater than 100 pounds. If the sample mean is far less than 100 pounds, he will conclude that the population mean is less than 100 pounds.

We can use sampling theory to establish a range for the sample mean ($\overline{X}$). If the sample mean falls inside this range, we will accept the hypothesis. If the mean falls outside the range, we will reject it, as suggested in the previous paragraph.

We will have a single random sample of four observations that we assume to be from a normally distributed population with a standard deviation (σ) of 0.90 pounds. Furthermore, if the manufacturer's hypothesis happens to be correct, the population mean (μ) will be 100 pounds.

The statement on page 176 and the discussion following it now come into play. The mean ($\overline{X}$) of the particular four observations that the manufacturer will select belongs to a statistical population composed of sample means from all possible random samples of four observations each. If the hypothesis and the foregoing assumptions are true, then the distribution of sample means has three characteristics:

1. It is normally distributed.
2. It has the same mean ($\mu_{\overline{x}}$) as does the parent population (μ). This is 100 pounds, by hypothesis.
3. It has a standard deviation ($\sigma_{\overline{x}}$), called the standard error of the sample mean, that can be found from the standard deviation of the parent population ($\sigma = 0.90$ pounds) and the sample size ($n = 4$). Specifically,

$$\sigma_{\overline{X}} = \frac{\sigma}{\sqrt{n}} = \frac{0.90}{\sqrt{4}}, \quad \text{or } 0.45 \text{ pounds.}$$

If the sample mean ($\overline{X}$) of the four observations made by the manufacturer did, in fact, come from the distribution just described, that mean should not be very far from the population mean ($\mu_{\overline{x}}$). Since the distribution of sample means is normal and we know the values of its mean and standard error, we can use Appendix A to assess the probability that $\overline{X}$ will lie within a given number of standard errors of $\mu_{\overline{x}}$. For example, the probability is 0.99 that a sample mean selected randomly from this population will be within 2.58 standard errors of 100 pounds. Since the standard error is 0.45 pounds, 99 percent of the sample means based on four observations will lie between 98.84 and 101.16 pounds. The lower limit is found by subtracting the product of 2.58 and 0.45 from 100 pounds, and the upper limit is the result of adding that product to 100 pounds. This situation is illustrated in Figure 10.1.

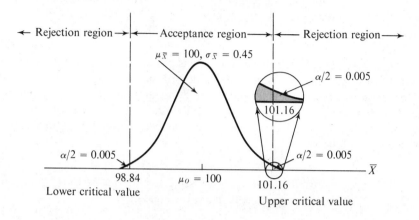

Null hypothesis: $\mu = 100$
Alternate hypothesis: $\mu < 100, \mu > 100$
Level of significance: $\alpha = 0.01$

Figure 10.1   Hypothesized Distribution of $\overline{X}$ and Test Criteria; Two-Tailed Test

In Figure 10.1 the sample size and the assumptions determine the shape of the sampling distribution of the mean and the value of its standard error. The hypothesized value of the population mean determines the mean of the population of sample means. *We will give the hypothesized mean the symbol μ_0.* Then the previous statement in symbols is $\mu_{\overline{X}} = \mu_0$. The statement that μ_0 is equal to 100 pounds is called the *null hypothesis*; it is subject to being nullified if evidence from the sample is unfavorable to the hypothesis.

If the null hypothesis is true, the probability is 0.99 that the particular sample mean ($\overline{X}$) the manufacturer obtains from his four random observations will lie between a *lower critical value* of 98.84 pounds and an *upper critical value* of 101.16 pounds. We will call the interval between these values the *acceptance region*, because the null hypothesis will be accepted if $\overline{X}$ falls within it. In this case the filling machine will be left as it is. The *rejection regions* are at either end of the acceptance region in Figure 10.1. The null hypothesis will be rejected if $\overline{X}$ falls in either of these and the filling machine will be adjusted.

When the null hypothesis is true and given the acceptance region in Figure 10.1, the probability is 0.99 that the null hypothesis will be accepted because 99 percent of the values of $\overline{X}$ arising from random samples of four observations will fall within the region. Alternatively, there is a probability of 0.01 of rejecting the null hypothesis *when it is true*. The combined area in the two tails lies in the rejection region and represents the probability of rejecting the null hypothesis when it is correct—a conclusion that constitutes one type of decision error. (Later in the chapter we will discuss decision errors.) The probability of rejecting the null hypothesis when it is true is called the *level of significance* for the given hypothesis test. This conditional probability of rejection is designated with the symbol α. For the example, α is 0.01. It follows that $1 - \alpha$ is the probability of accepting the null hypothesis when it is true. In the example, $1 - \alpha$ is 0.99.

Given the test procedure just described, suppose the cement manufacturer finds the sample mean ($\overline{X}$) of his random sample of four observations to be 98.3 pounds. This value falls in the lower rejection region. Consequently, the evidence is that the filling machine is out of adjustment. The population mean fill weight appears to be less than 100 pounds.

We can now summarize the discussion to this point. Our objective is to perform a two-tailed test of a hypothesis about the value of a population mean (μ). The test is based on three assumptions: independent random sampling, a normally distributed population, and a known value for σ, the population standard deviation. In addition, the sample size (n) to be used in the test has already been decided. We can perform the two-tailed test as follows:

1. Form the null hypothesis and the alternative to it by assigning a numerical value to μ_O in the statements

$$H_O : \mu = \mu_O,$$
$$H_A : \mu < \mu_O \quad \text{or} \quad \mu > \mu_O,$$

where H_O is the null hypothesis and H_A is the alternative statement to be accepted if H_O is rejected.

2. Select a value between 0.90† and 0.99 for the probability of accepting H_O if it is true. Set this probability equal to $1 - \alpha$ and solve for α, the probability of rejecting H_O if it is true. This value (α) is the level of significance.

3. From Appendix A find the magnitude of the standard normal deviate (z) that cuts off a tail area of $\alpha/2$. Call this magnitude $z_{\alpha/2}$ and ignore its algebraic sign.

Rules

4. Find the standard error of the mean ($\sigma_{\overline{X}}$) where $\sigma_{\overline{X}} = \sigma/\sqrt{n}$ and σ and n are known.

5. Find the upper (V_U) and lower (V_L) critical values that bound the acceptance region from the expression

$$\blacksquare \qquad \mu_O \pm z_{\alpha/2} \cdot \sigma_{\overline{X}}; \qquad \blacksquare \qquad (10.1)$$

that is,

$$V_U = \mu_O + z_{\alpha/2} \cdot \sigma_{\overline{X}} \quad \text{and} \quad V_L = \mu_O - z_{\alpha/2} \cdot \sigma_{\overline{X}}.$$

6. Find the mean of the random sample ($\overline{X}$), and accept the null hypothesis ($H_O: \mu = \mu_O$) if $\overline{X}$ is in the acceptance region, that is, $V_L \leq \overline{X} \leq V_U$. Otherwise, reject H_O and accept H_A; that is, accept $\mu < \mu_O$ if $\overline{X} < V_L$ or accept $\mu > \mu_O$ if $\overline{X} > V_U$.

† The level of significance must be small enough to prevent excessive rejection errors, yet large enough to keep the length of the acceptance region from being excessive. It is customary to choose values of α in the range from 0.01 to 0.10 for most applications.

When the hypothesis test just described is compared with the interval-estimation procedure described on pages 191 and 192, two major differences are apparent. As can be seen by comparing Equation (9.1) and Expression (10.1) or Figures 9.1 and ... on the sample mean ($\overline{X}$), while an acceptance ... st is centered on the hypothesized value of the ... is as illustrated for our cement sack example in ... figure shows the 0.99 confidence interval estimate ... on the sample mean of 98.3 the manufacturer ... e acceptance region for the hypothesis test for

$H_0: \pi = .2$

$\alpha = .05$

$n = 100, r = 10$

$Z_{.025} = 1.96$

$Z_{.05} = 1.645$

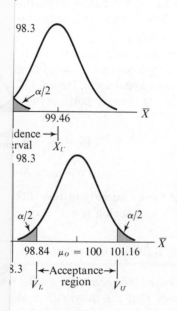

Figure 10.2 ...fidence ...erval and Acceptance Region Compared

The second difference between the two procedures involves the meaning of α. In the case of interval estimates, α is the extent to which we lack confidence in the estimate; in hypothesis tests, α is the probability of rejecting the null hypothesis when it is true. As with interval estimates, one of the values in Table 9.1 is commonly selected for use in a hypothesis test.

So long as the parent population is normally distributed and so long as the value of its standard deviation (σ) is known, the sampling distribution of $\overline{X}$ will be precisely normal for any specified value of n, the sample size. As a result, the procedure described in conjunction with Equation (10.1) applies to both large and small samples— to any sample size whatsoever.

10.1

Unspecified Population Distribution; $n \geq 20$ Although we do not assume that the parent population is normally distributed in this case, we do know the value of its standard deviation (σ). Furthermore, as in interval estimation for this case, a sample of 20 or more observations and the central limit theorem provide reasonable assurance that we are working with a sampling distribution of $\overline{X}$ that is virtually normal in shape. Consequently, we have everything we need to make the procedure described in the previous section applicable. Hence, the rules incorporating Expression (10.1) should be used to conduct a two-tailed test of a hypothesis about the value of the population mean when we know the value of the population standard deviation (σ) and when the sample size (n) is at least 20.

Unspecified Population Distribution; $n < 20$ A sample in this size range is too small for us to rely on the central limit theorem's ability to produce a normal sampling distribution for sample means. As in the analogous case for interval estimation, we must turn to Chebyshev's inequality.

We know the values of the parent population's standard deviation (σ) and of the sample size (n). As before, these can be used to find the standard error of the mean ($\sigma_{\bar{x}}$) from the relationship $\sigma_{\bar{x}} = \sigma/\sqrt{n}$.

Chebyshev's inequality assures us that $1/K^2$ is the maximum probability that a randomly selected sample mean ($\bar{X}$) will fall more than K standard errors from the population mean (μ_O by hypothesis). Hence, $1/K^2$ can be set equal to α, the significance level for a two-tailed test of the hypothesis that $\mu = \mu_O$. Finally, the acceptance region extends from

$$V_L = \mu_O - K \cdot \sigma_{\bar{X}} \tag{10.2a}$$

to

$$V_U = \mu_O + K \cdot \sigma_{\bar{X}}. \tag{10.2b}$$

We are assured that the probability of rejecting the null hypothesis when it is true is less than α, or $1/K^2$, with such a test.

Unknown Population Standard Deviation

Normally Distributed Population; $n \geq 30$ Because the population is normally distributed, we know that the distribution of sample means is also normally distributed. We do not, however, know the value of the population standard deviation (σ). Hence, although we do know the sample size (n), we cannot find the standard error of the mean ($\sigma_{\bar{x}}$). Instead we must find the sample standard deviation (s) as defined in Equation (9.3) and form a point estimate of the standard error, $s_{\bar{x}}$, where $s_{\bar{x}} = s/\sqrt{n}$, as in Equation (9.4). Values of $s_{\bar{x}}$ arising from samples of 30 or more should be closely clustered around $\sigma_{\bar{x}}$, so we can substitute our sample value of $s_{\bar{x}}$ for $\sigma_{\bar{x}}$ with negligible damage.

After selecting the desired significance level (α), we find the standard normal deviate $z_{\alpha/2}$. Then we use the expression

$$\mu_O \pm z_{\alpha/2} \cdot s_{\bar{x}}, \tag{10.3}$$

where μ_O is the hypothesized value of the population mean, to fix the limits of the acceptance region and proceed from this point on just as we did before.

Normally Distributed Population; $n < 30$ For normally distributed populations, all of our hypothesis tests prior to this have depended upon the normality of the distribution of the sampling statistics

$$z = \frac{\bar{X} - \mu_O}{\sigma_{\bar{X}}}, \quad \text{and} \quad z = \frac{\bar{X} - \mu_O}{s_{\bar{X}}}.$$

When the population of X values is normally distributed, the first of these two statistics is distributed as the standard normal distribution for any sample size. But the second statistic conforms to the standard normal distribution reasonably well only if two conditions are satisfied. The population of X values must be normally distributed *and* there must be 30 or more observations in the sample. Of course, for both statistics to be distributed as described, the null hypothesis ($\mu = \mu_0$) must also be true.

When the sample consists of fewer than 30 observations and all other conditions described in the previous paragraph are met, the ratio

$$\frac{\overline{X} - \mu_O}{s_{\overline{X}}}$$

is no longer distributed as z, the standard normal deviate. Instead this ratio will be distributed as t. In other words, we are concerned with the statistic t, where

$$t = \frac{\overline{X} - \mu_O}{s_{\overline{X}}}.$$

The t distribution and use of Appendix B were discussed in Chapter 9, pages 197 ff.

In the current situation, we have a random sample of less than 30 observations (n) from a normally distributed population for which we do not know the standard deviation. To perform a two-tailed test of the hypothesis that the population mean is equal to a specified value ($\mu = \mu_0$), we first select the level of significance (α). Then we find the value of $t_{\alpha/2}$ from Appendix B and establish the limits of the acceptance region for the hypothesis from the expression

$$\blacksquare \quad \mu_O \pm t_{\alpha/2} \cdot s_{\overline{X}}. \quad \blacksquare \tag{10.4}$$

Next, we find the sample mean ($\overline{X}$). If it falls within the acceptance region, we accept μ_O as the value of μ. Otherwise we accept either a lower or a higher value, as described previously.

For example, a reading test is being considered by an organization for use in hiring new employees. Test scores are very much affected by the time it takes to do the test. The developer of the test claims that the mean time to complete the test is 15 minutes and that the distribution of reading times is nearly normal. Three prospective employees of the organization have taken the test. Their times were 17, 18, and 19 minutes. The sample mean ($\overline{X}$) is 18 minutes and the sample standard deviation (s) is found from Equation (9.3):

$$s = \sqrt{\frac{\sum [(X - \overline{X})^2]}{n - 1}} = \sqrt{\frac{(-1)^2 + (0)^2 + (+1)^2}{2}}, \quad \text{or 1 minute.}$$

Then

$$s_{\overline{X}}^2 = \frac{s^2}{n} = \frac{1}{3}, \quad \text{or 0.333,}$$

and

$$s_{\overline{X}} = \sqrt{0.333}, \quad \text{or 0.58 minutes.}$$

The significance level selected is 0.05. From this, $\alpha/2$ is 0.025. Since $n-1$ is 2, we enter Appendix B with two degrees of freedom and find that $t_{\alpha/2}$ is 4.303. The acceptance region is

$$\mu_0 \pm t_{\alpha/2} \cdot s_{\bar{X}}, \quad \text{or} \quad 15 \pm 4.303(0.58).$$

The upper critical value is 17.50, which is less than the sample mean $(\bar{X})$ of 18 minutes. The null hypothesis should be rejected. If the three reading times are representative of the organization's prospective employees (that is, random), these people seem to take more time on the average than did those reported by the developer of the reading test.

It could be argued that the population of reading times in the example just described may have considerable positive skewness. If this is so, the t test would not be appropriate theoretically.

Unspecified Population Distribution; $n \geq 50$ By now it probably is apparent that the reasoning presented in a given case in the chapter on interval estimation parallels that used in the analogous case in this chapter. Consequently, you can refer to the section beginning on page 200 to review the reasoning that applies here.

The sample standard deviation, s, is found by applying Equation (9.3), as before. Then the estimate of the standard error of the mean ($s_{\bar{X}}$) is found by using Equation (9.4). The acceptance region is

$$\mu_0 \pm z_{\alpha/2} \cdot s_{\bar{X}}.$$

Once this region is established, the remaining test procedure is the same as has been illustrated in the previous cases.

Unspecified Population Distribution; $n < 50$ The sample is too small to justify using both the central limit theorem and a sample estimate of the standard error of the mean to form an acceptance region. Either gather more data or abandon the test in its present form.

Summary

Testing statements about the mean value of a statistical population is another application of the characteristics of the distribution of sample means discussed in Chapter 8. They were applied previously in Chapter 9 in connection with interval estimation of the population mean.

Acceptance regions for two-tailed tests of hypotheses about the population mean bear a strong resemblance to confidence intervals applicable under the same sets of assumptions. There are two major differences, however. Acceptance regions are centered on the hypothesized value of the population mean rather than on the value of the sample mean. Secondly, for interval estimates, $(1 - \alpha)$ measures the degree of confidence, while in a hypothesis test, $(1 - \alpha)$ measures the probability of accepting the null hypothesis, given that it is true.

See Self-Correcting Exercises 10.A

Exercises

1. In past years, the mean score on a national reading examination has remained virtually constant at 80. The national distribution of scores always has marked negative skewness, and the standard deviation varies materially from year to year. This year a random sample of 100 scores has been selected to perform a two-tailed test of the hypothesis that the population mean score is 80. The sample mean is 83.5 and the sample standard deviation is 12.

   a. What are the correct equations for finding the limits of the acceptance region? Explain your choice.

   b. Perform the test described for a 0.05 level of significance.

2. In a large department store, the mean balance due on active credit accounts was $214 for this month a year ago. The standard deviation of balances due has remained essentially constant at $42 for several years. The credit manager wants to perform a two-tailed test, at the 0.01 level, of the hypothesis that the mean balance due has not changed. He has selected a random sample of 36 accounts and has found the sample mean to be $199.

   a. What are the correct equations for the limits of the acceptance region? Explain your choice.

   b. Complete the test and state your conclusion.

3. For the situation described in Exercise 2, the acceptance region was described in terms of values of the sample mean.

   a. What would the values of the acceptance region limits be if they were stated in terms of the standard normal variate, z, instead of the sample mean?

   b. What would the general procedure be if z, rather than $\bar{X}$, were used as the test statistic?

4. The shelves in the mail room of a mail order firm have been designed for a mean package length of 16 inches. The manager wants to test the hypothesis that the mean length is actually 16 inches. He selects a random sample of 6 packages and finds their lengths to be 13, 14, 14, 14, 17, and 18 inches, to the nearest inch. Assume that package lengths are normally distributed.

   a. Find the sample mean and standard deviation.

   b. Find the limits of the acceptance region for a two-tailed test of the hypothesis that the population mean is 16 inches. Use a significance level of 0.02.

   c. Should the null hypothesis be rejected? Why?

5. A coin operated machine that pours milk by the cup is set to deliver 8 ounces per cup. A random sample of 10 cups has a mean weight of 8.60 ounces and a standard deviation of 0.5 ounces. This sample is to be used in a two-tailed hypothesis test that the setting is correct. Should this hypothesis be rejected at the 0.05 level of significance?

6. A plant engineer has stated that the time to assemble a new product should be 12 minutes, on the average. It is known that assembly time distributions for very similar products have marked positive skewness and standard deviations which vary little from 0.6 minutes. A random sample of 16 assembly times has been collected to make a two-tailed test of the engineer's statement at the 0.01 level of significance. The mean of the sample is 10.3 minutes. What conclusion is justified by the test?

7. Last year, the state tourist bureau estimated that visitors spent an average of $138 in the state. This year a random sample of 100 visitors has spent an average of $134 and the sample standard deviation is $20. The population is known to be positively skewed. Perform a two-tailed test of the hypothesis that this year's population mean is $138. The level of significance is 0.02.

10.2   One-Tailed Tests

In this section we return to the first set of conditions discussed under two-tailed tests. We shall also use a modification of the cement manufacturing example. Recall that the machine that fills sacks is supposed to put 100 pounds of cement in each sack. We assume that a filling machine produces a normally distributed population of fill weights and that the standard deviation (σ) of this population is known to be 0.90 pounds.

A state inspector of weights and measures wants assurance that customers get at least 100 pounds of cement, almost without exception, when they purchase from the plant in question. He is not concerned if they get more than that.

If the mean fill being made by the machine (μ) is 100 pounds, then the weights will be normally distributed around this value. Half the sacks will contain somewhat less than 100 pounds of cement. This is unacceptable to the inspector. He wants the mean fill weight to be greater than 100 pounds so that the majority of sacks will weigh at least 100 pounds. The cement manufacturer is aware of the slight variability in fill weights and attempts to keep his machine set to deliver a mean fill of slightly more than 100 pounds. The purpose of the inspection is to check that this practice is being followed.

The test rationale begins with (1) the supposition that the mean fill is 100 pounds and (2) a decision to select a random sample of, say, 36 observations (n). If true, the assumptions concerning the population of fill weights will result in a sampling distribution of $\overline{X}$ for samples of 36 that is normal in shape and has a mean (μ_o) of 100 pounds and a standard error ($\sigma_{\overline{X}}$) of 0.15 pounds—that is,

$$\sigma_{\overline{X}} = \frac{\sigma}{\sqrt{n}} = \frac{0.90}{\sqrt{36}} ; \quad \text{or 0.15 pounds.}$$

The next step is to select a significance level. The value chosen by the inspector is 0.01 ($\alpha = 0.01$). In the one-tail test we look up the magnitude of z_α in Appendix A, rather than the magnitude of $z_{\alpha/2}$ as in the two-tailed test. For our current example, z_α is 2.33.

We now come to the key point in the rationale. The hypothesized population mean is 100 pounds, and the standard error of the mean is 0.15 pounds. The distribution of sample means is normal. Given these conditions, only once in 100 times will we get a random sample of 36 observations with a mean ($\overline{X}$) that falls 2.33 standard errors or more *above* 100 pounds, that is, above 100.35 pounds. If the inspector selects a random sample of 36 observations and finds that its mean ($\overline{X}$) is greater than 100.35 pounds, he can reject the hypothesis that the machine is set too low and accept the desired alternative hypothesis that it is delivering a mean fill of more than 100 pounds.

The situation is illustrated in Figure 10.3. The critical value, 100.35 pounds, is 2.33 standard errors above the hypothesized mean (μ_o) of 100 pounds. If the mean of 36 observations ($\overline{X}$) falls in the acceptance region, the inspector will require the manufacturer to adjust his machine to deliver a greater mean fill. If $\overline{X}$ falls in the rejection region, the inspector will conclude that the mean fill is greater than 100 pounds.

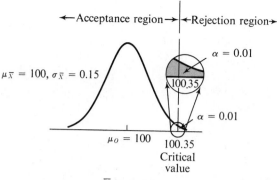

Null hypothesis: $\mu \leq 100$
Alternate hypothesis: $\mu > 100$
Level of significance: $\alpha = 0.01$

Figure 10.3 Hypothesized Distribution of $\bar{X}$ and Test Criteria; One-Tailed Test.

In this cement example the product is being tested to see whether it *exceeds* a minimum requirement (μ_0). This requirement is reflected in the *alternative* to the null hypothesis (H_A). Our complete hypothesis statement to test a *minimum* specification is

$$\text{Null hypothesis } (H_0): \quad \mu \leq \mu_0,$$
$$\text{Alternative hypothesis } (H_A): \quad \mu > \mu_0.$$

In other cases, to be satisfactory, products must not exceed some *maximum* requirement. Maximum specifications for impurities in foods and drugs are examples. One-tailed tests are appropriate here, too. But to reflect the requirement in H_A, the inequality signs must be reversed. The complete hypothesis statement to test a *maximum* specification is

$$\text{Null hypothesis } (H_0): \quad \mu \geq \mu_0,$$
$$\text{Alternative hypothesis } (H_A): \quad \mu < \mu_0.$$

We can now summarize the procedure for making a one-tailed test of a hypothesis about the population mean (μ). The test is based on three assumptions: independent random sampling, a normally distributed population, and a known value of σ, the population standard deviation. In addition, the sample size (n) has been decided. The one-tailed test procedure is given in the following rules.

Rules

1.   State the null hypothesis (H_0) and the alternative to it (H_A) by assigning a value to μ_0 in

$$H_0 : \mu \leq \mu_0, \quad \text{and} \quad H_A : \mu > \mu_0$$

or in

$$H_0 : \mu \geq \mu_0 \quad \text{and} \quad H_A : \mu < \mu_0.$$

Choose the first set of hypotheses for a *minimum* specification and the second set for a *maximum*.

2. Select the level of significance, α, and from Appendix A find the standard normal deviate (z) that cuts off a tail area of α. Call this z_α and ignore its sign.

3. Find the standard error of the mean ($\sigma_{\bar{X}}$).

4. Find the critical value (V) by substituting the proper values in

$$V = \mu_O + z_\alpha \cdot \sigma_{\bar{X}} \qquad (10.5a)$$

if H_O is of the form $\mu \leq \mu_O$
or in

$$V = \mu_O - z_\alpha \cdot \sigma_{\bar{X}} \qquad (10.5b)$$

if H_O is of the form $\mu \geq \mu_O$.

5. Find the sample mean ($\bar{X}$) and accept H_O if Equation (10.5a) applies and $\bar{X} \leq V$ or if Equation (10.5b) applies and $\bar{X} \geq V$. Otherwise accept the form of H_A that applies.

We will not describe one-tailed hypothesis tests concerning the population mean (μ) for the other sets of assumptions about the shape of the population distribution and our state of knowledge with regard to the population standard deviation (σ) covered under two-tailed tests. It should be clear from the case we did discuss in this section that only one change is involved to convert any two-tailed test of μ we discussed into a one-tailed test. We only need shift all the tail area defined by the level of significance (α) to one tail or the other to make the conversion.† This shift also causes a change in the number of standard errors that we must go out from the hypothesized mean (μ_O) to leave the required area in the appropriate tail of the sampling distribution.

10.2

Test Performance

10.3

In the previous portion of this chapter we selected sample sizes (n) and levels of significance (α) arbitrarily. Seldom in practice are the costs of error and observations so low that one can afford to set up tests by making these selections arbitrarily. Instead one can consider how test performance is affected by changes in sample size and significance level. This knowledge can then be used to construct tests to meet a desired performance criterion.

Types of Error

To examine the performances of alternative tests, we must begin by considering the types of error that can be made and the probabilities for making each type. First, we reason that a null hypothesis (H_O) must be either true or false. As a result of our sampling experiment, we can either accept or reject H_O. If we accept it when it is

† This cannot be done for tests involving the Chebyshev inequality.

true and reject it when it is false, we make correct decisions. On the other hand, if we reject the null hypothesis when it is true we make a *Type I error*. The maximum probability of making a Type I error is given the symbol "α," and it is this probability we set when we choose the level of significance for a test. In terms of our earlier hypothesis testing procedure, it should be pointed out that rejecting the null hypothesis (H_O) is equivalent to accepting the alternative (H_A).

If we accept the null hypothesis when it is false, we make a *Type II error*. The probability of making this type of error is given the symbol "β." The four possible situations are summarized in Table 10.1

Table 10.1   Possible Outcomes of Hypothesis Tests

	H_o *True*	H_o *False*
Accept H_o	Correct decision	Type II error Probability: β
Reject H_o	Type I error Probability: α	Correct decision

Error Probabilities for a One-Tailed Test

In our one-tailed test using the cement example, the null hypothesis was $\mu \leq 100$ pounds and the alternative was $\mu > 100$ pounds. The values of the standard deviation (σ) and sample size (n) were 0.90 pounds and 36 observations, respectively. The level of significance (α) was 0.01.

Figure 10.4(a) illustrates a situation *when the null hypothesis is true*. In this situation $\mu = \mu_o = 100$ pounds. The sampling distribution is centered on the hypothesized value of the mean. If a sample mean ($\overline{X}$) is selected that comes from the tail area above 100.35 pounds, a Type I error will result. The probability of a Type I error in this case equals the value of α exactly and is 0.01 when $\mu = 100$ pounds.

The null hypothesis also will be true if the population mean (μ) is any value less than 100 pounds. Should this be the case, the sampling distribution of $\overline{X}$ would be even farther to the left than it is in the upper half of the figure. Then the area to the right of the critical value will be less than α. Since the null hypothesis ($\mu \leq 100$) is still true, we can make only a Type I error. But now the probability of rejecting the hypothesis is less than 0.01. As we imagine the possibility of μ taking on values farther and farther to the left of 100 pounds, we can also visualize the tail area to the right of 100.35 approaching zero. The farther μ is to the left of μ_o the closer the probability of a Type I error is to zero.

Keep in mind that we do not know the value of μ at the beginning of a hypothesis test, at any time during the test, or after the test is completed. The sample information just improves our chances of accepting the correct alternative in the hypothesis statement. Even so, when we have completed the test, it is entirely possible that we have accepted the wrong alternative.

Figure 10.4(b) shows what could happen if the population mean (μ) were 100.20 pounds, without our knowledge. Since the critical value (100.35) is one standard error above μ, 84 percent of possible sample means ($\overline{X}$) fall below the critical value. Should

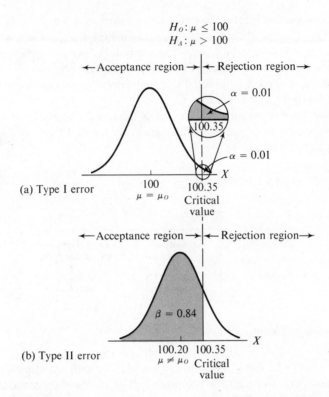

Figure 10.4   Types I and II Errors and Their Probabilities

our random sample of 36 observations yield a mean from this portion of the distribution, we would accept the null hypothesis that μ is no greater than 100 pounds when it is, in fact, 100.20 pounds. This is a Type II error. The probability of making this type of error (β) is 0.84 when μ is 100.20 pounds.

If μ were less than 100.20 pounds but more than 100 pounds, we would have an even greater probability of making a Type II error. The maximum value of β would result if μ were just infinitesimally greater than 100. In this event, very nearly 99 percent of the sampling distribution of $\overline{X}$ would be below the critical value and β would be just infinitesimally less than 0.99. In this case, the inspector is almost certain to accept the null hypothesis and to make the cement manufacturer increase the mean fill, even though that mean fill is already, if just barely, greater than 100 pounds.

If the population mean is somewhat greater than the critical value (100.35 pounds), the area in the sampling distribution that falls to the left of the critical value will be relatively small. Consequently β will be relatively small. The inspector is much more likely to conclude that μ is greater than 100 pounds (H_A) and not require that the mean fill be changed.

Table 10.2 presents several values that the population mean (μ) could have in the one-tailed test in this section. The type of error possible is shown for each potential value of μ listed. Measured in standard errors, the distance from μ to the critical value is found and used along with Appendix A to find the probability of error were μ to have the value shown.

Table 10.2   Potential Values of μ and Associated
Error Information

Potential value of μ	Type of error possible	α	β
99.90	I	.0014	
99.96	I	.0047	
100.00	I	.0100	
100.05	II		.9772
100.11	II		.9452
100.20	II		.8413
100.35	II		.5000
100.50	II		.1587
100.56	II		.0808
100.65	II		.0228
100.70	II		.0100

Operating Characteristic for a One-Tailed Test

Table 10.2 presents us with the possibility of describing how a test performs for all possible values of the population mean (μ). An operating characteristic function is one way to exhibit this performance.

	For a specified hypothesis test concerning the mean (μ) of a statistical
Definition	population, the **operating characteristic function** expresses the probability of accepting the null hypothesis (H_o) for any value of μ.

For our one-tailed test example, we can find the values of the operating characteristic function for the values of μ listed in Table 10.2. Then, by graphing these points and passing a smooth curve through them, we can visualize the function.

Table 10.3 lists the probabilities of accepting the null hypothesis for each of the given values of μ. Furthermore, in the right column, the table shows how to find the proper probability. If the value of μ is such that the null hypothesis (H_o) is true, the maximum probability that the hypothesis will be accepted is illustrated by all the area to the left of 100.35 in (a) of Figure 10.4. This area is equal to $1 - \alpha$. Since we have already found the values of α in Table 10.2, finding $1 - \alpha$ is simple. When μ is such that the null hypothesis (H_o) is false, a proper illustration is (b) of Figure 10.4. In this case also, the probability of accepting H_o is illustrated by all area to the left of H_o. In this circumstance a Type II error will be committed and β is the relevant probability.

The operating characteristic partially developed in Table 10.3 is graphed in Figure 10.5 along with the operating characteristic for the same test with the same level of significance but with a sample size of 81.

Look at the curve for a sample of 36 observations. This test gives the inspector a great deal of protection against leaving the machine set so that it delivers a mean fill of 100 pounds or less. If the true population mean is anything under 100.2 pounds, the

Table 10.3   Probabilities of Accepting H_o
for a One-Tailed Test

Potential value of μ	Probability of accepting H_o	Probability
99.90	0.9986	$1 - \alpha$
99.96	0.9953	$1 - \alpha$
100.00	0.9900	$1 - \alpha$
100.05	0.9772	β
100.11	0.9452	β
100.20	0.8413	β
100.35	0.5000	β
100.50	0.1587	β
100.56	0.0808	β
100.65	0.0228	β
100.70	0.0100	β

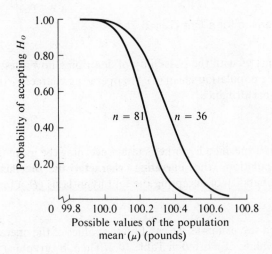

Figure 10.5   Operating Characteristics for a One-Tailed Test with Two Different Sample Sizes

probability is at least 0.84 that the inspector will accept the null hypothesis ($H_o : \mu \leq$ 100 and cause the manufacturer to adjust the mean fill upward. The probability that this action will be taken remains greater than 0.50 for values of μ all the way up to 100.35 pounds.

The curve for the larger sample size ($n = 81$) gives essentially the same high probability of having the mean fill raised as did the smaller sample, should it be 100 pounds or less. But the probability of taking this action drops off much more rapidly at higher values of μ. The manufacturer runs less risk of having to make an upward adjustment if he is already meeting the requirement that the mean fill (μ) be somewhat above 100 pounds. The larger sample provides a more sensitive test. On the other hand, the larger sample usually will cost more. In practice, the cost of error must be balanced off against the cost of sampling to arrive at a satisfactory sample size.

10.3

Operating Characteristic for a Two-Tailed Test

The first test we discussed in this chapter was a test of the null hypothesis (H_O) that $\mu = 100$ pounds against the alternative (H_A) that μ was either less than or greater than 100 in connection with a newly delivered cement-sack filling machine. The population standard deviation (σ) was 0.90 pounds, the sample size (n) was only 4, and the significance level (α) was 0.01.

To prepare an operating characteristic table and graph for this test we must, once again, find probabilities of accepting H_O for a wide range of possible values of the population mean (μ). Figure 10.6 illustrates the fact that finding these probabilities for a two-tail test entails finding the portion of a normal curve that falls between the two critical values. This contrasts with finding the normal curve area above (or below) a single critical value in the comparable one-tailed test.

Figure 10.6 also brings out another point of difference from the one-tailed test. For the one-tailed test illustrated in Figure 10.4, the null hypothesis ($H_O : \mu \le 100$) was true when the population mean (μ) was 100 pounds or *any value less than* 100

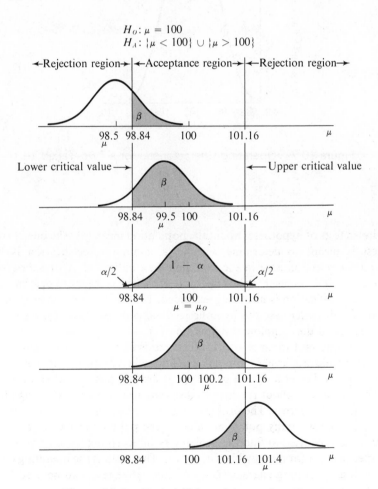

Figure 10.6   Probabilities of Accepting H_o

pounds. In the two-tailed test, Figure 10.6 shows that the null hypothesis ($H_o : \mu = 100$) is true only when μ takes on the single value of 100 pounds. Consequently, the inspector can make a Type I error only when the mean (μ) has the single value stated in the null hypothesis.

The operating characteristic for the two-tailed test we are discussing is shown in Figure 10.7. Although the curve is symmetrical with respect to μ_o (100 pounds) and tails off in both directions, it is *not* a normal distribution. Notice that the maximum height of the curve is equal to $(1 - \alpha)$ when the null hypothesis is true. For all other values of μ, the heights of the curve are values of β.

For a sample size greater than 4, the maximum height of the OC curve would still be 0.99 and the curve would be symmetrical with respect to μ_o. The width would be less, however. The larger the sample size, the less the width.

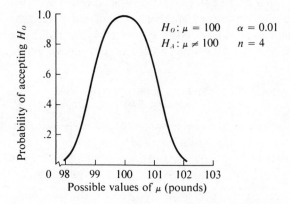

Figure 10.7   Operating Characteristic for a Two-Tailed Test

Summary

One-tailed tests of hypotheses about the population mean (μ) take one of two forms. If the test is meant to determine whether a minimum specification is being exceeded, all the rejection area in the sampling distribution of $\overline{X}$ specified by α (the significance level) is assigned to the upper tail. If the test is meant to determine whether a maximum specification is not being exceeded, all the rejection area is assigned to the lower tail. In either case the hypothesized value of the mean (μ_o) is taken as the central location of the sampling distribution of $\overline{X}$.

Changes in the performance of hypothesis tests in response to changes in sample size (n) and significance level (α) provide guidance when decisions are made concerning these two factors. Either a Type I or a Type II error is possible when drawing a conclusion from a sample about the hypothesis being tested. The level of significance (α) is reflected in Type I errors. The probability of a Type II error is called β. A value of α or β is associated with any potential value of the population mean (μ). These error probabilities can be used to find the probability of accepting the null hypothesis (H_o) for any specified test for all possible values of μ. The result is the operating characteristic for the test. Operating characteristics of alternative tests can be used to compare their performances.

See Self-Correcting Exercises 10.B

Exercises

1. A chemist in a pharmaceutical firm knows that the toxic dosage of an ingredient in a household product is 12 grains. He wants to make sure that the mean dosage in the lot currently being manufactured does not exceed 10 grains. The weights of this ingredient are normally distributed with a standard deviation of 0.2 grain. At the 0.01 level of significance, conduct the correct one-tailed test to determine whether the maximum specification is being exceeded if a sample of 25 observations has a mean dosage of 9.80 grains.

2. A firm makes a highly popular model of stereo FM radio that has a mean defect-free life of 2.3 years. A random sample of 9 of the most popular competitive model has a mean defect-free life of 1.8 years with a standard deviation of 0.7 years. The distribution from which the observations came is normal. At the 0.05 level of significance, test the hypothesis that will allow the firm to claim a longer mean life, if the test is successful.

3. A restaurant recorded mean daily sales of $326 during the summer season a year ago. The null hypothesis that mean daily sales during this year's summer season are no greater than last year's is to be tested at the 0.05 significance level. Daily sales during the summer season are normally distributed with a standard deviation of $10. If the mean this season should prove to be $330, the probability of rejecting the null hypothesis is to be 0.90.

   a. State the null hypothesis and the alternative to it.

   b. State an expression for the critical value such that the probability of a Type I error will be 0.05 when the population mean is $326.

   c. State an expression for the critical value for which the probability of a Type II error will be 0.10 when the population mean is $330.

   d. By equating the expressions in parts (b) and (c), find the sample size (n) and critical value (V) for the desired test.

4. For the test described in Exercise 1, find the probability of accepting the null hypothesis if the population mean is 10.0032 grains per dosage. What type of error is possible in this circumstance and what is the probability of making such an error?

5. For the test described in Exercise 1, find the probabilities of accepting the null hypothesis if the population mean is 9.9032 grains; if it is 9.8532 grains. What type of error is possible in these two circumstances and what are the probabilities of making these errors?

6. Refer to Exercise 2 in the first set of exercises for this chapter, page 211.

   a. If the population mean balance due on active credit accounts is now $234, which type of error is possible, and what is the probability of making such an error with the test described in that exercise?

   b. If the population mean balance due is $206, which type of error is possible and what is the probability of making such an error?

7. Will the operating characteristic curve for the testing procedure in Exercise 1 in this set have the same general appearance as those in Figure 10.5?

8. For the test procedure discussed in Exercise 6 in this set, what will be the height of the operating characteristic curve above the horizontal axis when μ is $206?

Glossary of Equations

$$V_U = \mu_O + z_{\alpha/2} \cdot \sigma_{\overline{X}}$$
$$V_L = \mu_O - z_{\alpha/2} \cdot \sigma_{\overline{X}}$$

The acceptance region for the sample mean ($\overline{X}$) in a two-tailed test of the hypothesis that the population mean is μ_O has an upper limit of V_U and a lower limit V_L. The level of significance is α. The value of the population standard deviation (σ) is known. The sample may be any size if the population is normally distributed. Otherwise the sample should contain at least 20 observations.

$$V_U = \mu_O + K \cdot \sigma_{\overline{X}}$$
$$V_L = \mu_O - K \cdot \sigma_{\overline{X}}$$

Chebyshev's inequality applies to find the upper and lower limits of the acceptance region for a two-tailed test about the value of the population mean when the sample size is less than 20, the population standard deviation is known, and the shape of the population distribution is not known. The level of significance is $1/K^2$.

$$V_U = \mu_O + z_{\alpha/2} \cdot s_{\overline{X}}$$
$$V_L = \mu_O - z_{\alpha/2} \cdot s_{\overline{X}}$$

When the population is normally distributed but the value of its standard deviation is not known and when the sample consists of 30 or more observations, the estimated standard error ($s_{\overline{X}}$) may be substituted for the true value ($\sigma_{\overline{X}}$) in the first set of critical value equations above to find the acceptance region. These equations also apply for a sample size of 50 or more when the shape of the population distribution is not known.

$$V_U = \mu_O + t_{\alpha/2} \cdot s_{\overline{X}}$$
$$V_L = \mu_O - t_{\alpha/2} \cdot s_{\overline{X}}$$

When the sample contains fewer than 30 observations and the population is normal but σ is unknown, the t distribution is used to find the acceptance region.

$$V = \mu_O + z_{\alpha} \cdot \sigma_{\overline{X}}$$

The single critical value for an upper one-tailed test, given normality and the value of the population standard deviation (σ), differs from V_U in the two-tailed test only in that z_{α} replaces $z_{\alpha/2}$. For a lower one-tailed test, a minus sign replaces the plus sign.

11

Other Tests and Estimates

Discussions of the past two chapters have centered on estimation and hypothesis tests for the mean of a single population—the subject of frequent questions in business and government. Now we want to examine some more tests and estimates used to answer further questions that often arise —for example, whether the population variance is a certain value, whether two populations have the same mean, and whether two populations have the same variance.

The general rationale underlying the statistical tests and estimates in this chapter is the same as that for Chapters 9 and 10. The interval-estimation and testing procedures for a certain population parameter are based upon some relevant statistic. The manner in which values of the statistic are distributed under specified sampling conditions is known. Specifically, the manner in which the sampling distribution relates to the parameter in question is known. We shall combine this knowledge of the distribution with concepts from probability theory to formulate the procedure.

Variance of a Single Population

One example in Chapter 10 involved a question about the mean fill delivered by a cement-sack filling machine. Another, equally important question to industrial quality control concerns the value of the variance for the population of fill weights. Not only must the mean fill be satisfactory, but also the variability of fill weights must be reasonably small. A set of two cement sacks that weigh 100 pounds each has a mean weight of 100 pounds per sack. So does a set in which one sack weighs 25 pounds and the other weighs 175 pounds.

A firm that specializes in conducting national direct mail advertising campaigns employs a large number of typists to prepare the material to be mailed. In the past it has checked random samples of each typist's work in its efforts to keep mean errors per 1000 words at a low level. Recently, in reaction to a customer complaint about several errors in a single advertisement, controls have been instituted to hold down the variability in each typist's error rate.

A random sample of 11 different letters prepared by a certain typist has a mean ($\overline{X}$) of 32 errors per 1000 words and a sample standard deviation (s) of 3.50 errors per 1000 words. Hence, the sample variance (s^2) is 12.25.

Experience has shown it is reasonable to expect a variance (σ^2) of not over 8.50 in error rates. The supervisor wants to test the null hypothesis that this typist has a basic error-rate variance of 8.50 or less against an alternative hypothesis that the typist's underlying error-rate variance is greater than 8.50. As we shall see, for the sample evidence described ($s^2 = 12.25$, $n = 11$), this null hypothesis must be accepted at the 0.05 level of significance.

11.1   The Chi-Square Distribution

The test statistic needed to conduct a one-tailed hypothesis test about the population variance (σ^2) just described is called chi-square (χ^2, pronounced keye, to rhyme with eye). To test hypotheses about the population variance or make interval estimates of it, we must assume that the population from which the sample came is normally distributed. In the case of the typist we must assume that for different typing specimens picked at random, the number of errors per 1000 words is normally distributed. This is not unreasonable if each specimen contains a minimum of several hundred words.

We are now ready to define and discuss the chi-square statistic and its distribution.

Definition

Consider all possible samples, each of which has the same number of observations (n), selected randomly from a normally distributed population for which the variance is σ^2. Suppose that the sample variance (s^2), where

$$s^2 = \frac{\sum (X - \overline{X})^2}{n - 1},$$

is calculated for each sample. Then the statistic

$$\blacksquare \quad \chi^2 = \frac{(n - 1)s^2}{\sigma^2} \quad \blacksquare \tag{11.1}$$

will have a chi-square (χ^2) distribution.

Just as Appendix B shows a different t distribution for each number of degrees of freedom, Appendix C gives a different chi-square distribution for each number of degrees of freedom (d.f.). Recall from the discussion of t distributions in Chapter 9 that a sample of n observations has $n - 1$ degrees of freedom. The same is true for the chi-square statistic in Equation (11.1).

The left column of Appendix C lists the degrees of freedom for 23 different chi-square distributions. The column headings are areas under the distributions, or probabilities, cumulated from the left. The body of the table contains values of chi-square that cut off the areas given in the column headings.

The graph at the top of Appendix C gives an example of how to read the table. There the chi-square distribution is shown for five degrees of freedom. The probability is 0.90 that a value of chi-square (χ^2) selected at random from this distribution will be less than 9.24. Only 10 percent of the time will values of χ^2 selected randomly from this distribution exceed 9.24. The value 9.24 appears in the table under $\chi^2_{.90}$ in the row for five degrees of freedom.

All normal distributions and all t distributions are symmetrical with respect to the mean. As shown in Figure 11.1, chi-square distributions are skewed to the right;

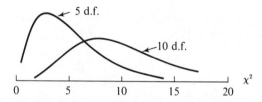

Figure 11.1   Two Chi-Square Distributions

they have positive skewness. The greater the degrees of freedom, the smaller the positive skewness. The mean of any chi-square distribution is equal to its degrees of freedom, and the variance is twice the degrees of freedom. If we let the letter m represent degrees of freedom, then $\mu = m$ and $\sigma^2 = 2m$ for a chi-square distribution. As m grows large without limit, the chi-square distribution approaches a normal distribution with $\mu = m$ and $\sigma^2 = 2m$. It is hardly safe, however, to assume normality when there are fewer than 100 degrees of freedom.

Hypothesis Test

For our direct mail advertising example, we have a sample of 11 observations (n) from a population of errors per 1000 words that is assumed to be essentially normally distributed. The sample variance (s^2) is 12.25.

We want to test the null hypothesis that the population variance (σ^2) is no more than 8.50 ($\sigma^2 \leq 8.50$) against the alternative that it is greater than 8.50. The level of significance (α) is 0.05.

Suppose the population variance (σ^2) is 8.50. Also assume that we calculate the values of s^2 for all possible random samples of 11 observations from this population. For each value of s^2 we find the associated value of $(n-1)s^2/\sigma^2$, where we know that $n-1$ is 10 and σ^2 is 8.50. From Equation (11.1) we know that these last quantities will form a chi-square distribution with 10 degrees of freedom. Appendix C tells us that 95 percent of this distribution lies below a chi-square (χ^2) value of 18.31. This is the critical value for our hypothesis test.

We now want the value of the χ^2 statistic for the sample we selected. It is

$$\chi^2 = \frac{(n-1)s^2}{\sigma^2} = \frac{10(12.25)}{8.50}, \quad \text{or } 14.41.$$

Since 14.41 is less than 18.31, the critical value, we must accept the null hypothesis that the typist is performing within the limit set for variability in error rates.

The majority of hypothesis tests about the population variance are upper one-tailed tests, in which the hypothesis statement is of the form

$$H_O : \sigma^2 \leq \sigma_O^2;$$

$$H_A : \sigma^2 > \sigma_O^2.$$

We can summarize our discussion of this test that the population variance (σ^2) is no more than a specified limit.

Statement | For a normally distributed population, the critical value (V) for an upper one-tailed hypothesis test concerning the population variance (σ^2) at the α level of significance is

$$V = \chi^2_{1-\alpha, m}, \tag{11.2}$$

where $(1 - \alpha)$ is the area below the value of χ^2 in the chi-square (χ^2) distribution with m degrees of freedom. The value of m is one less than the number of sample observations.

The sample value of χ^2 is formed from Equation (11.1) and compared with the value of V from Equation (11.2). The null hypothesis is accepted if the sample value is equal to or less than V. Otherwise the alternative hypothesis is accepted.

Note that in Equation (11.2) we stated the critical value for the hypothesis test in terms of chi-square rather than the sample variance. Given the equation

$$\chi^2_{1-\alpha} = \frac{(n-1)s^2}{\sigma^2},$$

we can solve for the critical value of the sample variance, s_c^2:

$$s_c^2 = \frac{(\sigma^2)\chi^2_{1-\alpha}}{n-1}.$$

Analogously, in Chapter 10 we could have found critical values in terms of the standard normal deviate, z, or t instead of the sample mean. Below expression (10.1) we have

$$V_U = \mu_O + (z_{\alpha/2})\sigma_{\bar{x}},$$

where V_U is the upper critical value of the sample mean. From this equation we get

$$z_{\alpha/2} = \frac{V_U - \mu_0}{\sigma_{\overline{X}}},$$

and an equally valid alternative procedure is to use the observed sample mean, $\overline{X}$, to find

$$z = \frac{\overline{X} - \mu_0}{\sigma_{\overline{X}}},$$

which is then compared with the tabled values $z_{\alpha/2}$ and $z_{1-\alpha/2}$. For convenience, from here on we will state critical values in terms of the statistics tabled in the appendices.   11.1

Interval Estimates

A purchasing agent for a large firm is considering buying 100-watt light bulbs from a certain supplier. The mean length of life for this type of bulb has already been estimated. Now the purchasing agent wants an estimate of the variability in the lengths of lives for bulbs of this type. A sample of 31 bulbs (n) has a sample standard deviation (s) of 40 hours in the lives of the bulbs. The sample variance (s^2) is, therefore, 1600.

Let the value of the confidence coefficient be 0.95. Then to form the 0.95 confidence interval estimate of the population variance (σ^2) for the life distribution, we will assume that the lives of bulbs are normally distributed and return to Equation (11.1). We know that $(n-1)$ is 30 and that s^2 is 1600 in this equation. For a chi-square distribution with 30 degrees of freedom, Appendix C shows that 2.5 percent $(\alpha/2)$ of the values are less than 16.79, and 97.5 percent $(1-\alpha/2)$ of the values are less than 46.98. Suppose we substitute these values in Equation (11.1) and solve for the resulting values of σ^2:

$$\chi^2 = \frac{(n-1)s^2}{\sigma^2};$$

$$\left\{ \begin{array}{c} 16.79 = \dfrac{30\,(1600)}{\sigma^2}, \\[2mm] \text{from which} \\[2mm] \sigma_U^2 = 2859; \end{array} \right. \quad \text{and} \quad \left\{ \begin{array}{c} 46.98 = \dfrac{30\,(1600)}{\sigma^2}, \\[2mm] \text{from which} \\[2mm] \sigma_L^2 = 1022. \end{array} \right.$$

Then the 0.95 confidence interval estimate of the population variance is 1022 to 2859. We can also express this in terms of the standard deviation. The 0.95 confidence interval estimate of σ is 31.9 hours to 53.5 hours.

The general procedure can now be stated.

	Given a confidence coefficient and a sample of n observations, to form an interval estimate of the population variance (σ^2) from the sample variance (s^2), find the values of $\chi^2_{\alpha/2}$ and $\chi^2_{1-\alpha/2}$ for $m = n - 1$ degrees of freedom. Then the upper limit of the interval is

Statement

$$\blacksquare \qquad \sigma^2_U = \frac{(n-1)s^2}{\chi^2_{\alpha/2,\,m}}, \qquad \blacksquare \qquad (11.3a)$$

and the lower limit is

$$\blacksquare \qquad \sigma^2_L = \frac{(n-1)s^2}{\chi^2_{1-\alpha/2,\,m}}. \qquad \blacksquare \qquad (11.3b)$$

The population must be assumed to be normally distributed.

Because chi-square distributions are not symmetrical about the mean, $\chi^2_{\alpha/2}$ is not as far below the mean as $\chi^2_{1-\alpha/2}$ is above it. In our light bulb example, the sample has 31 observations, or 30 degrees of freedom. The chi-square distribution with 30 degrees of freedom has a mean of 30. In this distribution $\chi^2_{.025} = 16.79$, and this value is 13.21 units below the mean. On the other hand, $\chi^2_{.975}$ is 46.98, 16.98 units above the mean.

Difference in Two Population Means

11.2   Up to now our discussion of statistical inferences has been limited to those about a single population parameter. Now we will consider inferences about a set of two population parameters.

We first take up the fairly common problem of testing whether or not two statistical populations have the same mean. Selection of the correct test procedure depends upon whether or not the standard deviations of the two populations are known.

Population Standard Deviations Known

An automatic drilling machine is set to drill holes one inch deep in engine blocks coming down an assembly line. At any single setting, the machine is known to produce a normal distribution of hole depths with a standard deviation of 0.01 inch. A similar machine farther down the line drills a narrower hole below the first hole, one-half inch deeper, for a total depth of one and one-half inches. The second machine produces a normal distribution of hole depths with a standard deviation of 0.02 inches. It is important that the smaller hole be a half inch deeper than the larger hole.

A gauge measures the depth of 25 holes drilled by the first machine and finds the sample mean to be 0.99 inches. The full depths of 20 holes are measured after the second machine is finished. Their mean depth is 1.51 inches. Is it reasonable to assume that the mean depth of the inner hole is 0.50 inches? We can summarize the information in a table.

	Shallow hole	*Deep hole*
Standard deviation	$\sigma_1 = 0.01$ in.	$\sigma_2 = 0.02$ in.
Sample size	$n_1 = 25$	$n_2 = 20$
Mean depth	$\overline{X}_1 = 0.99$ in.	$\overline{X}_2 = 1.51$ in.

We can frame the problem as a hypothesis test. Suppose we call the population mean depths for the shallow and deep holes μ_1 and μ_2, respectively. Then the requirement that the deep hole be 0.50 inch deeper than the shallow one, on the average, can be reflected in the null hypothesis:

$$H_O : \mu_2 - \mu_1 = 0.50 \text{ inch.}$$

Presuming that detecting an error in either direction is important, we form the two-tailed alternative

$$H_A : \mu_2 - \mu_1 < 0.50, \quad \text{or} \quad \mu_2 - \mu_1 > 0.50.$$

For independent random samples of n_1 and n_2 observations from normally distributed populations with means μ_1 and μ_2 and variances σ_1^2 and σ_2^2, the differences in sample means, $\overline{X}_2 - \overline{X}_1$, will be normally distributed. Furthermore, the distribution of differences in sample means will have as its mean $(\mu_{\overline{X}_2 - \overline{X}_1})$ and standard deviation $(\sigma_{\overline{X}_2 - \overline{X}_1})$

$$\mu_{\overline{X}_2 - \overline{X}_1} = \mu_2 - \mu_1$$

and

$$\sigma_{\overline{X}_2 - \overline{X}_1} = \sqrt{\frac{\sigma_2^2}{n_2} + \frac{\sigma_1^2}{n_1}}. \tag{11.4}$$

The standard deviation of this population of differences in sample means is called the *standard error of the difference between means*. It is then true that the statistic z will be distributed as the standard normal deviate, where

$$z = \frac{(\overline{X}_2 - \overline{X}_1) - (\mu_2 - \mu_1)}{\sigma_{\overline{X}_2 - \overline{X}_1}}.$$

For our example,

$$z = \frac{(1.51 - 0.99) - (0.50)}{\sqrt{\dfrac{0.0004}{20} + \dfrac{0.0001}{25}}} = \frac{0.02}{0.0049}, \quad \text{or 4.08 standard errors.}$$

The difference in sample means (0.52) exceeds the specification of 0.50 inch by more than four standard errors of the difference between sample means of $n_1 = 25$ and $n_2 = 20$ observations. If the level of significance (α) is 0.01, then from Appendix A, $z_{\alpha/2} = 2.58$ and our sample result is significant at well beyond this level. Since the difference between 1.51 and 0.99 is greater than 0.50, we accept H_A, the alternative to the null hypothesis. In practice, we would conclude that $\mu_2 - \mu_1 > 0.50$ and that the small holes are deeper than one-half inch on the average.

The formal summary can now be stated.

Statement

Consider two normal distributions with means μ_1 and μ_2 and standard deviations σ_1 and σ_2. Also consider a sample mean of $\overline{X}_1$ based on n_1 observations from the first population and a sample mean of $\overline{X}_2$ based on n_2 observations from the second. The difference in sample means can be incorporated in the statistic

$$\blacksquare \qquad z = \frac{(\overline{X}_1 - \overline{X}_2) - (\mu_1 - \mu_2)}{\sigma_{\overline{X}_1 - \overline{X}_2}}, \qquad \blacksquare \qquad (11.5)$$

where z is distributed as the standard normal deviate and $\sigma_{\overline{X}_1 - \overline{X}_2}$ is as defined in Equation (11.4).

11.2

Several examples in Chapters 9 and 10 showed how a standard normal deviate can be used in both one-tailed hypothesis tests and confidence interval estimates. Since the procedures with the z statistic from Equation (11.5) are identical to those used earlier, we will not discuss them further.

Population Standard Deviations Unknown

Large Samples   When the populations sampled are normally distributed, we can use sample estimates of the population variances as substitutes for σ_1^2 and σ_2^2 in Equation (11.4) provided that each of the two samples has 20 or more observations in it.

Suppose we have a sample of $n_1(n_1 \geq 20)$ observations from a normal population and suppose this sample has a mean of $\overline{X}_1$ and a variance of s_1^2. A second normal population has yielded an independent random sample of $n_2(n_2 \geq 20)$ observations with a mean of $\overline{X}_2$ and a variance of s_2^2. We want to test the null hypothesis that the difference between the two population means is equal to a certain value. We find $\overline{X}_1 - \overline{X}_2$, the difference in sample means. We also find the standard error of the difference as estimated from the sample variances; that is,

$$s_{\overline{X}_1 - \overline{X}_2} = \sqrt{\frac{s_1^2}{n_1} + \frac{s_2^2}{n_2}}. \qquad (11.6)$$

Then, the statistic

$$z = \frac{(\overline{X}_1 - \overline{X}_2) - (\mu_1 - \mu_2)}{s_{\overline{X}_1 - \overline{X}_2}} \qquad (11.7)$$

will follow an essentially normal distribution if each of the two samples is large. From this point on, hypothesis testing procedures and interval estimation procedures are identical to those discussed for standard normal deviates in the two preceding chapters.

Small Samples   The only procedure presented here assumes that the two populations in question are normally distributed and their variances are identical. A test applicable to the latter assumption (the F test) is the final test discussed in this chapter.

A health club director has developed two reducing diets for women. Two weeks ago he started a random sample of five women on the first diet and 11 women on the second. Today, when the women's individual weight losses were checked, he made up the following table.

	Diet A	Diet B
Sample size	$n_1 = 5$	$n_2 = 11$
Mean weight loss	$\bar{X}_1 = 4.2$ pounds	$\bar{X}_2 = 5.6$ pounds
Sample standard deviation	$s_1 = 0.7$ pounds	$s_2 = 0.9$ pounds

The health director has no preconceptions about which diet should produce the greater weight loss. Consequently, he wants to make a two-tailed test of the null hypothesis that $\mu_1 - \mu_2 = 0$, where μ_1 and μ_2 are the means of the two populations of weight losses. The alternative hypothesis, then, is either $\mu_1 > \mu_2$ or $\mu_2 > \mu_1$. He selects the 0.05 level of significance.

We begin with assumptions that the two populations of weight losses are normally distributed and have the same variances ($\sigma_1^2 = \sigma_2^2$). There is no highly sensitive test for these two assumptions with samples as small as these. On the other hand, the two distributions are not likely to be either heavily skewed or radically different in their variances. Under these circumstances, the test we are about to describe is safe because it is not highly sensitive to moderate departures from the two assumptions we have made.

The first step is to calculate a pooled estimate of the common population variance (σ^2) from the two sample estimates (s_1^2 and s_2^2). We find this estimate ($\bar{s}^2$) by using the degrees of freedom as weights and taking a weighted mean of the sample variances, which gives

$$\bar{s}^2 = \frac{(n_1 - 1)s_1^2 + (n_2 - 1)s_2^2}{n_1 + n_2 - 2}. \tag{11.8}$$

For our example,

$$\bar{s}^2 = \frac{4(0.49) + 10(0.81)}{14},$$

or

$$\bar{s}^2 = 0.7186,$$

from which

$$\bar{s} = 0.848 \text{ pounds.}$$

The second step is to find the standard error of the difference between sample means. The procedure is to use Equation (11.6) with $\bar{s}^2$ substituted for s_1^2 and s_2^2. After simplifying, the result is

$$\bar{s}_{\bar{X}_1 - \bar{X}_2} = \bar{s} \sqrt{\frac{1}{n_1} + \frac{1}{n_2}}. \tag{11.9}$$

A bar is placed over s on the left side to distinguish it from the left side of Equation (11.6). For our example,

$$\bar{s}_{\bar{X}_1 - \bar{X}_2} = 0.848 \sqrt{\frac{1}{5} + \frac{1}{11}} = 0.848\sqrt{0.2909}, \quad \text{or } 0.457.$$

If the populations are normally distributed with equal variances, then the statistic

$$t = \frac{(\bar{X}_1 - \bar{X}_2) - (\mu_1 - \mu_2)}{\bar{s}_{\bar{X}_1 - \bar{X}_2}}$$

is a t distribution with $n_1 + n_2 - 2$ degrees of freedom. For our example, $\mu_1 - \mu_2 = 0$ because of the null hypothesis, so

$$t = \frac{(4.2 - 5.6)}{0.457}, \quad \text{or } -3.06,$$

with $5 + 11 - 2$, or 14, degrees of freedom. Since the level of significance (α) is 0.05 for the two-tailed test, the critical values from Appendix B are -2.145 and $+2.145$. On the basis of this experiment we reject the null hypothesis of equal population means. Diet B apparently produces a greater weight loss than Diet A.

We can summarize the test procedure just described in the following statement:

Statement	Assume that two normally distributed populations have equal variances ($\sigma_1^2 = \sigma_2^2$). A random sample of n_1 observations is selected from one population and a random sample of n_2 is selected from the other. The difference in sample means ($\bar{X}_1 - \bar{X}_2$) can be incorporated in the statistic $$t = \frac{(\bar{X}_1 - \bar{X}_2) - (\mu_1 - \mu_2)}{\bar{s}_{\bar{X}_1 - \bar{X}_2}}, \qquad (11.10)$$ which is distributed as t with $n_1 + n_2 - 2$ degrees of freedom. The standard error $\bar{s}_{\bar{X}_1 - \bar{X}_2}$ is defined in Equations (11.8) and (11.9).

To form an interval estimate of the difference between the population means ($\mu_1 - \mu_2$), we would proceed very much as we did in the situation described in conjunction with Equation (9.7). If the confidence coefficient is $(1 - \alpha)$, we would find $t_{\alpha/2}$ for $n_1 + n_2 - 2$ degrees of freedom. Then from the sample we would find $\bar{X}_1 - \bar{X}_2$ and $\bar{s}_{\bar{X}_1 - \bar{X}_2}$. Finally the $(1 - \alpha)$ confidence interval estimate of $\mu_1 - \mu_2$ is

$$(\bar{X}_1 - \bar{X}_2) \pm (t_{\alpha/2})\bar{s}_{\bar{X}_1 - \bar{X}_2}.$$

Summary

Control of variability in output is often as important in industrial processses as is control of the mean output. The chi-square (χ^2) distribution plays a major role in testing a hypothesis about or making a confidence interval estimate of the variance (σ^2) of a normally distributed population. As with t distributions, there is a different chi-square distribution for every differing number of degrees of freedom. Chi-square distributions, however, are not symmetrical with respect to their means as are t distributions. Instead, chi-square distributions have positive skewness. The fewer the degrees of freedom, the greater the skewness. Because of skewness, two values of chi-square must be found to form a $(1 - \alpha)$ confidence interval estimate of σ^2.

Frequently it is important to test hypotheses about or make interval estimates of the difference in the means of two populations ($\mu_1 - \mu_2$). In the three examples discussed, all assumed both populations to be normally distributed. The example of the drilling machines applies for any sample sizes when the values of the population standard deviations are known. When the two population variances are not known, a large and a small sample procedure apply. For samples of 20 or more, sample estimates of the population variances are substituted for the unknown population values. In the small-sample case, the difference in means can be expressed in terms of a statistic that follows a t distribution.

See Self-Correcting Exercises 11.A

Exercises

1. In a chi-square distribution with 7 degrees of freedom, what value will be exceeded with probability 0.05? 0.95?

2. A machine that is set to cut wires to a length of 50 feet is known to cut with a standard deviation of 0.3 inches. A random sample of 11 lengths is to be measured and the sample standard deviation, s, will be calculated. What is the probability that the sample standard deviation will be less than 0.171 inches?

3. In a lumber yard, the standard deviation of the amounts shown on sales tickets is thought to be $3.74. For a random sample of 15 sales tickets, the sum of squared deviations from the sample mean is 392. At the 0.025 level of significance, test the null hypothesis that the population variance is no greater than $(3.74)^2$.

4. A salesman claims that his machine will automatically weigh and pack one-pound packages of ground beef with "negligible" variability. A random sample of 16 packs is found to have a sample standard deviation of 0.2 ounces. What is the 0.95 confidence interval estimate of the population variance? Of the standard deviation?

5. Two types of automobile tires have been tested by exposing them to the same driving conditions for an equal number of months. Four tires of the first type lasted a mean of 85.25 weeks with a variance of 25. Five tires of the second type lasted a mean of 88.00 weeks with a variance of 5.5. Assume normal populations with equal variances. The first type is to be used unless the second type is shown to be definitely superior. At the 0.05 level of significance, conduct a one-tailed hypothesis test that will show whether or not the second type is superior.

6. Two types of truck engines are being compared for mileage between major tune-ups. The truckers' trade association selects a sample of 100 trucks with engine A and 121 trucks with engine B. The test results were as follows:

| | *Mileage* | |
	Engine A	*Engine B*
Sample mean	24,800	25,100
Sample variance	1,210,000	1,742,400

With these large samples, assume that the sample variances are equal to the population variances. Conduct a two-tailed test of the hypothesis that the two population means are equal. Use 0.01 as the level of significance.

Matched Pairs

An oil company executive believes that his firm's newly formulated "regular" gasoline will yield more miles per gallon than the older type. To test his belief he selects one pair of automobiles from each of five weight classes. He then selects one member of each pair at random and assigns it to the old type of gasoline. The other member of the pair uses the new type. As we shall see, using matched pairs eliminates much variability attributable to weight differences and thereby improves sensitivity of the test to differences in mileage.

Each automobile makes several runs of a designated course at 60 miles per hour, after which the miles per gallon of gasoline consumed on all the runs are calculated. The data are shown in Table 11.1 along with the differences (d) in mileage between the old and new types. The mean mileages for each type as well as the difference in means (1.8 miles per gallon) are shown.

Table 11.1   Mileages from two Types of Gasoline

| | *Automobile weight* | | | | | |
Gasoline type	*Very heavy*	*Heavy*	*Medium*	*Light*	*Very light*	*Mean*
New	13.3	13.4	15.7	17.6	19.0	15.8
Old	12.5	12.5	13.7	14.9	16.4	14.0
New minus old (d)	0.8	0.9	2.0	2.7	2.6	1.8

Regardless of what the data show, the executive reasons that, if the new gasoline is no better than the old, the mean of the population of differences (μ_d) should be zero or negative. Since he is interested in establishing that, at the minimum, the new gasoline produces more miles per gallon than the old, he forms the hypothesis statements

$$\text{Null hypothesis } (H_O): \quad \mu_d \leq 0,$$

$$\text{Alternative hypothesis } (H_A): \quad \mu_d > 0;$$

he would, of course, prefer to accept H_A.

We now want to think about how sample values of paired differences (d) and means of paired differences ($\bar{d}$) are distributed; that is, we want to study the sampling distributions of d and $\bar{d}$ for a moment. Consider the population of all values of paired differences that could result from the test. This population of d values will have some mean (μ_d) and some standard deviation (σ_d), both of which are unknown. Next consider n samples selected at random from the population of d values and consider the mean of each sample ($\bar{d}$). (Note that $\bar{d}$ is 1.8 for our example.) If we assume that the original population of d values is normally distributed, then the distribution of values of $\bar{d}$ for samples of n differences will also be normally distributed. This sampling distribution will have a mean of μ_d and a standard error of $\sigma_{\bar{d}}$, or $\sigma_d/\sqrt{n}$.

We do not know the value of σ_d, the standard deviation of the population of differences. But our situation with respect to this population is identical to that discussed in conjunction with the t distribution and Equation (9.6), page 197. We can estimate the population standard deviation (σ_d) from the sample by finding

$$s_d = \sqrt{\frac{\sum (d - \bar{d})^2}{n - 1}}.$$
(11.11)

Then we can estimate the standard error of the sampling distribution of mean differences ($\sigma_{\bar{d}}$) by finding

$$s_{\bar{d}} = \frac{s_d}{\sqrt{n}}.$$
(11.12)

Finally, the statistic

$$t = \frac{(\bar{d} - \mu_d)}{s_{\bar{d}}}$$

will form a t distribution with $n - 1$ degrees of freedom.

The calculations necessary to find the value of $s_{\bar{d}}$ for our gasoline mileage example appear in Table 11.2. In this table note that n is 5, the number of *differences*, and not 10,

Table 11.2 The Standard Error of The Mean Difference ($s_{\bar{d}}$)

d	$d - \bar{d}$	$(d - \bar{d})^2$	
			$s_d^2 = \dfrac{\sum (d - \bar{d})^2}{n - 1}$
0.8	-1.0	1.00	
0.9	-0.9	0.81	$= \dfrac{3.30}{4}$, or 0.825;
2.0	0.2	0.04	
2.7	0.9	0.81	
2.6	0.8	0.64	$s_{\bar{d}}^2 = \dfrac{s_d^2}{n} = \dfrac{0.825}{5}$, or 0.165;
Total 9.0	0.0	3.30	
$\bar{d} = 1.8$		$n = 5$	$s_{\bar{d}} = \sqrt{0.165}$, or 0.406 miles per gallon.

the number of mileage observations. The value of the test statistic is

$$t = \frac{\bar{d} - \mu_d}{s_{\bar{d}}} = \frac{1.8 - 0}{0.406}, \quad \text{or } 4.43.$$

There are four degrees of freedom associated with the value of t. For a 0.01 level of significance and an upper tail test, we see from Appendix B that $t_{.99}$ is 3.747. The executive can reject the null hypothesis and accept the alternative that at 60 miles per hour the new gasoline delivers more miles per gallon than does the old.

The test procedure is summarized in the following statement.

Statement

Consider a set of n matched pairs. One treatment (A) is applied to one member of each pair. Another treatment (B) is applied to the other member. Assignment of members to treatments is made at random. Within each pair, the response to B is subtracted from A to obtain a sample of n differences (d). If these differences form a normal population with a mean of μ_d, then the statistic

$$\blacksquare \quad t = \frac{\bar{d} - \mu_d}{s_{\bar{d}}} \quad \blacksquare \tag{11.13}$$

belongs to a t distribution with $n - 1$ degrees of freedom. In this equation $\bar{d}$ is the mean of the n sample differences and $s_{\bar{d}}$ is as defined in Equations (11.11) and (11.12).

Instead of using the matched-pairs test for the gasoline-mileage problem, the oil company executive could have used the approach we described in the section just prior to this one. He could have paid no attention to automobile weight. Rather, he could have selected ten automobiles at random, assigned five of them the new gasoline at random, and let the remaining five use the old gasoline. Had he done this, it is highly unlikely that the new gasoline would have demonstrated its mileage superiority.

Under the circumstances described, the t test discussed in conjunction with Equation (11.10) would be far less sensitive than the one we performed using Equation (11.13). All of the mileage variability attributable to differences in vehicle *weight* would remain in the denominator of Equation (11.10), whereas matching on weight removes this variability from the denominator of Equation (11.13). Although the symbols differ, the numerators of the two equations produce identical values for identical data. When there really is a treatment difference, we can expect numerators under the two approaches to be the same but the denominator in Equation (11.13) to be smaller. As a result we can expect larger t values from the matched-pair approach. This of course presumes that variation in the factor used to match the pairs (weight in this case) would cause variability in the response (mileage) if it were left uncontrolled.

Even with the matched-pairs approach, we could have used a technique that might be expected to be even more sensitive to differences in mileage from the two gasolines. Instead of using a pair of different automobiles in each weight class, we could have used the same automobile with each type of gasoline. Then differences other than weight that cause variations in mileage (such as age and condition of engines and running gear), would have been controlled also. Whether the old or the new fuel is to be used first in such a test should be decided by a random process to minimize the possibility of a systematic carry-over from one treatment to the succeeding one. For example, suppose that under random selection five autos that started the test with the old gasoline and then switched to the new achieved better mileages when the new fuel was in use. It might later be found that the old fuel left a residue that was largely responsible for the better performance with the new fuel. We chose to use different automobiles in each treatment pair because this is the only guarantee we have that an unknown systematic factor would not produce any mileage difference.

Difference in Two Population Variances

Earlier we used an example of reducing diets in discussing a small sample t test for a difference in population means [Equation (11.10)]. The population variances, although unknown, were assumed to be equal. This assumption can be tested, as we shall see next. The test assumes that the two populations are normally distributed and is effective even if there are mild departures from normality.

The t test for a difference in the mean weight losses attributable to two diets was based on samples of 5 (n_1) and 11 (n_2) observations. The sample standard deviations were 0.7 pounds (s_1) and 0.9 pounds (s_2). The two populations of weight losses were assumed to be normally distributed.

Recall from our discussion for finding s (Equation 9.3, page 196), that the sum of squared deviations from the sample mean $(\overline{X})$ is divided by one less than the sample size $(n - 1)$, rather than by n. The result is s^2, the sample variance. Dividing by $n - 1$ makes s^2 *an unbiased estimator of σ^2*, the population variance.

Suppose we have one random sample from each of two populations that have equal variances $(\sigma_1^2 = \sigma_2^2 = \sigma^2)$. Then the sample variance from either of the populations $(s_1^2$ or $s_2^2)$ will be an unbiased estimator of the common population variance. Now suppose we form a ratio of the two sample variances by selecting one of them at random as the numerator and the other as the denominator. The ratio thus formed, say s_1^2/s_2^2, will have a probability distribution.

We can put the point of our discussion as follows.

Statement

Consider the variances (s_1^2) of random samples of n_1 observations from a normally distributed population, and the variances (s_2^2) of similar samples of n_2 observations from another normally distributed population with the same variance (σ^2) as the first. Define the ratio of ordered pairs of sample variances as the statistic F, where

$$F = \frac{s_1^2}{s_2^2}.$$ (11.14)

The sampling distribution of this statistic (all possible values of the statistic from samples subject to the given conditions) is called an F distribution.

As in the cases of the t and chi-square (χ^2) distributions, there is a family of F distributions. For every different pair of numbers $(m_1$ and $m_2)$ designating the degrees of freedom for the numerator and denominator in Equation (11.14), there is a different F distribution. The F distribution for a ratio of s_1^2/s_2^2 where $m_1 = 8$ and $m_2 = 14$ is a different distribution from the one in which $m_1 = 14$ and $m_2 = 8$.

For our example of the two diets, suppose we toss a coin to decide which sample variance to put in the numerator of the F ratio and suppose the result is

$$F = \frac{s_1^2}{s_2^2} = \frac{0.7^2}{0.9^2}, \quad \text{or } 0.60,$$

where s_1^2 is based on a sample of five and s_2^2 on a sample of 11 observations. Suppose we want to make a two-tailed test of the hypothesis that $\sigma_1^2 = \sigma_2^2$ for these two populations.

Appendix D identifies different F distributions by showing the degrees of freedom (m_1) for the numerator of an F ratio along the top of the table and the degrees of freedom (m_2) for the denominator along the sides. For each set $\{m_1, m_2\}$, a pair of numbers appears in the body of the table. The top number is the value of F such that 5 percent of the values in the distribution are greater. The bottom number is the value of F such that 1 percent of the values in the distribution are greater.

For our example, m_1 is 4 and m_2 is 10. For 4 degrees of freedom in the numerator and 10 in the denominator, Appendix D lists F values of 3.48 and 5.99. Suppose two normal populations with equal variances are sampled repeatedly. Suppose variances for samples of five (s_1^2) are found for the first population, and variances for samples of 11 (s_2^2) are found for the second population and paired to form sample values of the F ratio s_1^2/s_2^2. Then only 5 percent of these ratios will be larger than 3.48 and only 1 percent will be larger than 5.99.

We are making a two-tailed test of the null hypothesis that $\sigma_1^2 = \sigma_2^2$ for our weight-loss populations. To form an acceptance region we need an upper critical value (F_U) that leaves $\alpha/2$ in the upper tail of the relevant F distribution and a lower critical value (F_L) that leaves $\alpha/2$ in the lower tail, where α is the level of significance. With Appendix D we have two choices; $\alpha/2$ can be either 0.05 or 0.01 for the upper tail. Hence, the level of significance (α) can be either 0.10 or 0.02. Suppose we choose 0.02 for α. Then the upper critical value, F_U, is 5.99.

We now must find the lower critical value, F_L, such that 1 percent of the F distribution with 4 and 10 degrees of freedom lies below F_L. To do so we take advantage of a fortunate property of F distributions, which can be stated in the relationship

$$F_{\alpha/2, m_1, m_2} = \frac{1}{F_{1-\alpha/2, m_2, m_1}}. \tag{11.15}$$

This equation states that the value of F which cuts off the portion $\alpha/2$ in the *lower* tail of the distribution with m_1 degrees of freedom in the numerator and m_2 in the denominator is the reciprocal of the value of F which cuts off the portion $\alpha/2$ in the *upper* tail of *another* distribution with m_2 degrees of freedom in the numerator and m_1 in the denominator. Note carefully the reversal in the order of the degrees of freedom on the two sides of Equation (11.15).

For our example we want $F_{.01, 4, 10}$. Equation (11.15) tells us that to get it, we must first look up $F_{.99, 10, 4}$. The value of this latter quantity, 14.54, is found in Appendix D. Then,

$$F_{.01, 4, 10} = \frac{1}{F_{.99, 10, 4}} = \frac{1}{14.54}, \quad \text{or } 0.069,$$

and the acceptance region for our test is $0.069 \le F_{4, 10} \le 5.99$. Since our sample F value is 0.60, we must accept the null hypothesis of equal population variances at the 0.02 level of significance.

	To perform a two-tailed test of the null hypothesis that the variances of two normally distributed populations are equal ($\sigma_1^2 = \sigma_2^2$), find the ratio of sample variances (s_1^2/s_2^2), one from each population, and call this ratio F. Accept the null hypothesis if F falls within the acceptance region defined by upper and lower critical values, F_U and F_L, on the relevant F distribution. The proper F distribution is based on $m_1 = n_1 - 1$ degrees of freedom for s_1^2 and $m_2 = n_2 - 1$ degrees of freedom for s_2^2. The critical values for F are
Statement	

$$F_U = F_{1 - \alpha/2, \, m_1, \, m_2} \qquad\qquad (11.16a)$$

and

$$F_L = F_{\alpha/2, \, m_1, \, m_2}. \qquad\qquad (11.16b)$$

Appendix D gives values of F_U for levels of significance (α) of 0.02 and 0.10. Values of F_L are found from Equation (11.15).

One-tailed tests of hypotheses concerning the equality of two normal population variances can make use of Appendix D without modification for 0.05 and 0.01 levels of significance. All that is necessary is to frame the question so that an upper one-tailed test is appropriate.

11.3

Summary

To test for a difference between the means of two populations, a matched-pairs test can provide much greater sensitivity than the t test without matching, described in the first half of this chapter. However, the increased sensitivity will be present only if the pairs are matched by controlling a source of variability that could otherwise obscure differences in means.

The F ratio, that is, the ratio of sample variances from two normal populations, can be used to test the hypothesis that the two population variances are equal in value. F distributions, like t and chi-square (χ^2) distributions, depend upon the degrees of freedom in sample observations. Although this causes some inconvenience in using F tables, the principle of leaving tail probabilities each equal to half the level of significance still applies for a two-tailed test.

See Self-Correcting Exercises 11.B

Exercises

1. A physician is studying the effectiveness of a reducing regimen. "Before and after" figures have been obtained from a random sample of 10 adult males from 40 to 50 years of age. The data are (in pounds):

253 and 248,	210 and 211,
203 and 195,	185 and 180,
194 and 192,	194 and 196,
239 and 236,	225 and 214,
187 and 175,	176 and 169,

At the 0.01 level of significance, test the null hypothesis that adult males weigh at least as much after the regimen as they did before. Assume normal populations.

2. Several changes designed to improve productivity have been made in the assembly department of an electronics manufacturing firm. "Before and after" data on number of assemblies completed in a day's time have been gathered for 8 of the assemblers:

Assembler	1	2	3	4	5	6	7	8
Before	37	42	28	32	38	35	29	34
After	40	40	32	35	41	37	36	38

Find the 0.95 confidence interval estimate of the mean number of additional assemblies made per day after the change. Assume normal population.

3. For an F distribution with 12 degrees of freedom in the numerator and 28 degrees of freedom in the denominator, above what F value will one percent of the distribution lie?

4. For an F distribution with 9 degrees of freedom in the numerator and 24 degrees of freedom in the denominator, below what F value will five percent of the distribution lie?

5. A random sample of 10 observations is selected from a normally distributed population of length measurements and the sample variance (s_1^2) is formed. The sample variance (s_2^2) is formed for a second random sample of 17 measurements from the same population. The ratio of the first sample variance to the second can be expected to lie in what interval with probability 0.98?

6. Using the data given in Exercise 2, test the hypothesis that the "before" and "after" population distributions have equal variances. Use 0.10 as the level of significance.

7. Should the larger of two sample variances always be placed in the numerator of the F ratio to test the null hypothesis that the two population variances are equal?

Glossary of Equations

$$\chi^2 = \frac{(n-1)s^2}{\sigma^2}$$

The statistic formed by multiplying the sample variance (s^2) from a normal population by one less than the sample size ($n-1$) and dividing that product by the population variance (σ^2) belongs to a chi-square (χ^2) distribution with $n-1$ degrees of freedom.

$$\sigma_U^2 = \frac{(n-1)s^2}{\chi_{\alpha/2,m}^2}$$

$$\sigma_L^2 = \frac{(n-1)s^2}{\chi_{1-\alpha/2,m}^2}$$

The upper (σ_U^2) and lower (σ_L^2) limits of a $(1-\alpha)$ confidence interval estimate of the population variance for a normal population are found by dividing the product of the sample variance (s^2) and its degrees of freedom ($m = n - 1$) by the values of χ^2 that leave $\alpha/2$ of the distribution in each tail.

$$z = \frac{(\overline{X}_1 - \overline{X}_2) - (\mu_1 - \mu_2)}{\sigma_{\overline{X}_1 - \overline{X}_2}}$$

The difference in sample means from two normally distributed populations can be transformed to the standard normal statistic (z) by subtracting the difference in population means and dividing by the standard error of the difference in sample means. This statistic is used as a basis for hypothesis tests about and interval estimates of the difference in population means; it applies when the two population variances are known or have been estimated from samples of more than 20.

$$t = \frac{(\overline{X}_1 - \overline{X}_2) - (\mu_1 - \mu_2)}{\tilde{s}_{\overline{X}_1 - \overline{X}_2}}$$

This statistic belongs to a t distribution and is used for tests and estimates concerning the difference in population means when the populations are normal. The statistic is based on the assumption that the two populations have equal variances. The denominator is a pooled estimate of the population variance.

$$t = \frac{\overline{d} - \mu_d}{s_{\overline{d}}}$$

The mean difference ($\overline{d}$) of a sample set of differences from matched pairs can be used to test hypotheses about the difference in two population means (μ_d). If the differences (d) are normally distributed, then the sample-mean difference can be transformed into a statistic distributed as t by this equation. The denominator is the estimated standard deviation of the sampling distribution of the variable $\overline{d}$.

$$F_{m_1,m_2} = \frac{s_1^2}{s_2^2}$$

The ratio of two sample variances, one from each of two identically normally distributed populations, constitutes a statistic that follows an F distribution. The degrees of freedom are $m_1 = n_1 - 1$ and $m_2 = n_2 - 1$, when n_1 and n_2 are the sample sizes for the numerator and denominator of the variance ratio.

$$F_{\alpha/2, m_1, m_2} = \frac{1}{F_{1-\alpha/2, m_2, m_1}}$$

To find the value of F below which lies $\alpha/2$ of the F distribution with m_1 and m_2 degrees of freedom, take the reciprocal of the value of F *above* which lies $\alpha/2$ of the F distribution with the degrees of freedom reversed.

12

Statistical Inference for Classification Data

In this chapter we turn our attention to statistical inferences for nominal scale data. In many situations, the only sense in which characteristics are "measured" is on an all or none basis. Each unit of observation is simply categorized as possessing a characteristic of interest or not possessing it. Each toaster assembled on a production line may be classed as functioning or not functioning; paintings may be classed as for sale or not for sale; families may be classed as purchasers or nonpurchasers of our brand, and so forth. Even where there are more than two classes, the nature of "measurement" is still an all or nothing proposition *with respect to each class*. Consider the classification "state of birth." There are 50 states in the United States, and every individual born in the United States either does or does not possess the characteristic "born in Arizona," and so it is for each of the other 49 states.

Nominal scales (or categories) serve only to identify or label units of observation. A sample of such data would be summarized by a count, or *enumeration*, of the number of sample observations belonging in each class. The problem of inference is to employ the sample data to make a statement about a parameter—usually the proportion, or relative frequency, of observations in a particular class for the population. From the number of families found to purchase our brand in a survey of 200 families, we want to make an inference about the proportion of families in the entire market area who purchase our brand. Another common problem is that of making an inference from sample data about association between two classification variables. Sponsors of the market survey may want to know whether their brand has greater acceptance among some ethnic groups than among others. Here we would have a sample tabulation of families who purchase our brand and those who do not, classified by ethnic group; we would then need a way of using this data to make an inference about association between patronage and ethnicity in the population.

This chapter deals first with methods of inference for a single categorical variable where the data are contained in a simple one-way frequency, or enumeration, table.

Then it deals with inference about association where the data are contained in a two-way classification table. Methods of handling multiple samples of a single categorical variable are also treated. All the methods in this chapter are approximations that are appropriate when certain minimal requirements about the number of sample observations are met.

Methods Using the Normal Approximation to the Binomial

12.1   In Chapter 6 we introduced the binomial sampling distribution of the number of successes (r) in a sample of n independent draws from a binary population with a given probability of success (π) on a single draw. When the sample size (or number of draws) is adequately large, binomial probabilities are closely approximated by normal probabilities. Figure 12.1 shows the sampling distributions of r/n for $n = 4$, 8, and 16, and $\pi = 0.20$ and 0.50. The graphs for $\pi = 0.20$ show skewness decreasing with increasing n. Also present is a definite suggestion of the normal distribution shape as n increases—most evident in the $\pi = 0.50$ graphs. Of course, binomial probability distributions are discrete, while the normal is a continuous function. But as n is increased,

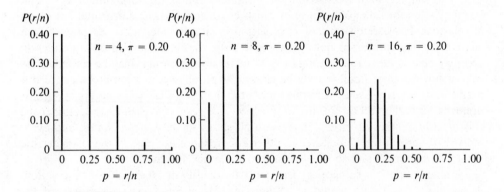

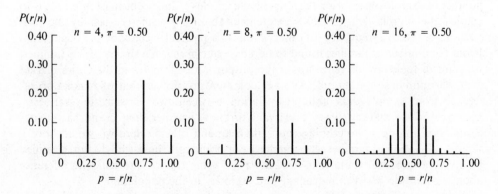

Figure 12.1   Selected Binomial Sampling Distributions of r/n

the ordinates for the possible values of r/n are more numerous and closer together, giving a suggestion of continuity. Mathematicians have shown that *the limit of the binomial distribution as n increases is a normal distribution for any value of π*. This generalization is like the central limit theorem in that it assures us that the normal shape is a better approximation as n increases for any specified π, but it does not give us an operating rule. The rule of thumb used by many practitioners follows.

Rule	The normal distribution provides reasonable approximations to binomial probabilities when the values of $n\pi$ and $n(1 - \pi)$ both equal or exceed 5.

This means, for example, that if $\pi = 0.10$, we need to have a sample size of at least 50, while for $\pi = 0.50$ a sample size of 10 would be sufficient to use the normal approximation.

The normal distribution is continuous, while the binomial distribution yields probabilities only for integer values of r, or corresponding values of r/n. This means that when using the normal distribution to approximate binomial probabilities, we must make some adjustment. We show this adjustment now in an example of estimating binomial probabilities from a normal distribution.

Consider the binomial probabilities for $r = 0, 1, 2, \ldots, 16$ when $n = 16$ and $\pi = 0.50$, which can be found in Appendix E. They are graphed in Figure 12.2. In (a) of the figure, we see that $P(r = 8) = 0.1964$ and $P(r = 11) = 0.0667$. These are the probabilities we will estimate. In (b) of Figure 12.2 the binomial probabilities are cumulated

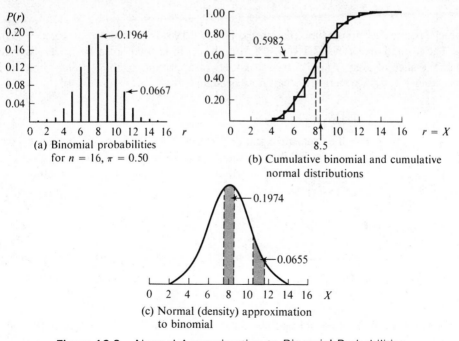

(a) Binomial probabilities
for $n = 16$, $\pi = 0.50$

(b) Cumulative binomial and cumulative
normal distributions

(c) Normal (density) approximation
to binomial

Figure 12.2   Normal Approximation to Binomial Probabilities

and we see that $P(r \leq 8)$ is approximately 0.60. The smooth cumulative curve is that of a normal distribution with a mean of 8 and a standard deviation of 2. We use these values because they are the mean and standard deviation of the binomial distribution of r when $n = 16$ and $\pi = 0.50$. Recall from Chapter 6 that

$$\mu_r = n\pi = 16(0.50) = 8.0,$$

$$\sigma_r = \sqrt{n\pi(1 - \pi)} = \sqrt{16(0.50)(0.50)} = \sqrt{4.0} = 2.0.$$

Notice that the cumulative normal distribution very nearly bisects the discrete steps of the cumulative binomial distribution. Consequently the 60th percentile of the matching cumulative normal is not 8.0 but 8.5. Using X for a continuous normally distributed variable with $\mu = 8.0$ and $\sigma = 2.0$, we would calculate this from

$$z_{.60} = 0.25, \quad \text{that is, } F'(z = 0.25) = 0.5987 \simeq 0.60,$$

$$X_{.60} = \mu + z_{.60}\sigma = 8.0 + 0.25(2.0) = 8.5.$$

Note that if we find the normal probability $P(X \leq 8.5)$ minus $P(X \leq 7.5)$, we will approximate the cumulative binomial step from $P(r \leq 7)$ to $P(r \leq 8)$, and this step is $P(r = 8)$. This is

$$z(8.5) = \frac{8.5 - 8.0}{2.0} = 0.25,$$

$$z(7.5) = \frac{7.5 - 8.0}{2.0} = -0.25;$$

$$P(7.5 \leq X \leq 8.5) = F'(z = 0.25) - F'(z = -0.25) = 0.5987 - 0.4013,$$

$$P(7.5 \leq X \leq 8.5) = 0.1974,$$

which is quite close to the binomial probability of 0.1964 for $P(r = 8 | n = 16, \pi = 0.50)$.

The normal area for $P(7.5 \leq X \leq 8.5)$ is shown in (c) of Figure 12.2. Here we see that the slice of area from 7.5 to 8.5 is used to approximate the "point" binomial probability at 8. A general rule can now be stated.

Rule

> The binomial probability for any integer r is approximated by the normal probability in the interval from $r - 0.50$ to $r + 0.50$. The appropriate normal distribution is one having $\mu = n\pi$ and $\sigma = \sqrt{n\pi(1 - \pi)}$.
>
> ■
> $$P(r | n, \pi) \simeq F'\left(z = \frac{r + 0.50 - n\pi}{\sqrt{n\pi(1 - \pi)}}\right)$$
> $$- F'\left(z = \frac{r - 0.50 - n\pi}{\sqrt{n\pi(1 - \pi)}}\right). \qquad ■ \quad (12.1)$$

In our example with $n = 16$ and $\pi = 0.5$, we would find the normal approximation to the binomial probability for $r = 11$ by

$$P(r = 11 \mid n = 16, \pi = 0.5) \simeq F' \left[z = \frac{11 + 0.5 - 16(0.5)}{\sqrt{16(0.5)(0.5)}} \right] - F' \left[z = \frac{11 - 0.5 - 16(0.5)}{\sqrt{16(0.5)(0.5)}} \right]$$

$$\simeq F'(z = 1.75) - F'(z = 1.25);$$

$$P(r = 11 \mid n = 16, \pi = 0.5) \simeq 0.9599 - 0.8944 = 0.0655.$$

This compares favorably with the actual binomial probability of 0.0667.

12.1

Hypothesis Tests for a Single Proportion

12.2

In Chapter 10 we encountered hypothesis tests for a single mean where the test statistic was the standard normal z statistic. Hypothesis tests for proportions using the normal approximation to the binomial sampling distribution of r are carried out in similar fashion. For example, suppose that 49 football games will be played this year between members of two leagues, A and B. Although the league commissioners believe they have achieved balance between the leagues, they wish to use the season's results to test their belief. They are willing to run a risk of 0.05 of discarding their belief when it is really correct. The commissioners believe that the underlying probability of a League A team winning a game is 0.50. This is their null hypothesis. If this hypothesis were true, then the number of wins for League A teams in 49 trials (games) would have a binomial probability distribution with $\pi = 0.50$ and $n = 49$. Since $n\pi$ and $n(1 - \pi)$ well exceed 5, the normal approximation to the binomial can be safely used. We would have:

Null hypothesis: $\pi = 0.50$
Alternative hypothesis: $\pi < 0.50$ or $\pi > 0.50$
Significance level: $\alpha = 0.05$

Using the normal distribution, $z_{\alpha/2} = z_{.025} = -1.96$ and $z_{1-\alpha/2} = z_{.975} = 1.96$. The observed result will lie in the lower-tail rejection region if the entire "slice" of probability corresponding to $P(r)$ lies to the left of (below) $z = -1.96$. This will be true if z for $r + 0.5$ is less than -1.96. The observed r will lie in the upper-tail rejection region if the entire "slice" of probability for $P(r)$ lies to the right of (above) $z = 1.96$. For this to happen the z value for $r - 0.5$ must exceed 1.96. Look again at Figure 12.2c to see how these "slices" extend from $r - 0.5$ to $r + 0.5$. We can now state the hypothesis testing procedure.

Statement

In a two-tailed test for a single proportion using the normal approximation to the binomial, the observed result will lie in the lower-tail rejection region if

$$\frac{r + 0.5 - n\pi}{\sqrt{n\pi(1 - \pi)}} < z_{\alpha/2},$$

and in the upper-tail rejection region if

$$\frac{r - 0.5 - n\pi}{\sqrt{n\pi(1 - \pi)}} > z_{1-\alpha/2}.$$

In our example, with $\alpha = 0.05$, the rejection region is $z < -1.96$ and $z > 1.96$. Suppose that League A teams win 30 games. Since our observed value, $r = 30$, exceeds $n\pi = 49(0.5) = 24.5$, we need only check against the upper-tail rejection region. We would have

$$\frac{30 - 0.5 - 49(0.5)}{\sqrt{49(0.5)(0.5)}} = \frac{5.0}{3.5} = 1.43.$$

We cannot reject the hypothesis that $\pi = 0.50$. The evidence is consistent with the commissioners' belief that they have achieved balance between the leagues.

One-tailed tests using the normal approximation to the binomial can be summed up in the following procedure.

Statement

When the rejection region for a hypothesis test of a single proportion is entirely in the lower tail, reject the null hypothesis if

$$\frac{r + 0.5 - n\pi}{\sqrt{n\pi(1 - \pi)}} < z_\alpha.$$

When the rejection region is entirely in the upper tail, reject the null hypothesis if

$$\frac{r - 0.5 - n\pi}{\sqrt{n\pi(1 - \pi)}} > z_{1-\alpha}.$$

Consider a marketer who has enjoyed 40 percent brand preference in a certain market in the past but has recently made strenuous efforts to increase his market share. A random sample of 1350 consumers in the market produces 594, or 44.0 percent, who prefer his brand. Test the null hypothesis, using $\alpha = 0.10$, that his brand-preference share has not increased.

Null hypothesis: $\pi \leq 0.40$
Alternative hypothesis: $\pi > 0.40$

Rejection region: Upper tail; reject null hypothesis if $\dfrac{r - 0.5 - n\pi}{\sqrt{n\pi(1 - \pi)}} > z_{.90}$.

$$\frac{r - 0.5 - n\pi}{\sqrt{n\pi(1 - \pi)}} = \frac{594 - 0.5 - 1350(0.40)}{\sqrt{1350(0.40)(0.60)}} = \frac{53.5}{18.0} = 2.97.$$

$$z_{1-\alpha} = z_{.90} = 1.28; \quad \text{reject null hypothesis.}$$

12.2

Confidence Interval for a Proportion

Suppose the marketer in the example above wanted to state the 0.95 confidence interval for the true proportion of consumers preferring his brand. In Chapter 6 we saw that the parameters of the sampling distribution of a proportion are

$$\mu_p = \pi,$$

$$\sigma_p = \sqrt{\frac{\pi(1 - \pi)}{n}}.$$

The marketer has observed a particular value of $p = r/n$, namely, $p = 594/1350 = 0.44$. He has conducted one experiment of 1350 independent draws among the many possible experiments (or samples of $n = 1350$) contemplated in the sampling distribution of p. The fact that the expected value of the sample proportion is the population proportion ($\mu_p = \pi$) means that p is an unbiased point estimate of π. The sample proportion is, in fact, the best (minimum variance) unbiased estimate of the population proportion.

If we knew σ_p and if we knew that the sampling distribution of p was normal (at least to an approximation) regardless of what the actual value of π was, then the interval $p \pm 1.96\ \sigma_p$ would establish a symmetrical confidence interval with a confidence coefficient of 0.95. But a little reflection reveals several problems. First, to know σ_p we would have to know π. Clearly, we are trying to estimate π *because* we do not know π. Second, the shape of the sampling distribution depends also on π. Recall that as π departs from 0.50, the binomial distribution of r (and of $p = r/n$) becomes skewed. Not only does σ_p depend on π, but the shape of the sampling distribution depends on π.

The upshot is that there is no simple and exactly correct algebraic method for setting confidence intervals for proportions when the sample size is small. If the sample size is large enough, an approximation will give reasonably accurate confidence intervals. Therefore, we use an estimate of σ_p and restrict the application to sample sizes large enough for the error in this estimate to be of little consequence. The sample sizes are large enough so that a normal sampling distribution is a satisfactory approximation regardless of what values π could reasonably take on.

Statement

A large sample procedure for setting a $1 - \alpha$ symmetrical confidence interval for a fixed but unknown proportion of successes (π) in a binary population based on observing r successes in n independent draws from the population ($r/n = p$, the sample proportion) is

$$p + z_{\alpha/2} \sqrt{\frac{p(1 - p)}{n}} \quad \text{to} \quad p + z_{1-\alpha/2} \sqrt{\frac{p(1 - p)}{n}}. \qquad (12.2)$$

To apply this large sample procedure, the following minimum sample sizes are recommended.

Lesser of p or $1 - p$	0.01	0.02	0.05	0.10	0.20	0.30	0.40	0.50
Minimum sample size (n)	10,000	5000	2000	900	300	150	60	30

The marketer observed $p = 0.44$ in a sample where $n = 1350$, well above the recommended minimum sample size. For a 0.95 confidence interval for π, the lower limit is

$$0.44 + (-1.96)\sqrt{\frac{0.44(0.56)}{1350}} = 0.44 - 1.96(0.0135) = 0.4135$$

and the upper limit of the interval is

$$0.44 + 1.96\sqrt{\frac{0.44(0.56)}{1350}} = 0.44 + 1.96(0.0135) = 0.4665.$$

If one encounters situations in which the sample sizes are below those suggested for this large sample approximation, classical methods given in Reference 1 can be employed. Alternatively, a Bayesian method illustrated in Chapter 18 of this text can be used.

Summary

For testing a hypothesis about a population proportion based on a sample comprised of independent draws from the population, a normal approximation can be used as long as $n\pi$ and $n(1 - \pi)$ equal or exceed 5. For setting confidence limits for a population proportion, the sample size requirement for an approximation employing the normal distribution is a more stringent one because the method involves several elements of approximation.

See Self-Correcting Exercises 12.A

Exercises

1. Find the normal approximations to the following binomial probabilities and compare the results with the true binomial probabilities from Appendix E.

   a. $P(r \geq 12 \,|\, n = 16, \pi = .5)$,   b. $P(r < 7 \,|\, n = 14, \pi = .6)$,
   c. $P(r = 6 \,|\, n = 16, \pi = .5)$.

2. Consider the outcomes of samples of size 100 taken from a process which is producing 10 percent defective products. Use the normal approximation to the binomial probability distribution to find

   a. $P(r < 4)$,   b. $P(r = 8)$,   c. $P(r > 14)$.

3. During the past year, a company enjoyed a brand-preference proportion in a certain market of 0.30. A sampling of current brand preference is to be taken. If α is set at 0.10, express the rejection region as a level of z

   a. for a test of the null hypothesis that the firm's current brand preference in the market is 0.30, and

b. for a test of the null hypothesis that its current brand preference is not greater than 0.30.

4. A random sample of 805 consumers in the situation of Exercise 3 yields 265 who prefer the company's brand. Carry out the test in (b) of that exercise.

5. In a random sample of 100 students, 64 agree with recent criticisms of selections of speakers for an "ecological awareness" program. Test the hypothesis, at $\alpha = 0.05$, that student attitude is evenly divided on the issue.

6. Calculate the 95 percent confidence interval for the proportion of all students agreeing with the criticisms in Exercise 5. Is the sample size adequate for the procedure?

7. In a random sample of 400 hourly wage employees in a certain area, 53 percent worked less than 36 hours the preceding week. Test the hypothesis that the median hours worked was 36.

8. Establish the 90 percent confidence interval for the proportion of all employees working less than 36 hours in Exercise 7.

9. In an acceptance sampling procedure a random sample of 144 parts was taken from each large lot of parts and the lot quality declared unacceptable if eight or more defectives were found. What is the probability under this scheme of

a. declaring a lot with 4 percent defectives unacceptable?

b. declaring a lot with 10 percent defectives acceptable?

10. A random sample of 900 persons in a scientific manpower registry revealed 27 who met certain specifications for a space exploration team. Is the sample size sufficient for setting confidence limits using the procedure of Formula (12.2)?

11. Is the sample size in Exercise 10 sufficient for testing the hypothesis (using the normal approximation) that the population proportion is 0.02? How about 0.04?

12. Carry out the hypothesis tests suggested in Exercise 11, using $\alpha = 0.05$.

Tests for a Single Categorical Variable

12.3

Tests of hypotheses involving proportions can be carried out by using the chi-square distribution. While we will illustrate the method first for a test involving a twofold classification, the real usefulness of the chi-square approach is that it can be extended to tests involving a manyfold classification.

Recall our example of inter-league games in which League A won 30 of 49 games played. We were concerned with testing the hypothesis that the underlying probability of a League A win was 0.50—that is, the leagues were evenly matched. In the chi-square approach we deal with two sets of frequencies. The first is the *actual outcome* —League A won 30 games and League B won 19 games. The second set of frequencies is the *expected outcomes under the hypothesis being tested*. If the probability of a League A win was indeed 0.50, then the expected outcome in an experiment of 49 games would be 24.5 wins for League A, or in binomial terminology, $\mu_r = n\pi$. The expected number of wins for League B is also 24.5, or $n(1 - \pi)$. We have, then, a table of actual frequencies and a corresponding table of expected frequencies under the null hypothesis. We show this below, introducing the symbol f for actual and g for expected frequencies.

League	Actual wins *f*	Expected wins *g*
A	30	24.5
B	19	24.5
Total	49	49.0

We can now think of our hypothesis test problem as one of evaluating the probability of observing an outcome at least as extreme (that is, different from the expected result under the null hypothesis) as our sample result. If the probability of such a result is small—that is, less than a predesignated α level—we reject the null hypothesis. Otherwise the null hypothesis is accepted.

The probability of a result at least as different from the expected outcome is evaluated by means of a statistic that has a chi-square sampling distribution.

Statement | When a table of actual frequencies has r rows and one column, and a corresponding set of frequencies expected under a null hypothesis has been derived, the following statistic has an approximate chi-square sampling distribution with $r - 1$ degrees of freedom under the condition that the null hypothesis is true.

$$\chi^2 = \sum_i \left[\frac{(f_i - g_i)^2}{g_i} \right].$$ (12.3)

In our example, χ^2 is calculated from

i	f_i	g_i	$f_i - g_i$	$(f_i - g_i)^2 / g_i$
1	30	24.5	5.5	1.235
2	19	24.5	−5.5	1.235
				$2.470 = \chi^2$

Our original table had two rows, so we consult the chi-square table for $r - 1 = 1$ degree of freedom. There we find $\chi^2_{.90} = 2.71$. The calculated χ^2 statistic of 2.470 is our measure of discrepancy between actual and expected frequencies. Since $2.470 < 2.71$, our χ^2 measure has a probability of greater than 0.10 of being exceeded. Thus, with any predesignated alpha level of 0.10 or less, we would have to accept the hypothesis of equal balance between the leagues. For example, at $\alpha = 0.05$, the critical value of χ^2 for one degree of freedom is 3.84. The rejection region is then $\chi^2 > 3.84$. Since $2.470 < 3.84$, we cannot reject the hypothesis. When an α level has been predesignated, the rejection region for a chi-square test of a null hypothesis will be $\chi^2 > \chi^2_{1-\alpha}$.

Extension to Manyfold Classification

The chi-square approach is extremely useful because it can be extended to a many-fold classification. For example, suppose 150 randomly selected school children are asked to indicate a preference among three flavors of gum. Their preferences are tabulated below.

	Number preferring
Flavor A	62
Flavor B	46
Flavor C	42
Total	150

The sponsor of the study is interested in whether the flavors enjoy equal preference among all school children. Given this null hypothesis, the expected frequencies and the calculation of the chi-square measure of discrepancy are

i	f_i	g_i	$(f_i - g_i)$	$(f_i - g_i)^2/g_i$
1	62	50	12.0	2.88
2	46	50	−4.0	0.32
3	42	50	−8.0	1.28
				$4.48 = \chi^2$

There are three rows in the table, so the degrees of freedom are $3 - 1 = 2$. Consulting this row of the chi-square table, we find $\chi^2_{.90} = 4.61$. To reject the hypothesis of equal preference on the basis of the evidence above would entail a risk of greater than 0.10 of rejecting a true hypothesis.

One is not restricted to testing hypotheses involving equal expected frequencies. Suppose the sponsor's advertising manager believes that 50 percent of all school children prefer flavor A, 25 percent prefer flavor B, and 25 percent prefer flavor C. If this hypothesis were true, the expected number preferring A in a sample of 150 children would be $0.50(150) = 75$, and the expected frequencies for B and C would be $0.25(150) = 37.5$. The calculation of chi-square would be as follows.

i	f_i	g_i	$(f_i - g_i)$	$(f_i - g_i)^2/g_i$
1	62	75	−13.0	2.253
2	46	37.5	8.5	1.927
3	42	37.5	4.5	0.540
				$4.720 = \chi^2$

The probability of a discrepancy as large as $\chi^2 = 4.720$, given the truth of the advertising manager's belief, is only slightly less than 0.10. The hypothesis that preferences were in a 2-1-1 ratio would be acceptable at $\alpha = 0.05$ but not if α were set as high as 0.10. The equal-preference hypothesis and the advertising manager's belief are about equally acceptable, given the sample outcome. This is indicated by the comparable values obtained for the chi-square measure of discrepancy.

12.3

12.4   Tests for Association

A frequently encountered situation is illustrated by the following example. A test was run on the appeal of a new magazine cover format among a random sample of 300 subscribers to the magazine. The magazine staff questioned whether preference for the new design over the old would be the same for male and female subscribers. The sample results are in the following table.

Sex	Prefer new	Prefer old	Total
Men	122	38	160
Women	88	52	140
Total	210	90	300

If we analyzed these data as a population (in the manner of Chapter 3), we would say that preference for the new cover format was associated with sex of subscriber. About 76 percent of the men interviewed prefer the new cover, while only about 63 percent of the women interviewed prefer the new cover. However, the data are a sample—not a population. We should check to see if an inference that association exists between preference and sex in the population is justified. The method followed is to test the null hypothesis that no association exists—that is, that preference and sex are independent in the population. If this hypothesis can be rejected at a low α risk level, then we are justified in concluding that some association exists in the population—that is, the proportion preferring the new cover is not the same for all male subscribers as for all female subscribers.

The key to the chi-square test of independence is the calculation of expected frequencies under the hypothesis of independence. In the present case we reason that if men had the same preference as women for the new cover, then the best estimate of that preference level is the proportion preferring the new cover among all persons interviewed. That proportion is $210/300 = 0.70$. Then we reason that if that proportion were true for men, the expected number of men preferring the new cover in a sample of 160 men would be $0.70(160) = 112$. In like manner the expected number of women preferring the new cover, *if the null hypothesis were true*, is $0.70(140) = 98$. The expected numbers of men and women preferring the old cover are $0.30(160) = 48$ and $0.30(140) = 42$. We now have a table of expected frequencies and a table of actual frequencies. Both tables have two rows and two columns in the present case. The tables are constructed below, including the symbols that we will employ.

Actual frequencies   Expected frequencies

$122 = f_{11}$	$38 = f_{12}$	$160 = f_{1.}$
$88 = f_{21}$	$52 = f_{22}$	$140 = f_{2.}$

$210 = f_{.1}$ $90 = f_{.2}$ $300 = f_{..}$

$112 = g_{11}$	$48 = g_{12}$	$160 = g_{1.}$
$98 = g_{21}$	$42 = g_{22}$	$140 = g_{2.}$

$210 = g_{.1}$ $90 = g_{.2}$ $300 = g_{..}$

Here, we use row-column subscripts for frequencies in the body of the table. The actual frequencies are f_{ij}, with i indexing the row and j the column location of a frequency. The "dot" notation found in the various totals is a convenient way to indicate various sums. The dot indicates the element summed over. Thus $210 = f_{.1}$ is the total frequency for the first column, that is, summed over rows; $140 = f_{2.}$ is the total frequency for the second row (summed over columns); and $300 = f_{..}$ is the grand total frequency (summation over rows and columns).

Note that because the marginal totals for the expected frequencies are the same as for the actual frequencies, we need only determine one expected frequency before the rest can be obtained by subtraction. However, a general formula for expected frequencies in tests of independence from $r \times c$ tables can be expressed as

$$\blacksquare \quad g_{ij} = \frac{(f_{.j})(f_{i.})}{f_{..}}. \quad \blacksquare \tag{12.4}$$

For example, we found above

$$g_{11} = \frac{(f_{.1})(f_{1.})}{f_{..}} = \frac{(210)(160)}{300} = 112.$$

We can now state the chi-square test of independence for an $r \times c$ table.

Statement | Given an $r \times c$ frequency table representing a random sample from a statistical population, the following statistic has an approximate chi-square sampling distribution with $(r - 1)(c - 1)$ degrees of freedom under a null hypothesis of independence between the variables represented by the rows and columns.

$$\blacksquare \quad \chi^2 = \sum_{ij} \left[\frac{(f_{ij} - g_{ij})^2}{g_{ij}} \right]. \quad \blacksquare \tag{12.5}$$

The calculation for our example is carried out in tabular fashion below.

ij	f_{ij}	g_{ij}	$(f_{ij} - g_{ij})$	$(f_{ij} - g_{ij})^2/g_{ij}$
11	122	112	10	0.893
12	38	48	−10	2.083
21	88	98	−10	1.020
22	52	42	10	2.381
				$6.377 = \chi^2$

The degrees of freedom are $(2-1)(2-1) = 1$, and $\chi^2_{.975}$, for one degree of freedom, is 5.02. Therefore, a discrepancy as great as $\chi^2 = 6.377$ (or greater) between actual frequencies and frequencies expected under independence would occur with probability less than 0.025. We can conclude that association exists between preference and sex with less than a 0.025 risk of having done so when the variables are in fact independent.

Chi-square tests of independence from two-way tables may involve tables of larger dimension than 2×2. As an example Table 12.1 shows a tabulation of frequency of newspaper purchase by educational level of head for 411 families randomly selected from all families in a model cities area.

Table 12.1   Frequency of Newspaper Purchase by Educational Level

	Educational level			
Newspaper purchase	*0–6th grade*	*7–9th grade*	*10th grade and above*	*Total*
Daily	42	47	96	185
Once or twice/week	24	32	35	91
Occasionally	17	11	12	40
Never	59	21	15	95
Total	142	111	158	411

We will test the null hypothesis that frequency of newspaper purchase is independent of educational level among all families in the model cities area. If this hypothesis were true, the proportion of persons in each educational class who purchased newspapers daily would be the same. Our best estimate of that proportion from the sample is $185/411 = 0.450$. Given this, the expected frequency for daily purchases in the 0–6th grade class is $0.450(142) = 63.9$; for daily purchases in the 7–9th grade class it is $0.450(111) = 50.0$; and for daily purchases in the 10th grade and above class it is $0.450(158) = 71.1$. These calculations are, in general, $g_{ij} = (f_{.j})(f_{i.})/f_{..}$, and have been effectively incorporated into generalized computer programs for chi-square tests. The actual and expected frequencies, using our system of symbols, are shown in Table 12.2. The calculation of χ^2 is shown in Table 12.3.

The basic table is four rows by three columns; that is, $r = 4$ and $c = 3$. The degrees of freedom for chi-square are $(4-1)(3-1) = 6$. The degrees of freedom are the number of independent opportunities for difference between f_{ij} and g_{ij} that contribute to the total chi-square measure. A way of counting these independent opportunities in tests of independence from $r \times c$ tables is as follows. Given that the marginal totals in the g_{ij} table are predetermined to be equal to those in the f_{ij} table, once a row has $c-1$ expected frequencies entered, the final entry can be obtained by subtraction from the marginal row total. Also, once a column has $r-1$ entries of g_{ij}, the final entry can be obtained by subtraction. In a 4×3 table this would mean that once $3 \times 2 = 6$ appropriate g_{ij} values are entered, the remainder can be obtained by subtraction from marginal totals. In general, the product of $(r-1)(c-1)$ reflects the independent opportunities for difference, or degrees of freedom, in chi-square tests for association in two-way tables.

Table 12.2 Actual versus Expected Frequency of Newspaper Purchase by Educational Level

Actual (f_{ij})

i \ j	1	2	3	
1	42	47	96	185
2	24	32	35	91
3	17	11	12	40
4	59	21	15	95
	142	111	158	411

Expected (g_{ij})

i \ j	1	2	3	
1	63.9	50.0	71.1	185
2	31.4	24.6	35.0	91
3	13.8	10.8	15.4	40
4	32.8	25.7	36.5	95
	141.9	111.1	158.0	411

$$g_{ij} = \frac{(f_{\cdot j})(f_{i\cdot})}{f_{\cdot\cdot}}; \quad \text{e.g., } g_{32} = \frac{(f_{\cdot 2})(f_{3\cdot})}{f_{\cdot\cdot}} = \frac{(111)(40)}{411} = 10.8$$

Table 12.3 Calculation of χ^2 for a 4×3 Table

ij	f_{ij}	g_{ij}	$(f_{ij} - g_{ij})$	$(f_{ij} - g_{ij})^2/g_{ij}$
11	42	63.9	−21.9	7.506
12	47	50.0	−3.0	0.180
13	96	71.1	24.9	8.720
21	24	31.4	−7.4	1.744
22	32	24.6	7.4	2.226
23	35	35.0	0.0	0.000
31	17	13.8	3.2	0.742
32	11	10.8	0.2	0.004
33	12	15.4	−3.4	0.751
41	59	32.8	26.2	20.928
42	21	25.7	−4.7	0.860
43	15	36.5	−21.5	12.664
				$56.325 = \chi^2$

For 6 degrees of freedom, $\chi^2_{.995} = 18.55$. Our χ^2 value is 56.325, one that would hardly ever occur if there were no association between frequency of newspaper purchase and educational level. We conclude that there is association between newspaper purchase frequency and educational level among all families in the area. Having concluded that there is association in the population, we can go ahead to use the sample data to estimate this pattern. We have assurance now that in looking at the association in the sample we are not looking at a pattern that could readily have been produced by sampling from a population in which no association prevailed. Figure 12.3 shows the percentage distributions of purchase frequency within each educational class in a way designed to bring out the pattern of association in the sample data. We see that the percentage of families purchasing daily increases with increasing educational level and the percentage of families never purchasing a newspaper decreases markedly as educational level increases. The null hypothesis of independence that we rejected for the population would be represented on this type of graph as a series of horizontal lines for each class of newspaper purchase frequency.

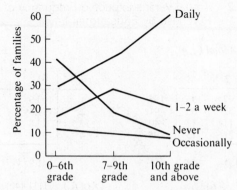

Figure 12.3   Percentage Distributions of Newspaper Purchase Frequencies by Educational Level—Model Cities Families

12.4

Multiple Samples—Single Categorical Variable

In the section on tests for a single categorical variable we studied the chi-square test for actual frequencies versus expected frequencies in a table with r rows and one column. In the example of inter-league win-loss record, there were only two rows (or classes) in the table. The chi-square test was, then, a test of a hypothesis about a binomial parameter, π. If there are more than two classes, as in the chewing gum flavors example, there are more than two parameters (true proportions) involved. These are sometimes called the multinomial (as opposed to binomial) parameters and are frequently identified as $\pi_1, \pi_2, \ldots, \pi_r$, where r is the number of classes involved.

In both of the earlier examples only one sample was involved. It often happens that we want to test whether on not several samples came from the same binomial or multinomial population. For example, suppose five persons are each asked to obtain 40 interviews for a sample survey of student opinions on a university campus. It is known that 50 percent of all students are lower division (freshmen and sophomores), 30 percent are upper division, and 20 percent are graduates and unclassified students. The distribution of each interviewer's sample is given below.

	Interviewer				
Class	A	B	C	D	E
Lower division	17	24	28	16	20
Upper division	7	6	8	11	14
Graduate and unclassified	16	10	4	13	6
Total	40	40	40	40	40

A relevant question is whether the five samples can be regarded as *random* samples from the population with $\pi_1 = 0.50$, $\pi_2 = 0.30$, and $\pi_3 = 0.20$. If this is not an acceptable hypothesis at a reasonable α level, say 0.05, then we may have reservations about whether the selection processes used by the interviewers were unbiased and consequently whether the opinions revealed by the survey should be taken as a fair representation of overall student opinions.

If the hypothesis about random selection were true, the expected number of lower division students in any sample of 40 would be 0.50(40) = 20, the expected number of upper division students would be 0.30(40) = 12, and the expected number of graduates and unclassified students would be 0.20(40) = 8. We then have five different samples with actual frequencies to compare with these expected frequencies. Below we show χ^2 calculated for each sample and then summed for the five samples.

		f_i					χ^2				
i	g_i	A	B	C	D	E	A	B	C	D	E
1	20	17	24	28	16	20	0.45	0.80	3.20	0.80	0.00
2	12	7	6	8	11	14	2.08	3.00	1.33	0.08	0.33
3	8	16	10	4	13	6	8.00	0.50	2.00	3.12	0.50

$$\chi^2 = 10.53 + 4.30 + 6.53 + 4.00 + 0.83 = 26.19$$

There are two degrees of freedom in the comparison of actual and expected frequencies for each sample. For comparison over five samples there are ten degrees of freedom. The critical level of chi-square for ten degrees of freedom with $\alpha = 0.05$ is $\chi^2_{.95} = 18.31$. With an actual chi-square of 26.19 we should suspect the selection procedures, as indicated earlier.

The samples can be individually tested by means of their own chi-square values. Such a series of tests would ask the same question of each interviewer's sample in turn, whereas our original question was whether the *five* samples could be regarded as *five* random samples from the underlying population. For tests of the individual samples at $\alpha = 0.05$, the critical value is $\chi^2_{.95} = 5.99$ for two degrees of freedom. Interviewers A and C would be suspect on the basis of individual tests. Interviewer A apparently has an underlying predisposition to select graduates and unclassified students and interviewer C tends to favor lower division students. Notice that the test we originally performed simply adds together the chi-square values and the degrees of freedom for the several samples to form a total chi-square and total degrees of freedom for the collection of five samples.

Relation to Test for Association

The question posed in the last section was whether the five interviewer samples could be regarded as random samples from a population with $\pi_1 = 0.50$, $\pi_2 = 0.30$, and $\pi_3 = 0.20$. Another type of question is often relevant. In this case it is the question whether the five samples can be regarded as random samples from populations with the same π_1, the same π_2, and the same π_3—whatever the values of π_1, π_2, and π_3. This amounts to asking whether the selection processes—biased or not—are uniform among interviewers.

In formal terminology this new question asks whether several samples came from *the same* population, while the earlier question was whether several samples came from *a specified* population. To answer this new question we must estimate the binomial or multinomial parameters from the data and compare the resulting frequencies expected under the null hypothesis against actual frequencies. The null hypothesis that the samples come from the same population amounts to a statement that the population proportions by categories are independent of (do not vary according to) the sample. In our example, independence between class of student and interviewer would mean that the underlying probabilities of selection of a given class of student are the same for all interviewers. The data are:

	Actual (f_{ij})								Expected (g_{ij})				
i \ j	1	2	3	4	5	$f_{i.}$		i \ j	1	2	3	4	5
1	17	24	28	16	20	105		1	21.0	21.0	21.0	21.0	21.0
2	7	6	8	11	14	46		2	9.2	9.2	9.2	9.2	9.2
3	16	10	4	13	6	49		3	9.8	9.8	9.8	9.8	9.8
$f_{.j}$	40	40	40	40	40	200							

Since $f_{.j}$ is the same for each interviewer, the set of expected frequencies for each interviewer (column) is the same. Remember that $g_{ij} = (f_{.j})(f_{i.})/f_{..}$. The calculation of chi-square proceeds as before.

$$\chi^2 = \sum_{ij} \left[\frac{(f_{ij} - g_{ij})^2}{g_{ij}} \right] = 19.29.$$

The degrees of freedom that apply are now for the test of independence from an $r \times c$ table, namely $(r - 1)(c - 1)$. This is $(3 - 1)(5 - 1) = 8$ degrees of freedom, and $\chi^2_{.95}$ is 15.51. We may reject the hypothesis of independence and conclude that the selection probabilities are not the same among interviewers for a given class of respondents.

One of the dimensions in $r \times c$ tables often can be regarded as designating separate samples. Usually this will be the dimension corresponding to the independent variable. In the earlier example of newspaper purchase frequency versus educational level, we could regard the data as representing three samples—one from each class of educational level. If a hypothesis were proposed that the proportion purchasing daily (π_1) was a specified value regardless of educational level, and similarly specific values of π_2, π_3, and π_4 prevailed regardless of educational level, the appropriate test would be the several-samples test described at the beginning of this section. The chi-square would have $3(4 - 1) = 9$ degrees of freedom, as opposed to $(4 - 1)(3 - 1) = 6$ degrees of freedom in the test for independence. The difference in degrees of freedom reflects the fact that in the several-samples test the hypothesized multinomial parameters are not derived from the sample data. In the test for independence they are effectively estimated from the sample data.

Cautions in the Use of Chi-Square

Chi-square tests from a 2×1 table for a single sample, a $2 \times c$ table for c samples, and a 2×2 table for association between two classification variables are each related to the binomial probability distribution. A "correction factor" is sometimes introduced that increases the accuracy with which the chi-square statistic approximates underlying binomial probabilities. This correction is to reduce the absolute magnitudes of difference between actual and expected frequencies by 0.5 before squaring in the calculation of the χ^2 statistic. In our formulas for χ^2 we did not include this correction factor. The larger the expected frequencies the less will be the inaccuracy introduced by omitting the correction factor. Larger expected frequencies correspond with larger sample sizes, other things being equal.

For three types of tests the probabilities approximated by the chi-square statistic are related to multinomial, rather than binomial, probabilities. These tests are: $r \times 1$ single-sample tests where r is greater than 2, $r \times c$ many-sample tests with c samples and r greater than 2, and $r \times c$ tests for association in other than 2×2 tables. The proper correction factor by which to reduce the absolute difference between actual and expected frequencies is not 0.5, and a simple generally applicable rule-of-thumb is not available. Generally, the larger the dimension of the table and the larger the expected frequencies, the more accurate will be the uncorrected value of chi-square. In dealing with tables of the type and dimension indicated, most authorities recommend introducing no correction factor.

In both of the sets of situations mentioned, the continuous chi-square distribution is being used to approximate probabilities that are related to discrete distributions. The discrete distributions (binomial and multinomial) arise because the random variables in the tables are counts rather than continuous measurements. The larger the expected numbers for the counts—which is to say the larger the expected frequencies— the better are the continuous approximations. A conservative rule of thumb employed is that each of the expected frequencies should be at least 5 before chi-square tests are employed at all.† When this criterion is not met, one recourse is to exact probabilities determined from binomial or multinomial probability formulas. Except for the 2×1 table for a single sample, we have not described these exact methods in this text. They may be found in advanced texts such as Reference 2.

Summary

Chi-square tests can be applied to enumerations of sample data by classes of a single categorical variable to test a hypothesis about the underlying probabilities (or population proportions) for each class. There may be one sample, two samples, or many samples. The chi-square statistic can also be used to test a hypothesis of independence from an enumeration of sample data by classes of two categorical variables. There will be two or more classes of each of the two categorical variables. When there are two or more samples, the data table for a test of a single categorical variable looks just like the data table for a test of independence between two categorical variables. The critical distinction is that in a test of a single categorical variable the expected frequencies are

---

† When adjacent cells represent adjacent classes of an implied ordinal or interval scale, the cells may be combined in order to achieve the required level of expected frequency. The degrees of freedom are then determined from the number of cells after grouping.

determined from proportions that are not derived from the table, while in a test of independence between two categorical variables the expected frequencies are derived entirely from the sample data.

The classification of chi-square tests provided in Table 12.4 is intended to serve as an aid in recognizing data situations and as a reference for the proper degrees of freedom to employ in connection with the calculated value of chi-square. In the case of a test of proportions for classes of a single categorical variable we assume the data have been arranged so that these classes are identified by the row headings of the table. Thus the number of columns in the table will correspond to the number of samples involved.

Table 12.4   Degrees of Freedom for Chi-Square Tests

	Number of columns (samples or classes)		
Situation	1	2	*More than* 2
Tests for specified proportions			
Binomial (two classes)	1	2	c
Multinomial (more than two classes)	$r - 1$	$(r - 1)2$	$(r - 1)c$
Tests for independence			
Two classes (rows)	N.A.	1	$1(c - 1)$
More than two classes (rows)	N.A.	$r - 1$	$(r - 1)(c - 1)$

In all cases of tests for specified proportions, the degrees of freedom are $(r - 1)c$, where r is the number of classes of the categorical variable and c is the number of samples. For tests for independence, the degrees of freedom are $(r - 1)(c - 1)$, where r is the number of classes of the "row" variable and c the number of classes of the "column" variable.

See Self-Correcting Exercises 12.B

Exercises

1. In a random sample of 1000 elderly people in a large city, 460 indicated that previous savings were a source of current income. Use chi-square to test the hypothesis that the true proportion for the population is 0.50. Use $\alpha = 0.01$.

2. Establish the 99 percent confidence interval for the true proportion in Exercise 1. Does this result appear consistent with your result for the chi-square test?

3. An equipment manufacturer changed the service features connected with his guarantee. A random sample of guarantee holders under the old and under the new system produced the following figures:

	Old	New
Satisfied with service	145	155
Dissatisfied	55	45

Can the manufacturer conclude that there is any association between satisfaction and the change in service features? Explain.

4. A random sample of families in four neighborhoods produced the following data on new arrivals. Test the hypothesis that the proportion of families who took up residence in the past year is the same in all four neighborhoods.

| | Neighborhood | | | |
Lived in neighborhood	1	2	3	4
Less than 1 year	24	2	19	15
One year or more	126	78	71	65

5. A manufacturer of a meat seasoning feels that the spiciness of his seasoning is properly positioned when consumers are divided evenly among those who want more spice, those who want less spice, and those who consider the spiciness "about right." Recently, taste tests were conducted in three cities with the following results:

Reaction	City A	City B	City C
Want more spice	17	27	25
About right	21	17	22
Want less spice	22	16	13

Can the data be regarded as three samples from a population meeting the manufacturer's desires?

6. Using the data in Exercise 5, test the hypothesis that the three samples come from the same multinomial population.

7. Test the data for each individual city in Exercise 5 against the theoretical population desired by the manufacturer.

Glossary of Equations

$$P(r\,|\,n, \pi) \simeq F'\!\left(z = \frac{r + 0.5 - n\pi}{\sqrt{n\pi(1 - \pi)}}\right)$$

$$- F'\!\left(z = \frac{r - 0.5 - n\pi}{\sqrt{n\pi(1 - \pi)}}\right)$$

The normal approximation to a discrete binomial probability is found by finding the probability in a unit interval centered on r for a normal distribution whose mean and standard deviation match those of the given binomial distribution.

$$p + z_{\alpha/2}\sqrt{\frac{p(1 - p)}{n}} \quad \text{to}$$

$$p + z_{1-\alpha/2}\sqrt{\frac{p(1 - p)}{n}}$$

A large sample normal approximation to the $1 - \alpha$ confidence interval for a fixed but unknown proportion of successes in a binary population based on observing r successes in n independent draws from the population ($p = r/n$, the sample proportion).

$$\chi^2 = \sum_i \left[\frac{(f_i - g_i)^2}{g_i} \right]$$

The chi-square statistic with $r - 1$ degrees of freedom for testing a hypothesis that the actual frequencies (f_i) in r classes of a single categorical variable were generated by independent draws from a population leading to a set of expected frequencies (g_i).

$$g_{ij} = \frac{(f._j)(f_i.)}{f..}$$

The expected frequency in the ijth cell of a two-way $(r \times c)$ table under a hypothesis of independence between the row and column variables is the product of the marginal totals corresponding to the cell location divided by the total frequency for the entire table.

$$\chi^2 = \sum_{ij} \left[\frac{(f_{ij} - g_{ij})^2}{g_{ij}} \right]$$

The chi-square statistic with $(r - 1)(c - 1)$ degrees of freedom for testing a hypothesis that the actual frequencies in a two-way $(r \times c)$ classification table were generated by independent draws from a population in which the row and column variables were independent—that is, not associated.

13

Analysis of Variance

In Chapter 11 we discussed how to test whether the observed difference between two sample means indicates a difference between the two population means or is merely a sampling fluctuation. In this chapter we will consider a general class of techniques for simultaneously testing for differences among several population means. The general class of techniques is called *analysis of variance*. We will discuss just two of the many types: one-way and two-way analysis of variance. In one-way analysis there is only one *treatment*—that is, one basis for distinguishing among populations. In our first example below, the treatment is poster advertising. In two-way analysis of variance, observations are classified simultaneously on two bases, or treatments. Within any treatment there are several *levels*. In the first example four posters, or levels, are used.

General Description

In this section we will examine the underlying logic on which analysis of variance is based. We begin by describing an example. Then we illustrate the notation for analysis of variance and the explanation which follows by referring to the example. The context for the explanation will be one-way analysis of variance. Essentially the same sort of reasoning applies to two-way analysis of variance.

Suppose a marketing executive for a chain of small convenience stores is considering ways of advertising a certain product. He has four different posters prepared, for display near the product, that he wants to test for differences in effectiveness. On the first day of the test, a different poster is displayed near the product in each of four markets, and the numbers of product sold that day are recorded. On the second day, each poster is moved to a market not involved in the first day of the experiment, and the same type of data are collected. On the third and final day, the procedure is repeated at a third set of four markets.

There are good reasons for using a different set of markets every day. If the same market were used for the same poster during all three days of the experiment, we could not be certain whether it was the posters or the markets that were the source of any differences in sales. If posters were rotated among the same set of markets on successive days, the possibility of an effect from one poster carrying over to its successor could not be ruled out.

The sales results from the experiment appear in Table 13.1. The mean daily sales

Table 13.1   Number of Packages Sold

| Day | Poster | | | |
	A	B	C	D
1	55	32	38	47
2	43	42	35	48
3	43	40	38	55
Total	141	114	111	150
Grand total				516
Mean	47	38	37	50
Grand mean				43

from poster D, 50, is higher than the mean for any other poster. The mean sales for posters B and C are very close and well below mean sales for the other two posters. The mean for poster A is fairly close to that for poster D. The purpose of the analysis of variance to be described is to test whether these differences in sample means reflect differences in the means of the statistical populations from which they came.

Notation and Terminology

In order to discuss analysis of variance effectively, we need a way to designate observations and the several types of means we must refer to. We also need a standard set of terms. In Table 13.2, the data of Table 13.1 have been transformed to the standard notation we will use.

In one-way analysis of variance there can be only one treatment, but it may be investigated at any number of levels greater than one. The symbol k is used to designate the number of levels. In our example, the use of posters constitutes the treatment and there are four levels of this treatment, that is, four types of poster. Hence, k is 4.

We elected to make the same number of independent observations within all treatment levels. Thus, we are conducting a *balanced* experiment with *replications*. The symbol r designates the number of such replications. For our example, $r = 3$.

In general, an observation is designated symbolically as X_{ij}, where i is the row index and j is the column index. For our example, X_{32} is 40, for instance. The manner of designating the various totals, means, and numbers of observations is indicated in the lower portion of Table 13.2. An example of each type of total and mean can be had by referring to the comparable position in Table 13.1.

Table 13.2   Standard Notation

Replication	Treatment-level			
	1	2	3	4
1	X_{11}	X_{12}	X_{13}	X_{14}
2	X_{21}	X_{22}	X_{23}	X_{24} the X_{ij}
3	X_{31}	X_{32}	X_{33}	X_{34}
Level totals	$\sum_i X_{i1}$	$\sum_i X_{i2}$	$\sum_i X_{i3}$	$\sum_i X_{i4}$
Grand total				$\sum_i \sum_i X_{ij}$
Level means	$\bar{X}_{\cdot 1}$	$\bar{X}_{\cdot 2}$	$\bar{X}_{\cdot 3}$	$\bar{X}_{\cdot 4}$
Grand mean				$\bar{X}_{\cdot \cdot}$
Number of observations per level	$n_{\cdot 1}$	$n_{\cdot 2}$	$n_{\cdot 3}$	$n_{\cdot 4}$
Total number of observations	n			
Number of levels	k			
Replications	r			

The Ratio of Variances

In Table 13.1, we saw that the mean sales for each of the four posters $(\bar{X}_{\cdot j})$ are 47, 38, 37, and 50. The grand mean $(\bar{X}_{\cdot \cdot})$ is 43 packages a day. We can establish a null hypothesis by stating that the four sales population distributions for the four posters have equal means—that is,

$$\mu_{\cdot 1} = \mu_{\cdot 2} = \mu_{\cdot 3} = \mu_{\cdot 4},$$

where the $\mu_{\cdot j}$ represent the daily sales population means for the four types of poster. We want to test this hypothesis against the alternative that the population means are not all equal, that at least one of the means differs from the others.

In addition to the assumption that (1) the null hypothesis is true, we must also assume that (2) all four sales populations are *normally distributed* and (3) all four populations *have the same variance*. In other words, our three assumptions result in our having four normally distributed sales populations for which $\mu_{\cdot 1} = \mu_{\cdot 2} = \mu_{\cdot 3} = \mu_{\cdot 4}$ and $\sigma_{\cdot 1}^2 = \sigma_{\cdot 2}^2 = \sigma_{\cdot 3}^2 = \sigma_{\cdot 4}^2$. It follows that we can consider the 12 sales observations as coming from the *same population*. For simplicity we will use μ and σ^2 for the parameters of this population. Furthermore, it follows that we can view Table 13.1 as being a subdivision of the random sample of 12 observations into four samples comprised of three observations apiece.

The rationale of our test will be to compare the variability among the four treatment level means to the average variability within the treatment levels. We can expect the ratio of the variability among means to the variability within levels to be much smaller if the four sales populations have the same mean than if their means differ widely.

From the assumptions just described, the four sample means (47, 38, 37, and 50) belong to the sampling distribution of the mean for $n_{\cdot j} = 3$. From our previous work, we know that the sampling distribution to which the observed values of the column means (the $\bar{X}_{\cdot j}$) belong is a normally distributed statistical population that has a mean and variance of μ and $\sigma^2/n_{\cdot j}$, where $n_{\cdot j}$ is equal to 3.

Although we do not know the value of the population variance for the sampling distribution of the mean ($\sigma_{\bar{X}_{.j}}^2 = \sigma^2/n_{.j}$), we can form an unbiased estimate of it from the sample information.

$$s_{\bar{X}_{.j}}^2 = \frac{\sum (\bar{X}_{.j} - \bar{X}_{..})^2}{k - 1}. \tag{13.1}$$

For our example,

$$s_{\bar{X}_{.j}}^2 = \frac{(47 - 43)^2 + (38 - 43)^2 + (37 - 43)^2 + (50 - 43)^2}{4 - 1},$$

or

$$s_{\bar{X}_{.j}}^2 = 42.$$

We know that $s_{\bar{X}_{.j}}^2$ is an unbiased estimator of $\sigma_{\bar{X}_{.j}}^2$. We also know that $\sigma_{\bar{X}_{.j}}^2$ is $\sigma^2/n_{.j}$. Hence, if we multiply $s_{\bar{X}_{.j}}^2$ by the sample size $(n_{.j})$, we will have an unbiased estimator of σ^2, the population variance. In symbols this estimator is

$$(n_{.j})s_{\bar{X}_{.j}}^2 \tag{13.2}$$

In our example,

$$(n_{.j})\, s_{\bar{X}_{.j}}^2 = 3(42). \quad \text{or } 126.$$

Because there are four levels in the experiment, three degrees of freedom are associated with this estimate of population variance.

If we happened to know the value of the population variance (σ^2), we could use the ratio

$$\frac{(n_{.j})s_{\bar{X}_{.j}}^2}{\sigma^2} \tag{13.3}$$

as the basis for a test of our original null hypothesis ($H_0 : \mu_{.1} = \mu_{.2} = \mu_{.3} = \mu_{.4}$). When the hypothesis is true, the numerator of the ratio $(n_{.j}s_{\bar{X}_{.j}}^2)$ is an estimate of σ^2, and the ratio will have an expected value of one. On the other hand, if the four population means are markedly different, the dispersion of the four sample means typically will be much greater than $\sigma_{\bar{X}_{.j}}^2$. In turn, the numerator of the test ratio, Equation (13.3), will be much greater than the denominator, σ^2, and the value of the ratio will be much greater than one. In this manner, *we can use a ratio of variances to test for a difference in means.* This is the central concept at the heart of all analysis of variance procedures.

Almost never, however, do we know the value of the population variance (σ^2) when we want to test for a difference in several population means. To construct a test for the differences in several population means we must use sample estimates of population variance.

We have assumed that the sales population distributions for the four posters in our experiment are identical normal distributions and are, therefore, all the same distribution. As a consequence, their variances are assumed to be equal to each other and to that of the composite population ($\sigma^2 = \sigma_{.1}^2 = \sigma_{.2}^2 = \sigma_{.3}^2 = \sigma_{.4}^2$).

Now consider the sample of three observations from the sales population for poster A. The variance of this sample $(s_{.1}^2)$ is also an unbiased estimator of the population variance (σ^2) if our assumption about variances stated in the previous paragraph is correct. In terms of our present notation,

$$s_{\cdot 1}^2 = \frac{\sum (X_{i1} - \overline{X}_{\cdot 1})^2}{n_{\cdot 1} - 1} = \frac{(55 - 47)^2 + (43 - 47)^2 + (43 - 47)^2}{3 - 1},$$

or

$$s_{\cdot 1}^2 = 48.$$

Similarly, for the other posters, $s_{\cdot 2}^2$, $s_{\cdot 3}^2$, and $s_{\cdot 4}^2$ are also unbiased estimators of σ^2. We can form a pooled, unbiased estimator (s_p^2) of the population variance (σ^2) by averaging these four sample values. When there are an equal number of observations in all samples,

$$s_p^2 = \frac{\sum s_{\cdot j}^2}{k}, \tag{13.4a}$$

or

$$s_p^2 = \frac{1}{k}\left\{\sum_j \left[\frac{\sum_i (X_{ij} - \overline{X}_{\cdot j})^2}{n_{\cdot j} - 1}\right]\right\}. \tag{13.4b}$$

For our example the estimate is

$$s_p^2 = \frac{1}{4}\left[\frac{(55 - 47)^2 + (43 - 47)^2 + (43 - 47)^2}{2} + \cdots\right.$$
$$\left. + \frac{(47 - 50)^2 + (48 - 50)^2 + (55 - 50)^2}{2}\right]$$

$$= \tfrac{1}{4}(48 + 28 + 3 + 19),$$

or

$$s_p^2 = 24.5.$$

Since there are two degrees of freedom in each of the four samples, a total of eight degrees of freedom is associated with this estimate.

The pooled sample variance (s_p^2) is an unbiased estimate of σ^2 based solely upon the dispersion within each of the four independent samples. The four independent sample means ($\overline{X}_{\cdot j}$) constitute the reference points from which the respective dispersions are measured. Consequently, the pooled sample variance within treatment levels is not at all affected by differences among the four sample means.

Given the assumptions, we now have two independent estimates of the population variance (σ^2). The first estimate, $n_{\cdot j}s_{\overline{X}_{\cdot j}}^2$, is based on the dispersion of the four sample means ($\overline{X}_{\cdot j}$) about the grand sample mean ($\overline{X}_{\cdot\cdot}$). The second estimate, s_p^2, is based on the dispersion of the sample observations (X_{ij}) about their respective treatment-level means ($\overline{X}_{\cdot j}$).

Suppose we form a ratio by placing the estimate of σ^2 based on variance among poster means ($n_{\cdot j}s_{\overline{X}_{\cdot j}}^2$) in the numerator and the estimate based on average variance of sales within a poster type (s_p^2) in the denominator. Then we can expect the value of the ratio to be smaller when the sales populations are identical normal distributions for all four posters than when the means of these populations differ.

Finally, we note that the ratio we are discussing satisfies all the requirements of the F statistic as defined in conjunction with Equation (11.14) when the assumptions we have made are true. As a result, we can make the following statement.

Statement

Consider an experiment to test for differences among the means of k normally distributed populations with the same variance (σ^2). Under the null hypothesis of no differences among means, two independent estimates of σ^2 are $n_{\cdot j}s_{\bar{X}\cdot j}^2$ and s_p^2 as defined in conjunction with Equations (13.2) and (13.4). Define the ratio of the first of these to the second as an F statistic, that is,

$$F = \frac{n_{\cdot j}s_{\bar{X}\cdot j}^2}{s_p^2}.$$ (13.5)

The sampling distribution of this statistic is an F distribution with $k-1$ degrees of freedom in the numerator and $n-k$ degrees of freedom in the denominator.

For our example, we have already found that the numerator has the value 126 with three degrees of freedom and the denominator has the value 24.5 with eight degrees of freedom. The value of the test statistic is therefore

$$F = \frac{126}{24.5}, \quad \text{or } 5.14.$$

From Appendix D we see that the value of $F_{.95}$ for three and eight degrees of freedom is 4.07. At the 0.05 level of significance, we can reject the null hypothesis that the long-run mean numbers of sales from the four posters are equal. There is too much variability among the four sample means ($\bar{X}_{\cdot j}$) as compared with the variability within the four samples for us to accept the hypothesis that all four samples come from populations with the same mean.

13.1

13.2   Partitioning Total Variability

Additional insight into one-way analysis of variance is provided by an alternative approach. Returning to Tables 13.1 and 13.2, we can begin by focusing on the sum of squared deviations of all 12 of the original observations (X_{ij}) from their grand mean ($\bar{X}_{\cdot\cdot}$). We will call this quantity the total sum of squares (SST) and define it symbolically as

$$SST = \sum_j \sum_i (X_{ij} - \bar{X}_{\cdot\cdot})^2.$$ (13.6)

For our example,

$$SST = [(55-43)^2 + (43-43)^2 + (43-43)^2] + [(32-43)^2 + (42-43)^2 + (40-43)^2]$$
$$+ [(38-43)^2 + (35-43)^2 + (38-43)^2]$$
$$+ [(47-43)^2 + (48-43)^2 + (55-43)^2]$$
$$= 574.$$

We could go on to find the variance for this sample of 12 observations by dividing SST by 11, the number of degrees of freedom present. But SST is also a measure of total variability for the entire set of observations and, as we shall see, we can partition SST into a portion that measures variability among treatment level means and another portion that measures variability within treatment levels.

The deviation of each original observation from the grand mean $(X_{ij} - \overline{X}_{..})$ can be separated into two portions as follows

$$X_{ij} - \overline{X}_{..} = X_{ij} - \overline{X}_{.j} + \overline{X}_{.j} - \overline{X}_{..}$$
$$= (X_{ij} - \overline{X}_{.j}) + (\overline{X}_{.j} - \overline{X}_{..}). \tag{13.7}$$

For example, for the observation $X_{14} = 47$ in Table 13.1,

$$47 - 43 = (47 - 50) + (50 - 43).$$

Adding and subtracting the column mean $(\overline{X}_{.4})$ does not change the value of any given deviation. It does, however, allow us to partition a total deviation into two portions. The first portion $(X_{ij} - \overline{X}_{.j})$ represents the deviation of an observation from its column mean and the second portion $(\overline{X}_{.j} - \overline{X}_{..})$ represents the deviation of the column mean from the grand mean.

We can substitute the partitioned deviation from Equation (13.7) into the expression for the total sum of squares as given in Equation (13.6):

$$SST = \sum_j \sum_i (X_{ij} - \overline{X}_{..})^2$$
$$= \sum_j \sum_i [(X_{ij} - \overline{X}_{.j}) + (\overline{X}_{.j} - \overline{X}_{..})]^2.$$

By performing some algebraic operations, it can be shown that an equivalent relationship is

$$SST = \sum_j n_{.j}(\overline{X}_{.j} - \overline{X}_{..})^2 + \sum_j \sum_i (X_{ij} - \overline{X}_{.j})^2. \tag{13.8}$$

The importance of this result can be appreciated when it is expressed as follows:

Statement	The total sum of squares (SST) as defined in Equation (13.6) can be expressed as the sum of the two terms on the right side of Equation (13.8). The first term, $\sum_j n_{.j}(\overline{X}_{.j} - \overline{X}_{..})^2$ is a measure of the variability *among treatment-level means*. It will be called the "sum of squares among means" and will have the symbol SSA. The second term, $\sum_j \sum_i (X_{ij} - \overline{X}_{.j})^2$, is a measure of the variability *within treatment levels*. It will be called the "within sum of squares" and will have the symbol SSW. In symbols,

$$\blacksquare \qquad SST = SSA + SSW. \qquad \blacksquare \tag{13.9}$$

For the sum of squares among means (SSA) there are k column means, one for each treatment level. Among the k treatment-level means there are $k - 1$ degrees of freedom. We will divide the sum of squares among means (SSA) by the associated degrees of freedom $(k - 1)$ and define the result to be the mean square among means (MSA). In symbols,

$$MSA = \frac{SSA}{k - 1},$$

or

$$MSA = \frac{\sum_j n_{.j}(\overline{X}_{.j} - \overline{X}_{..})^2}{k - 1}. \tag{13.10}$$

This is an exact algebraic equivalent of the unbiased estimator of σ^2 defined in Equation (13.2). This estimate is based on the variance among the treatment-level means.

Similarly, we can divide the within sum of squares (SSW) by the degrees of freedom within columns $(n - k)$ and call the result the within mean square (MSW). In symbols

$$MSW = \frac{SSW}{n - k},$$

or

$$MSW = \frac{\sum_j \sum_i (X_{ij} - \overline{X}_{.j})^2}{n - k}. \tag{13.11}$$

This is an exact algebraic equivalent of the unbiased estimator of σ^2 defined in Equation (13.4). This estimate is based on the variance within treatment levels.

Given the null hypothesis and assumptions mentioned earlier, we can form the F ratio in Equation (13.5) from the mean squares:

$$F = \frac{MSA}{MSW}. \tag{13.12}$$

Then we can test the hypothesis of equal means for treatment level populations by referring to Appendix D as we did before.

One-Way Analysis of Variance

As early as Chapter 1 we found that calculating variances from deviations usually is very inconvenient unless we are working with carefully chosen examples. There and in subsequent chapters we developed convenient calculation relationships based on sums and sums of squares of the original observations.

Similar relationships convenient for computational purposes with real data exist for analysis of variance. The four relationships that apply to one-way analysis of variance are

$$■ \quad C = \frac{(\sum_j \sum_i X_{ij})^2}{n}, \quad ■ \tag{13.13a}$$

$$■ \quad SST = \sum_j \sum_i (X_{ij}^2) - C, \quad ■ \tag{13.13b}$$

$$■ \quad SSA = \sum_j \left[\frac{(\sum_i X_{ij})^2}{n_{.j}} \right] - C, \quad ■ \tag{13.13c}$$

and

$$\blacksquare \quad SSW = SST - SSA. \quad \blacksquare \qquad (13.13\text{d})$$

The first relationship, Equation (13.13a), is used to find a correction term for use in Equations (13.13b) and (13.13c). The correction term is found by summing all of the original observations, squaring the sum, and dividing by the total number of observations in the entire experiment. In Table 13.1, the grand total was found to be 516 sales and there were 12 observations in all. Hence,

$$C = \frac{(\sum_j \sum_i X_{ij})^2}{n} = \frac{(516)^2}{12},$$

or

$$C = 22{,}188.$$

The relationship for the total sum of squares (SST), Equation (13.13b), states that each observation must first be squared. Then the sum of all n squared observations is found and the correction term is subtracted from this result. For our example,

$$\sum_j \sum_i (X_{ij}^2) = (55^2 + 43^2 + 43) + \cdots + (47^2 + 48^2 + 55^2) = 22{,}762.$$

Then

$$SST = \sum_j \sum_i (X_{ij}^2) - C = 22{,}762 - 22{,}188,$$

or

$$SST = 574.$$

The relationship for the sum of squares among means (SSA), Equation (13.13c), states that the sum of observations for each treatment level must first be squared. Then this result must be divided by the number of observations for that treatment level. The resulting quotients for the separate levels are then added and the correction term is subtracted from this latter total. For our example, we refer to Table 13.1 and see that

$$\sum_j \left[\frac{(\sum_i X_{ij})^2}{n_{\cdot j}} \right] = \frac{141^2}{3} + \frac{114^2}{3} + \frac{111^2}{3} + \frac{150^2}{3} = 22{,}566.$$

Then

$$SSA = 22{,}566 - 22{,}188,$$

or

$$SSA = 378.$$

Finally, the relationship for the sum of squares within treatment levels (SSW), Equation (13.13d), states that SSW is the difference between SST and SSA. Hence, for our example,

$$SSW = SST - SSA = 574 - 378,$$

or

$$SSW = 196.$$

When the sums of squares have been calculated, they are entered in a table, the format for which has become virtually standard. Table 13.3 shows the general format

Table 13.3   Format for One-Way ANOVA

Source of variation	Sums of squares	Degrees of freedom	Mean squares	F
Treatment	SSA	$k-1$	MSA	MSA/MSW
Residual	SSW	$n-k$	MSW	
Total	SST	$n-1$		

for one-way analysis of variance (called ANOVA in many references). The general format applied to our example appears in Table 13.4. The resulting F ratio is, of course, the same as we obtained earlier and is significant at the 0.05 level.

Table 13.4   ANOVA for Poster Experiment

Source of variation	Sums of squares	Degrees of freedom	Mean squares	F
Posters	378	3	126	5.14
Residual	196	8	24.5	
Total	574	11		

We have conducted our explanation of one-way analysis of variance in terms of a balanced experiment, which, you will recall, has the same number of observations within all treatment levels. A similar rationale produces identical results for one-way experiments in which the numbers of observations within levels are not equal. Equations (13.6) through (13.13) and Table 13.3 apply to both types of experiment.

13.2

Summary

One-way analysis of variance provides a technique for testing whether several levels of a given treatment result in a different mean response for some variable under that treatment. The test assumes that the sample observations within the treatment levels are independent random selections from identical normal populations. The null hypothesis of no difference in mean response for the treatment levels is tested by using a ratio of the variability among sample treatment-level means to the pooled variability within levels. Standard calculation procedures and standard formats for presenting the results of a one-way analysis have been described.

See Self-Correcting Exercises 13.A

Exercises

1. In an automobile service station location study, 3 traffic observers are being trained to count and record the number of cars making right turns adjacent to prospective sites. The observers have counted simultaneously but independently on the same site for four 4-hour periods. The mean number of cars per hour is then found. These results appear in the following table:

Number of Right-Turn Autos per Hour

| | Observer | |
A	B	C
51	34	25
33	46	57
32	50	40
65	29	29

   Use the method described in connection with Equation (13.5) and a 0.05 level of significance to test whether the means are equal for the three observers.

2. Use shortcut Equations (13.13a) through (13.13d) and the format given in Table 13.3 to test for a significant difference in observer means in Exercise 1. Let α be 0.05.

3. Four different instructors gave the same quiz to their students. A random sample of three student papers was selected from each instructor's class to see whether there was a difference in mean performance. The scores as listed by instructor are:

| | Instructor | | |
A	B	C	D
3	10	6	7
6	6	6	6
3	8	3	8

   a. Find the mean scores for each instructor and find the grand mean.

   b. State each of the twelve scores as a deviation from the grand mean and then find the sum of squares of these deviations.

   c. As shown in Equation (13.7), express each deviation as found in part (b) as the sum of a deviation from the treatment mean and a deviation of the treatment mean from the grand mean.

   d. Square each deviation in (c). Then sum each type of squared deviation to show that SST is the sum of SSA and SSW as expressed in Equations (13.8) and (13.9). Complete the test for the equality of means. Let α be 0.05.

4. Use short-cut calculations as described in Equations (13.13a) through (13.13d) to test the hypothesis of no difference in means for the data in Exercise 3. Let α be 0.05.

   a. Compare the values of SST, SSA, and SSW with those found in Exercise 3.

   b. Find the value of F, complete the test, and compare your results with those in Exercise 3.

5. Three different makes of desk top electronic calculators to be used for classroom programming instruction are being evaluated for a volume purchase by a university. A standard set of problems is created that is representative of the uses to which the calculators will be

put. It is estimated that about 10 hours will be required to process the set of problems. Because of variations in the times manufacturers could make demonstrators available, only 12 operators of 15 assigned to the three calculators were able to complete the standard set of problems. The completion times, rounded to the nearest hour are:

	Calculator	
A	B	C
7	8	14
6	6	9
8	9	12
	11	10
	7	

a. At the 0.05 level, test the hypothesis that the mean completion times are the same for all three calculators. Use short-cut Equations (13.13) and the format in Table 13.3.

b. Based only on the outcome of this experiment, which calculator should the university choose?

6. What assumptions permit use of the F table in Appendix D to test the significance of the sample F ratio in Exercise 5? Are these assumptions likely to be met in the experiment described?

7. Why are all the F tests for these exercises upper one-tailed tests; that is, why test only for excessively large values of F?

13.3   Two-Way Analysis of Variance

As we have seen, one-way analysis of variance separates into two-parts the total variation in a set of observations assigned to different levels of a single treatment. One part of the variation is attributable to the treatment imposed by the experiment, which for our example was poster advertising. The other part is attributable to the residual variation within treatment levels.

Two-way analysis of variance involves two treatments instead of one, and total variation is separated into three parts: one part for each treatment and the remaining part for residual variation within combinations of treatment levels. When one of the two treatments is included only to increase the sensitivity of the experiment—that is, it is not of primary interest—the two-way analysis is called a *randomized block design*.

A Randomized Block Design

Suppose that the safety design group in an automobile manufacturing firm is investigating the effect of differences in dashboard light intensity on the reaction time of drivers. A series of night time driving emergencies is programmed on automobile driving simulators. Each of four drivers is subjected to low, medium, and high levels of light from the dashboard, in random sequences, and reaction times are noted as shown in Table 13.5.

Table 13.5   Reaction Times* Classified by Light Intensity and Driver

Driver	Intensity Low	Intensity Medium	Intensity High	Totals	Means
A	33	18	35	86	28.6
B	22	11	25	58	19.3
C	36	22	39	97	32.3
D	29	18	26	73	24.3
Totals	120	69	125	314	
Means	30.00	17.25	31.25		26.2

* in hundredths of a second

For these data we will first carry out in summary fashion a one-way analysis of variance. Then we will carry out a two-way analysis, so that we can compare the two. The hypothesis of central interest under both types of analysis is that the selected differences in light intensity do not affect mean driver reaction time.

For a one-way analysis, we use Equations (13.13) to obtain

$$C = \frac{(\sum_j \sum_i X_{ij})^2}{n} = \frac{314^2}{12}. \quad \text{or } 8216.33;$$

$$SST = \sum_j \sum_i (X_{ij}^2) - C = (33^2 + 22^2 + \cdots + 39^2 + 26^2) - C$$

$$= 9010 - 8216.33, \quad \text{or } 793.67;$$

$$SSA = \sum_j \left[\frac{(\sum_i X_{ij})^2}{n._j} \right] - C = \frac{120^2}{4} + \frac{69^2}{4} + \frac{125^2}{4} - C$$

$$= 8696.50 - 8216.33, \quad \text{or } 480.17;$$

$$SSW = SST - SSA = 793.67 - 480.17, \quad \text{or } 313.50.$$

The results are shown in Table 13.6. For two and nine degrees of freedom, $F_{.95}$ is 4.26 and $F_{.99}$ is 8.02. Consequently, the difference between treatment means is significant at the 0.05 level, and light intensity appears to affect reaction time.

Table 13.6   One-Way Analysis of Light Intensities

Source of variation	Sums of squares	Degrees of freedom	Mean squares	F
Light intensity	480.17	2	240.08	6.89
Residual	313.50	9	34.83	
Total	793.67	11		

In this case we might reason that, as with other physiological differences, we can expect basic mean reaction times to differ from one individual to another. If this is true, it follows that when we conduct a one-way analysis, the variability in mean reaction times among individuals is left in the residual term. Leaving this variability in the residual term tends to inflate the denominator of the F ratio and to reduce the experiment's sensitivity to differences in the effect of light intensity on reaction times.

We now go to a two-way analysis, in the hope of increasing the experiment's sensitivity to differences in light intensity. This form of two-way analysis proceeds in much the same manner as did the one-way analysis described earlier. The part of total variation attributable to differences in driver mean reaction times is separated from the other parts along with its degrees of freedom. We do this by finding the treatment-B sum of squares (SSB), where

$$SSB = q\left[\sum_i (\overline{X}_{i.} - \overline{X}_{..})^2\right],$$

and q is the number of levels for treatment B.

In this relationship, we are using the deviations of the *row* means ($\overline{X}_{i.}$) from the grand mean ($\overline{X}_{..}$) in Table 13.5 as another independent estimate of the population variance (σ^2), given the null hypothesis and the identical normal population assumption for all combinations of two treatment levels.

For computational purposes in two-way analysis of variance we only need to add one relationship to Equations (13.13) and to change the relationship for SSW in that set of equations. These changes are as follows:

$$\blacksquare \qquad SSB = \sum_i \left[\frac{(\sum_j X_{ij})^2}{n_{i.}}\right] - C \qquad \blacksquare \qquad (13.14a)$$

and

$$\blacksquare \qquad SSW = SST - SSA - SSB. \qquad \blacksquare \qquad (13.14b)$$

In the one-way analysis of the data in Table 13.5 we found that C is 8216.33, SST is 793.67, and SSA is 480.17. These carry over to the two-way analysis.

The relationship for the sum of squares among means for treatment B, Equation (13.14a), states that the sum of observations for each treatment level (for each row in Table 13.5) must first be squared. Then this result must be divided by the number of observations for that treatment level. The resulting quotients for the different levels are then added and the correction term is subtracted from this latter total. For our example, we refer to Table 13.5 and see that

$$\sum_i \left[\frac{(\sum_j X_{ij})^2}{n_{i.}}\right] = \frac{86^2}{3} + \frac{58^2}{3} + \frac{97^2}{3} + \frac{73^2}{3} = 8499.33.$$

Then

$$SSB = 8499.33 - 8216.33,$$

or

$$SSB = 283.00.$$

The relationship for the residual sum of squares (SSW), Equation (13.14b), states that SSW is the difference between the total sum of squares (SST) and the sums of squares for the two treatments (SSA and SSB). Hence,

$$SSW = 793.67 - 480.17 - 283.00,$$

or, for our example,

$$SSW = 30.50.$$

As in the case of one-way analysis, the sums of squares for a two-way analysis are entered in a table with a format that has become almost standard. Table 13.7 shows

Table 13.7 Format for Two-Way ANOVA

Source of variation	Sums of squares	Degrees of freedom	Mean squares	F
Treatment A	SSA	$k-1$	MSA	MSA/MSW
Treatment B	SSB	$q-1$	MSB	MSB/MSW
Residual	SSW	$(k-1)(q-1)$	MSW	
Total	SST	$n-1$		

this format. Note that the value of SSW in the table is found by using Equation (13.14b) for two-way analysis, rather than Equation (13.13d) for one-way analysis. Also note that the degrees of freedom associated with SSW are the number of columns less one multiplied by the number of rows less one. Equations (13.14) and the format in Table 13.7 apply only to a two-way analysis in which there is exactly one observation for every combination of treatment levels and treatment effects are independent of each other. Other formats exist for different two-way arrangements. Our example in Table 13.5 fulfills the requirement just stated. There are three levels of treatment A (light intensity) and four levels of treatment B (drivers), for a total of $3 \times 4 = 12$ observations.

The general format of Table 13.7 applied to our light-intensity experiment appears in Table 13.8. The F ratio for differences in light intensity (treatment A) is 47.3 with two

Table 13.8 Two-Way ANOVA for Light Experiment

Source of variation	Sums of squares	Degrees of freedom	Mean squares	F
Light intensity	480.17	2	240.08	47.3
Driver reaction time	283.00	3	94.33	18.6
Residual	30.50	6	5.08	
Total	793.67	11		

and six degrees of freedom. For these degrees of freedom, Appendix D shows that $F_{.99}$ is 10.92. The F ratio for differences in mean driver reaction times (treatment B) is 18.6 with three and six degrees of freedom, for which $F_{.99}$ is 9.78. Both treatments produce differences in reaction times at well beyond the 0.01 level of significance.

A comparison of Tables 13.6 and 13.8 is informative. First we note that the sums of squares and degrees of freedom for both light intensity (treatment A) and total variation do not change from one table to the other. Second, we see that the residual sum of squares in Table 13.6 (313.50) is partitioned in Table 13.8 to driver reaction time (treatment B, 283.00) and two-way residual (30.50). Similarly the nine degrees of freedom for residual variation in Table 13.6 are also partitioned and allocated to treatment B and two-way residual variation.

Note that a far greater portion of the residual sum of squares in Table 13.6 (313.50) has been assigned to treatment B in Table 13.8 (283.00) than the portion of the nine degrees of freedom associated with 313.50 that was assigned to treatment B. Only one-third of the degrees of freedom went to treatment B, while just over 90 percent of the one-way residual sum of squares was assigned to treatment B. As a result, the mean square residual for the two-way analysis is much smaller than the mean square residual for the one-way analysis. Since the residual mean square is the denominator for all F ratios and since the mean square for treatment A is unchanged, the sensitivity of the experiment has been increased greatly. In the one-way analysis, F was 6.89, while $F_{.95}$ and $F_{.99}$ were 4.26 and 8.02, respectively. In the two-way analysis, F for treatment A is 47.3, while $F_{.95}$ and $F_{.99}$ are 5.14 and 10.92, respectively. Even though the criterion F values from Appendix D have increased because of a loss of degrees of freedom, identifying treatment B as a significant contributor of variation has caused the sensitivity of the experiment to increase even more.

On the other hand, if treatment B were some other treatment completely unrelated to differences in reaction times, no increase in sensitivity could be expected from the two-way analysis as compared with the one-way analysis. The one-way residual sums of squares and degrees of freedom would tend to be partitioned in the same proportions. Consequently, the residual mean square in the two-way analysis would remain about the same as in the one-way analysis, and the F ratio for the light-intensity factor would not change materially.

Recall that a two-way analysis is referred to as a *randomized block design* when only one of the two treatments is of primary interest. As we have indicated immediately above and in connection with the matched-pairs analysis in Chapter 11, blocking will help only when the variability of the dependent variable within blocks tends to be smaller than its variability among blocks. In our example, reaction times for any one driver tend to be more homogeneous than reaction times among drivers.

As you may suspect, the blocking technique can be extended to include additional independent factors. We could, for instance, design in a fatigue factor, such as number of hours without rest for each driver. We could then analyze the effect of differences in light intensity adjusted for differences in mean driver reaction time and for differences in fatigue.

When there is primary interest in both treatments, we often refer to the analysis as being concerned with two main effects. When there is one observation per cell in the cross-classified table of treatment levels (as with our example), the analytical procedure is identical to the one we have described. Other arrangements of the observations are covered in the references mentioned in the following section on design of experiments.

13.3

Some General Comments

Design of Experiments

Analysis of variance is the basis for a major branch of statistics called *design of experiments*. We have discussed only two of the many ways in which experiments can be designed to elicit desired information from a minimal number of observations.

References 1 and 2 are devoted entirely to this subject. A major advantage to be gained from studying in this field is an acquaintance with the rather rigorous requirements of sound scientific design.

Violation of Assumptions

It has been pointed out that the F test for significance in analysis of variance rests on two assumptions about the population of values from which the observations in any given cell of the table of observations are selected. These populations are assumed to be normally distributed and to have equal variances. Reference 3a describes a test for the hypothesis that population variances within treatment levels are equal. In addition to these two assumptions there are two more. The observations are chosen randomly and independently, and the components of total variability are additive, to conform with our breakdown of components in the sample into a simple sum. Investigations of the effect of violating the first two of the four assumptions have shown that the F ratio is relatively insensitive to departures from normality but is more sensitive to departures from homogeneity of variance within cells.

Estimation

If the null hypothesis in an analysis of variance is rejected, the conclusion is that differences exist among at least some of the population means. But the test provides very little information as to which population differences are most likely to be responsible for the significance of the F ratio. The analysis of the many possible comparisons among means is a fairly extensive subject and we shall not discuss it. References 3b, 4, and 5 cover this material.

Summary

Analysis of variance techniques can be extended to include a second treatment. To do this, the residual sum of squares in the one-way analysis is decomposed into a sum of squares attributable to differences among the level means for the second treatment and a new, smaller sum of squares within combinations of treatment levels. The technique described here applies only when there is a single observation for each combination of treatment levels. This technique is called randomized block design if there is no primary interest in the second treatment. When there is such interest, the experiment has two main effects or treatments.

See Self-Correcting Exercises 13.B

Exercises

1. For the automobile service station location study described in the first exercise of the previous set, no significant differences were detected in mean traffic counts as made by three observers. Another possible source of variation is the time of day. Traffic tends to be heaviest in the early morning and late afternoon. The traffic count data arranged by time of day as well as by observer are:

Number of Right-Turn Autos per Hour

	Observer		
	A	B	C
5 a.m.—9 a.m.	51	46	40
9 a.m.—1 p.m.	33	34	29
1 p.m.—5 p.m.	32	29	25
5 p.m.—9 p.m.	65	50	57

   Find SSB, the sum of squares for the second treatment (time of day) in this experiment.

2. Combine the results from the first exercise in this set and the first exercise in the previous set to complete a standard analysis of variance as described in Table 13.7.

   a. Is either F ratio significant at the 0.05 level? At the 0.01 level?

   b. What conclusion is justified by the two-way experiment?

3. Analyze the poster data in Table 13.1 as a two-way classification in which the type of poster is the treatment and days are blocks. Use Table 13.7 as the format for the analysis.

4. In the poster experiment described in conjunction with Table 13.1, three different sets of four markets were used on successive days.

   a. In view of this procedure, why might one wish to make the two-way analysis in Exercise 3?

   b. What outcome would one reasonably expect for the F test on blocks?

5. At the end of Chapter 11, the first exercise was concerned with a matched-pairs experiment to test the effectiveness of a weight reducing regimen. Six pairs of weights were:

$$253 \text{ and } 248, \quad 239 \text{ and } 236, \quad 194 \text{ and } 196,$$
$$203 \text{ and } 195, \quad 187 \text{ and } 175, \quad 225 \text{ and } 214.$$

   A two-way analysis of variance will test whether matching was effective, as well as whether the regimen works.

   a. To reduce arithmetic, subtract 175 (the smallest value) from all observations. This will not affect the analysis.

   b. Perform the two-way analysis and interpret the results.

6. Using the matched-pairs method from Chapter 11, perform a t test on the 6 matched pairs in Exercise 5 immediately above. Is t significant at the 0.01 level? at the 0.05 level?

7. Refer to Appendices B and D.

   a. Compare $t^2_{.975,60}$ and $F_{.95,1,60}$

   b. Compare $t^2_{.995,13}$ and $F_{.99,1,13}$

   c. In general, $t^2_{1-\alpha/2}$ for m degrees of freedom and $F_{1-\alpha,1,m}$ are equal. In view of this, discuss the tests conducted in Exercises 5 and 6 above.

Glossary of Equations

$$F = \frac{n_{\cdot j} s_{\bar{X} \cdot j}^2}{s_p^2}$$

The ratio of an estimator of population variance based on differences among treatment-level means to an estimator based on variability within levels is distributed as F when the treatment level populations are identical normal distributions.

$$SST = \sum_j \sum_i (X_{ij} - \bar{X}_{\cdot\cdot})^2$$

The total sum of squares is the sum of the squared deviations of every observation in an analysis of variance from the grand mean of all the observations.

$$SST = \sum_j n_{\cdot j} (\bar{X}_{\cdot j} - \bar{X}_{\cdot\cdot})^2 \\ + \sum_j \sum_i (X_{ij} - \bar{X}_{\cdot j})^2$$

The total sum of squares can be partitioned into the sum of squared deviations of the treatment-level means from the grand mean and of squared deviations of observations from their respective treatment-level means. This equation applies in one-way analysis of variance.

$$SST = SSA + SSW$$

In one-way analysis of variance the total sum of squares equals the sum of squares among treatment-level means and the sum of squares within treatment levels.

$$MSA = \frac{\sum_j n_{\cdot j} (\bar{X}_{\cdot j} - \bar{X}_{\cdot\cdot})^2}{k - 1}$$

The mean square among treatment-level means is the sum of squares among means divided by the degrees of freedom among means.

$$MSW = \frac{\sum_j \sum_i (X_{ij} - \bar{X}_{\cdot j})^2}{n - k}$$

In one-way analysis of variance, the mean square within treatment levels is the sum of squares within levels divided by the degrees of freedom within levels.

$$F = \frac{MSA}{MSW}$$

The ratio of the mean square among treatment levels to the mean square within levels is distributed as F if all treatment populations are identical normal distributions.

$$C = \frac{(\sum_j \sum_i X_{ij})^2}{n}$$

The correction term for one-way or two-way analysis of variance is the square of the grand total of all observations divided by the number of such observations.

$$SST = \sum_j \sum_i (X_{ij}^2) - C$$

The total sum of squares is the sum of the squared observations less the correction term.

$$SSA = \sum_j \left[\frac{(\sum_i X_{ij})^2}{n_{\cdot j}} \right] - C$$

The sum of squares among means is found by summing the observations within each level, squaring these sums, dividing by the number of observations in the respective levels, summing the quotients, and subtracting the correction factor.

$$SSW = SST - SSA$$

In one-way analysis of variance, the sum of squares within levels is found by subtracting the sum of squares among means from the total sum of squares.

$$SSB = \sum_i \left[\frac{(\sum_j X_{ij})^2}{n_{i\cdot}} \right] - C$$

In two-way analysis of variance the sum of squares among treatment levels for the second treatment is found by summing the observations in each level, squaring, dividing by the number of observations, summing the quotients for all levels, and subtracting the correction factor.

$$SSW = SST - SSA - SSB$$

In two-way analysis of variance, the sum of squares within treatment-level combinations is found by subtracting the sums of squares among means for both treatments from the total sum of squares.

14

Regression and Correlation

In Chapter 3, concerned with association between two variables, the cross-classified table was used for detecting association when the set of observations constitutes a statistical population. In Chapter 12, when the set of cross-classified observations came from a sample, we used chi-square to test for association in the population. Now we want to discuss making inferences from a sample of paired observations through the methods of sample regression and correlation analysis. Before going on, you may wish to review the latter portion of Chapter 3, concerned with fitting a least squares regression line to a complete set of paired observations for descriptive purposes and measuring the closeness of that fit. Here, as with regression analysis for a population, both variables from which the observations come usually are measured in interval or ratio scales.

Sample Regression

A retail credit bureau in a large city wants to find the variables associated with the credit behavior of customers. In one phase of the study, the number of months customers have lived at their present residence is the independent variable (X). The analyst making the study has decided to make sample observations at X values of 10, 20, and 30 months.

The dependent variable (Y) is the number of delinquent accounts in a set of 100 customer accounts selected at random from all those with a specified length of residence. Four such sets, each consisting of 100 accounts, are selected for each of the three values of X. Then the number of delinquent accounts per hundred active accounts is counted for each of these 12 sets. The observations are recorded in Table 14.1

Table 14.1   Credit Customer Data

Months of residence (X)	Delinquent accounts per 100 active accounts (Y)			
10	7	9	8	7
20	6	7	9	4
30	7	3	2	5

The delinquency variable (Y), as the dependent variable, is free to vary in accordance with its population distribution. On the other hand, the residence variable (X), the independent variable, cannot vary from the values 10, 20, and 30 that were preselected by the investigator. This situation is typical of a sample regression analysis and distinguishes it from a sample correlation analysis. By contrast, both the dependent and independent variables are free to vary at random in a sample correlation analysis. We will have more to say about sample correlation later in this chapter.

One advantage of a sample regression analysis over a correlation analysis is illustrated by the data in Table 14.1. When a value of the independent variable (X) is preselected, several values of the dependent variable (Y) associated with that value of X can be selected. Later we shall see that this enables us to make an analysis of variance test that can detect significant departures from a straight-line relationship between the variables. For our example, we have chosen an equal number (4) of observations of Y for each value of X. It is not necessary to have the same number of Y values for all X values; it is only necessary to have more than one Y value for each of a minimum of three X values.

We have one sample set of 12 paired observations (X, Y). Our first objective is to make the best estimate we can of the functional relationship between these variables in the population.

As in Chapter 3, we begin with a scattergram of the data to determine whether it is reasonable to assume that the functional relationship is a straight line. We see in Figure 14.1 that there is no reason at this point to reject a straight line. Of course, we recognize that this linear assumption is only meant to apply within the observed range of the data: $10 \leq X \leq 30$.

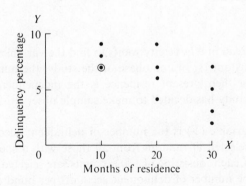

Figure 14.1   Scattergram for Credit Data.

The Sample Regression Line

To find that straight line which fits the sample data best, we use the same criterion for the sample that we used for the population in Chapter 3: the least squares criterion (see pages 65–68).

The general equation for a *sample* regression line that we shall use is

$$Y_c = a + bX, \tag{14.1}$$

where a is the Y intercept and b is the slope and the subscript c makes it clear that such values of Y are calculated, rather than observed. The lowercase c subscript signifies that sample data have been used, while capital C applies to a population. The sample Y intercept (a) corresponds in concept to its counterpart in the population, α. Similarly, the sample slope (b) corresponds to β [see Equation (3.5)]. Though a sample least squares regression line corresponds conceptually to a population least squares regression line, it is virtually impossible to select a random sample for which the intercept and slope of the sample line will be numerically identical to those of the population line. Sampling fluctuation makes such an occurrence extremely rare.

Using sample data, we would like to find numerical values for the slope (b) and intercept (a) that are, in some sense, the best estimates of their counterparts in the population (β and α). If we use exactly the same procedures for the sample that we did for the population in Chapter 3, we will have estimates (a and b) that are unbiased and have less sampling error than any other unbiased estimators.

Equations (3.7) provided one way to find the slope (β) and intercept (α) of the population least squares regression line ($Y_C = \alpha + \beta X$). If we use a and b in place of α and β in these equations, we will have the proper set for a sample:

$$\sum Y = an + b(\sum X), \tag{14.2a}$$
$$\sum (XY) = a(\sum X) + b(\sum X^2). \tag{14.2b}$$

The use of n rather than N also reflects the fact that a sample is involved. As in Chapter 3, we can find the necessary sums from our data and solve the equations simultaneously to obtain the proper numerical values for a and b.

Table 14.2 shows the 12 paired observations and the other calculations needed. In

Table 14.2   Sample Sums for Regression Line

X	Y	X^2	XY	Y^2
10	7	100	70	49
10	9	100	90	81
10	8	100	80	64
10	7	100	70	49
20	6	400	120	36
20	7	400	140	49
20	9	400	180	81
20	4	400	80	16
30	7	900	210	49
30	3	900	90	9
30	2	900	60	4
30	5	900	150	25
240	74	5600	1340	512

comparison with Table 14.1, Table 14.2 makes it clear that there are 12 values of the independent variable (X), rather than 3. This is, of course, necessary if we are to have one pair of values (X, Y) per observation for a total of 12 observations (n). Table 14.2 also gives $\sum Y^2$, although we will not need it until later.

After substitution in Equations (14.2), we have for our example

$$74 = 12a + 240b,$$
$$1340 = 240a + 5600b.$$

When solved simultaneously, they produce a value of 9.67 for the intercept (a) and -0.175 for the slope (b). The equation of the sample least squares regression line for our example is, therefore,

$$Y_c = 9.67 - 0.175X. \tag{14.3}$$

As an alternative to Equations (14.2) for finding the intercept and slope, Equations (3.8) can be adapted for use with samples. For sample data, the relationships are

$$b = \frac{n(\sum XY) - (\sum X)(\sum Y)}{n(\sum X^2) - (\sum X)^2} \tag{14.4a}$$

and

$$a = \frac{\sum Y - b(\sum X)}{n}. \tag{14.4b}$$

Direct substitution from Table 14.2 into these equations yields

$$b = \frac{12(1340) - 240(74)}{12(5600) - 240^2}, \quad \text{or} \quad -0.175$$

and

$$a = \frac{74 - (-0.175)(240)}{12}, \quad \text{or} \quad 9.67,$$

as before. The sample least squares regression equation is shown together with the data in Figure 14.2. At least for this sample, the negative slope indicates that the number of delinquent accounts per 100 active accounts declines as the length of residence increases.

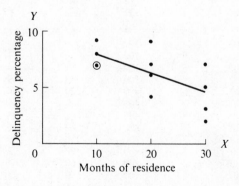

Figure 14.2   Sample Least Squares Regression Line.

The Sample Standard Error of Estimate

In Chapter 3 the standard error of estimate for the population of bivariate observations $(\sigma_{Y.X})$ was defined as the standard deviation of the values of the dependent variable (Y) from the least squares regression line [see Equation (3.9) and the related discussion]. The standard error of estimate is the standard deviation of the predictive errors. As such, it plays a major role in the process of making predictions by means of the least squares regression line.

The square of the sample standard error of estimate $(s_{Y.X}^2)$ represents the variance of the dependent variable (Y) about the regression line. It is the best estimate of its population counterpart $(\sigma_{Y.X}^2)$ in the sense that it is unbiased and has less sampling error than any other unbiased estimator. There is no simple unbiased estimate of the population standard error of estimate $(\sigma_{Y.X})$.

The sample standard error of estimate is

$$s_{Y.X} = \sqrt{\frac{\sum (Y - Y_c)^2}{n - 2}}, \tag{14.5}$$

where the values of $(Y - Y_c)$ are prediction errors of the sample regression line and n is the number of observations in the sample. In the denominator, decreasing the number of observations by 2 is necessary to make $s_{Y.X}^2$ an unbiased estimator of $\sigma_{Y.X}^2$.

Equation (14.5) makes clear the conceptual basis of the sample standard error of estimate $(s_{Y.X})$. But the form of the numerator on the right side of that equation is seldom convenient for computation. A more convenient form is

$$\sum [(Y - Y_c)^2] = \sum (Y^2) - a(\sum Y) - b(\sum XY). \tag{14.6}$$

For our retail credit example, we have $a = 9.67$ and $b = -0.175$. The sums we need are in Table 14.2. By substitution,

$$\sum [(Y - Y_c)^2] = 512 - 9.67(74) - (-0.175)(1340),$$

or

$$\sum [(Y - Y_c)^2] = \frac{187}{6}, \quad \text{or } 31.17$$

when corrected for rounding errors. Hence, for Equation (14.5),

$$s_{Y.X} = \sqrt{\frac{31.17}{10}}, \quad \text{or } 1.765.$$

The standard error of estimate is nearly 1.8 delinquent accounts per 100 active accounts. This value will be most useful when we discuss prediction and errors in prediction later in this chapter.

The Sample Coefficient of Determination

As we saw in Chapter 3 a relative measure of the strengths of regression functions is useful for comparing alternative functions. We also learned that the population coefficient of determination (ρ^2) is such a measure. The sample coefficient of determination (r^2) plays a role in the sample similar to that of ρ^2 in the population. The sample coefficient of determination is the portion of the variability in the dependent variable that is accounted for by the sample least squares regression with the independent variable. In symbols, the sample coefficient of determination is defined as follows:

$$r^2 = \left[1 - \frac{\sum (Y - Y_c)^2}{\sum (Y - \overline{Y})^2} \right], \tag{14.7}$$

where $\sum (Y - Y_c)^2$ is the sum of squared deviations about the regression line and $\sum (Y - \overline{Y})^2$ is the sum of squared deviations about the sample mean of the dependent variable.

For our example, we have already found from Equation (14.6) that the sum of squared deviations about regression, $\sum (Y - Y_c)^2$, is 187/6. To find the denominator of the fraction in Equation (14.7), we adapt Equation (1.3),

$$\sum (Y - \overline{Y})^2 = \sum Y^2 - \frac{(\sum Y)^2}{n}.$$

For the example,

$$\sum (Y - \overline{Y})^2 = 512 - \frac{74^2}{12}, \quad \text{or} \frac{334}{6}.$$

Then from Equation (14.7),

$$r^2 = \left(1 - \frac{187/6}{334/6} \right), \quad \text{or } 0.44.$$

In the sample, 44 percent of the variability in the number of delinquent accounts per hundred active accounts (Y) can be attributed to differences in the length of residence (X).

Hypothesis Tests

In regression analysis, the independent variable (X) must be known without error. On the other hand, the dependent variable (Y) is a random variable. So that inferences about population parameters can be made, three assumptions apply to values of the dependent variable:

1. For any given value of the independent variable (X) there is a population of values of the dependent variable (Y).

   At the designated X value, the mean of the dependent variable is obtained when that value of X is substituted in the population regression equation

$$Y_C = \alpha + \beta X.$$

   Call this mean $\mu_{Y \cdot X}$.

2. At the designated value of X, the population of Y values is normally distributed with a standard deviation equal to the population standard error of estimate ($\sigma_{Y \cdot X}$).

3. For *any* X value the population of Y values is normally distributed with a mean of

$$\mu_{Y \cdot X} = \alpha + \beta X$$

and a standard deviation of $\sigma_{Y \cdot X}$.

This set of assumptions is illustrated in Figure 14.3. Note there that the mean value

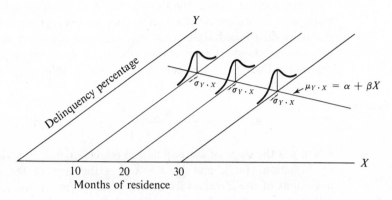

Figure 14.3   Linear Regression Model.

$(\mu_{Y \cdot X})$ of the dependent variable moves in accordance with the population regression equation as X changes. Also note that the conditional standard deviation of the Y values $(\sigma_{Y \cdot X})$ is the same for any value of X and is the population standard error of estimate. This figure shows the form of regression population from which our retail credit sample is assumed to come.

To gain understanding of what happens under sampling conditions, we can consider Figure 14.4. The sample least squares regression line for the data in Table 14.1

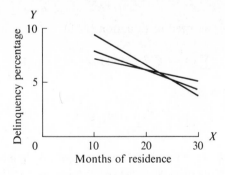

Figure 14.4   Regression Lines from Three Samples.

is shown along with sample regression lines from two other random samples, each composed of 12 observations from the same population. The entire regression line changes as we go from one sample to the next. Both the slope (b) and the intercept

14.1

(a) fluctuate under sampling conditions. From the model based on the conditions enumerated above, mathematical statisticians have found the sampling distributions for a and b and the other sample measures discussed earlier.

Perhaps the most often required test statistic based on the model discussed in this section is the one that applies to the slope of the sample regression line.

Statement

Consider the regression situation in which the conditional populations of Y values are centered on the population regression line, are normally distributed about that line, and have standard deviations equal to the standard error of estimate ($\sigma_{Y \cdot X}$). Then the slopes (b) of regression lines from independent random samples can be transformed to the statistic

$$\blacksquare \quad t = \frac{(b - \beta)\sqrt{\sum (X - \overline{X})^2}}{s_{Y \cdot X}}, \quad \blacksquare \qquad (14.8)$$

where β is the slope of the population regression line, $s_{Y \cdot X}$ is defined by Equation (14.5), and $\sum (X - \overline{X})^2$ is the sum of the squared deviations of the X values from their mean. This statistic forms a t distribution with $n - 2$ degrees of freedom.

The statistic in Equation (14.8) can be used to test a hypothesis about any pre-supposed value of the population slope (β). The most frequent test, however, is performed for the hypothesis that β is zero. If this hypothesis cannot be rejected, then we would be forced to conclude that there is no evidence of a linear regression relationship in the population.

For our example, we have already found that b is -0.175 and $s_{Y \cdot X}$ is 1.765. We can use our adaptation of Equation (1.3),

$$\sum (X - \overline{X})^2 = \sum X^2 - \frac{(\sum X)^2}{n},$$

to find the other value we need in Equation (14.8).

$$\sum (X - \overline{X})^2 = 5600 - \frac{240^2}{12}, \quad \text{or } 800.$$

Then Equation (14.8) becomes

$$t = \frac{[-0.175 - (0)]\sqrt{800}}{1.765}, \quad \text{or } -2.80,$$

with $12 - 2$, or ten degrees of freedom. This value of t is significant at the 0.02 level for a two-tailed test. Consequently, we reject the hypothesis that the slope (β) of the population regression line is zero at the 0.02 level of significance. This sample constitutes evidence for a negative slope in the population.

Tests of hypotheses about the intercept (α) and standard error of estimate ($\sigma_{Y \cdot X}$) also exist. The statistics appropriate for all of these tests can, of course, be used to make interval estimates too. These additional tests and estimates are discussed in References 1 and 2.

Summary

Sample regression analysis differs from sample correlation analysis in that values of the independent variable are known without error in regression, whereas in correlation they are subject to random variation. Techniques for finding the slope and intercept of the population least squares regression line apply to samples with only changes in symbols. The sample standard error of estimate requires some modification of its population counterpart to take account of degrees of freedom in the sample.

Point and interval estimates and hypothesis tests of the intercept, slope, and standard error of estimate are based on a regression model. The model assumes that, for any given value of the independent variable (X), the dependent variable (Y) is normally distributed and centered on the population least squares regression line. Furthermore, the standard deviations of all such populations are equal to the population standard error of estimate $(\sigma_{Y.x})$. A frequently used test is a t test of the hypothesis that the slope of the population regression line (β) has a value of zero—that is, there is no regression relationship in the population.

See Self-Correcting Exercises 14.A

Exercises

1. The following table shows the number of minutes that the caution light is on during a complete cycle of traffic lights at an intersection, and the number of accidents during a month at that intersection.

Intersection	Yellow light (Minutes/cycle) (X)	Accidents/month (Y)
1	1.0	11
2	1.0	12
3	1.2	12
4	1.2	13
5	1.2	15
6	1.4	12
7	1.4	14
8	1.8	15
9	1.8	16
10	1.8	16

a. Find the equation of the least squares sample regression line.

b. Use the least squares line to estimate the number of accidents to be expected from a setting of 1.6 minutes.

2. Find the standard error of estimate and the coefficient of determination for the sample in Exercise 1. Discuss the meaning of the coefficient of determination.

3. At the 0.005 level of significance, test the hypothesis that there is no linear relationship of accident rate to light timing.

4. A supervisor collected the following data on the number of days employees had been on a new job assembling a component and the average time spent per assembly. The employees were randomly selected.

Employee	Days on job (X)	Minutes per assembly (Y)
1	1	8
2	3	5
3	2	6
4	4	4
5	1	10
6	5	2
7	3	5
8	3	4

Find the equation of the least squares regression line for these observations.

5. What is the standard error of estimate for the sample in Exercise 4?

6. Do the data in Exercise 4 constitute evidence that there is a significant linear dependence of assembly time on learning time? Is the direction of the relationship plausible?

7. What is the coefficient of determination for the sample in Exercise 4?

8. Is there anything to be gained by comparing the standard errors of estimate found in Exercises 2 and 6? Discuss.

9. Is there anything to be gained by comparing the coefficients of determination in Exercises 2 and 7? Discuss.

14.2   Analysis of Variance in Linear Regression

When analysis of variance techniques are applied in linear regression, two types of results can be realized. First, analysis of variance can provide an algebraic equivalent to the test for the presence of regression described in connection with Equation (14.8). Second, if replication has been performed for three or more values of the independent variable, it is possible to test whether the regression relationship departs from linearity.

The data in Table 14.1 can be viewed in the context of a one-way analysis of variance. We can consider the length of residence (X) as the treatment and the values 10, 20, and 30 months as three treatment levels. Given multiple observations of the dependent variable (Y) at no less than two treatment levels, we can use analysis of variance to test the null hypothesis that there is no regression in the population. This is equivalent to the t test based on Equation (14.8). Given multiple observations of Y at no less than three treatment levels, not only can we test for the presence of regression in the population; we can also test whether the population regression relationship is likely to be nonlinear rather than linear (see pages 68–70).

Table 14.3 shows the data from Table 14.1 arranged more suitably for analysis of variance. The level means decline as length of residence increases. We want to test whether or not this indicates a linear trend in the population.

Table 14.3   Delinquent Accounts per Hundred

	Months of residence		
	10	20	30
	9	9	7
	8	7	5
	7	6	3
	7	4	2
Total	31	26	17
Grand total			74
Mean	7.75	6.50	4.25
Grand mean			6.17

In the manner developed in Chapter 13 [see Equation (13.7)], we can express the deviation of an observation of the delinquency variable (Y_{ij}) from the grand mean of the sample observations ($\overline{Y}_{..}$) as

$$Y_{ij} - \overline{Y}_{..} = (Y_{ij} - \overline{Y}_{.j}) + (\overline{Y}_{.j} - \overline{Y}_{..}),$$

where $\overline{Y}_{.j}$ is the treatment level mean. Alternatively, we can express the deviations $Y_{ij} - \overline{Y}_{..}$ in a manner that considers the effect of linear regression. Specifically, we can state that

$$Y_{ij} - \overline{Y}_{..} = (Y_{ij} - \overline{Y}_{.j}) + (\overline{Y}_{.j} - Y_c) + (Y_c - \overline{Y}_{..}). \qquad (14.9)$$

In Equation (14.9), Y_c is the value of the dependent variable calculated from the sample regression equation [Equation (14.1)] for the value of X which defines the treatment level. For example, consider the observation $Y_{21} = 8$ delinquencies in the left column of Table 14.3. As we see in that table,

$$\overline{Y}_{.1} = 7.75 \quad \text{and} \quad \overline{Y}_{..} = 6.17.$$

We know that the regression relationship is $Y_c = 9.67 - 0.175X$. When X is 10, therefore, Y_c will be 7.92. Hence,

$$8 - 6.17 = (8 - 7.75) + (7.75 - 7.92) + (7.92 - 6.17).$$

Similarly, each of the 12 deviations of Y values from their grand mean can be stated in terms of three components. These are the deviation of the observation from its treatment-level mean, the deviation of the treatment-level mean from the value on the sample regression line, and the deviation of the regression-line value from the grand mean of all the Y values.

When both sides of the relationship among components of deviations, Equation (14.9), are squared and these squared deviations are summed for all observations of the dependent variable, the result can be expressed as

Statement

$$SST = SSW + SSM + SSR, \tag{14.10}$$

where SST is the total sum of squared deviations and SSW is the sum of squared deviations within treatment levels, SSM is the sum of squared deviations of treatment level means from the values on the sample regression line, and SSR is the sum of squared deviations of regression line values from the grand mean.

The general format for our analysis of variance test appears in Table 14.4. There is always just one degree of freedom associated with SSR. Where k represents the

Table 14.4   Analysis of Variance for Linear Regression

Source of variation	Sums of squares	Degrees of freedom	Mean squares	F ratio
Linear regression	SSR	1	MSR	MSR/MSW
Nonlinearity	SSM	$k-2$	MSM	MSM/MSW
Within levels	SSW	$n-k$	MSW	
Total	SST	$n-1$		

number of treatment levels, there are $k - 2$ degrees of freedom associated with SSM. Finally, there are $n - k$ degrees of freedom associated with SSW.

The following relationships provide convenient computational equations for the sums of squares in Table 14.4. The relationships for SST and SSW are equivalent to those in Chapter 13 [see Equations (13.13)].

$$SST = \sum_i \sum_j (Y_{ij}^2) - \frac{(\sum_i \sum_j Y_{ij})^2}{n}; \tag{14.11a}$$

$$SSR = (r^2)SST; \tag{14.11b}$$

$$SSW = \sum_i \sum_j (Y_{ij})^2) - \sum_j \frac{(\sum_i Y_{ij})^2}{n_{\cdot j}}; \tag{14.11c}$$

$$SSM = SST - SSR - SSW. \tag{14.11d}$$

For our example, we can return to Table 14.2 for the necessary sums to substitute in Equation (14.11a),

$$SST = 512 - \frac{74^2}{12}, \quad \text{or } 55.67.$$

In connection with Equation (14.7), we found that the sample coefficient of determination (r^2) is 0.44. Hence,

$$SSR = 0.44(55.67), \quad \text{or } 24.50.$$

We can get the treatment-level totals from Table 14.3 to substitute in Equation (14.11c):

$$SSW = 512 - \left(\frac{31^2}{4} + \frac{26^2}{4} + \frac{17^2}{4}\right), \quad \text{or } 30.50.$$

Finally,

$$SSM = 55.67 - 24.50 - 30.50, \quad \text{or } 0.67.$$

The completed analysis of variance table for our example appears as Table 14.5. To test the null hypothesis of no linear regression in the population, we compare the

Table 14.5   ANOVA for Regression of Credit Delinquency and Length of Residence

Source of variation	Sums of squares	Degrees of freedom	Mean squares	F
Linear regression	24.50	1	24.50	7.23
Nonlinearity	0.67	1	0.67	0.20
Within levels	30.50	9	3.39	
Total	55.67	11		

sample F ratio 7.23 with $F_{.95}$ for one and nine degrees of freedom. We find that this value is 5.12. Hence, at the 0.05 level of significance, we accept the alternative hypothesis that a linear regression component is present in the population.

The second F ratio (0.20) is used to test the null hypothesis that the relationship between X and Y in the population is not nonlinear. From Appendix D, for one and nine degrees of freedom, the value of $F_{.95}$ is 5.12. Consequently, we must accept the null hypothesis of linearity.

When there is no evidence of nonlinearity, an alternative procedure is possible for the first of the two tests just described. Since there is no evidence of nonlinearity, we can combine the sums of squares and degrees of freedom for this source with those within levels to get a revised estimate of variability within treatment levels. This revision yields values for the sums of squares, degrees of freedom, and mean square of 31.17, 10, and 3.117, respectively. Then the revised F ratio is 24.5/3.117, or 7.86. For one and ten degrees of freedom, Appendix D shows a value of 4.96 for $F_{.95}$, and there is evidence of linear regression in the population. This latter test is identical to the t test of the null hypothesis that β is zero, which was described in conjunction with Equation (14.8).

14.2

14.3   Prediction

Our two major concerns so far in this chapter have been to find characteristics of the sample linear regression equation and test hypotheses about the population regression relationship. Another major concern in regression analysis is the problem of making point and interval predictions based upon the sample linear regression equation.

Point Prediction

For the credit data in Table 14.1 we found that the sample linear regression equation was $Y_c = 9.67 - 0.175X$. Suppose that we want to predict what the number of delinquent accounts per 100 will be when the length of residence is 14 months. In terms of our regression analysis, we want to know the value of Y when X takes on the value 14. We can predict the value of Y for this set of accounts by substituting 14 for X in the sample regression equation:

$$Y_c = 9.67 - 0.175(14), \quad \text{or } 7.22.$$

Our *point prediction* is that 7.22 accounts out of 100 with this length of residence will become delinquent.

Although it can be shown that Equation (14.3) produces the most likely single-number prediction for a stated value of X, from previous experience we know that we can place no confidence in Y taking on this exact value. For this reason we prefer prediction intervals for which we can state a known level of confidence.

Interval Prediction

When sampling, we have learned to expect successive sample estimates to fluctuate around the parameter being estimated. But when we examine the linear regression model described in conjunction with Figure 14.3, another source of fluctuation becomes evident in the population itself. The values of Y are dispersed about the population regression equation in accordance with a normal distribution centered on the regression equation with a standard deviation equal to the standard error of estimate ($\sigma_{Y \cdot X}$). Hence, even if we knew the population regression equation,

$$\mu_{Y \cdot X} = \alpha + \beta X,$$

we could not predict a specific value of Y without error. We would only know the mean value of Y at the designated value of X. This would, of course, be the value of $\mu_{Y \cdot X}$ found by substituting the given value of X in the population regression equation. Consequently, when we base a prediction of the value of Y on a sample regression equation, two sources of sampling error must be taken into account. As indicated in Figure 14.4, sample regression lines fluctuate about the population regression line. In addition, individual Y values are scattered above and below the sample regression line.

Suppose we have a random sample, as in Table 14.1. Then the $(1 - \alpha)$ prediction interval for an individual value of the dependent variable (Y) associated with a specific value of the independent variable (X') is

$$Y_c \pm (t_{\alpha/2})(s_{Y \cdot X}) \sqrt{1 + \frac{1}{n} + \frac{(X' - \overline{X})^2}{\sum (X - \overline{X})^2}} \, . \qquad (14.12)$$

In this expression, Y_c is the point estimate found by substituting the given value of the independent variable (X') into the sample least squares regression equation, and $t_{\alpha/2}$ has $n - 2$ degrees of freedom.

Suppose we let X' be 14 months. In the previous section we found that Y_c is 7.22 when 14 is substituted in the sample regression equation. Furthermore, $t_{.975}$ for ten degrees of freedom is 2.228, and we have already found that $s_{Y \cdot X}$ is 1.765. Therefore, the 0.95 prediction interval for Y, given that $X' = 14$ months, is

$$7.22 \pm 2.228(1.765) \sqrt{1 + \frac{1}{12} + \frac{(14 - 20)^2}{800}}, \quad \text{or} \quad 7.22 \pm 4.18.$$

Hence the interval runs from 3.04 to 11.40.

Consider the effect of the term under the radical that is farthest to the right in expression (14.12). If we elected to predict Y for a value of X' even farther from $\overline{X}$ than 14 is from 20, the numerator of this term would be larger. If X' were selected closer to the value of $\overline{X}$, this numerator would be smaller. Since all other values in the equation remain unchanged for a given sample, the effect of this term is to widen the prediction interval as the deviation of the selected value of the independent variable (X') from the sample mean $(\overline{X})$ increases. This extra width compensates for the greater variability of the extremes of fluctuating sample regression lines as compared with their central regions (see Figure 14.4).

Expression (14.12) applies when we want to predict a *single* Y value associated with a preselected value of the independent variable (X'). Sometimes, however, we want to predict the *average* Y value for the designated value of X'. As an example, on our credit study we may want to predict the number of delinquent accounts for a specified *single set* of 100 active accounts, all with 14 months of residence. On the other hand, we may want to predict the mean number of delinquent accounts per 100 active accounts for *all sets* of 100 accounts with 14 months of residence. In this latter case we could predict without error if we had the population regression relationship, $\mu_{Y \cdot X} = \alpha + \beta X$. We would substitute an X' value of 14 for X and the prediction would be the resulting value of $\mu_{Y \cdot X}$. Given only a random sample, however, we must form the prediction interval

$$\overline{Y}_c \pm (t_{\alpha/2})(s_{Y \cdot X}) \sqrt{\frac{1}{n} + \frac{(X' - \overline{X})^2}{\sum (X - \overline{X})^2}} \, . \qquad (14.13)$$

In this expression, all symbols are as defined for expression (14.12). The overbar in $\overline{Y}_c$ simply indicates that we are using the value calculated from the sample regression equation (7.22) to estimate a mean. Hence if we preselect an X' value of 14, our 0.95 prediction interval for $\mu_{Y \cdot X}$ is

$$7.22 \pm 2.228(1.765) \sqrt{\frac{1}{12} + \frac{(14 - 20)^2}{800}}, \quad \text{or } 7.22 \pm 1.41,$$

a much narrower interval than was our prediction for a single Y value. This reflects the fact that we are predicting the value of the mean of a distribution ($\mu_{Y \cdot x}$) rather than the value of a single observation in that distribution. Note that Expressions (14.13) and (14.12) differ only in the value "one" which constitutes the first term under the radical in Expression (14.12) but does not appear in Expression (14.13).

14.4   Correlation

The population correlation coefficient, ρ, was introduced in Chapter 3 as a dimensionless measure of the degree of association between two variables that are linearly related. For descriptive purposes in statistical populations, ρ was presented as an alternative to the coefficient of determination (ρ^2). In Chapter 3, where only statistical populations were under consideration, it was not necessary to distinguish between correlation analysis and regression analysis. On the other hand, when we are dealing with random samples rather than with populations, a basic distinction must be made between correlation analysis and regression analysis.

In sample regression analysis, one way to proceed is to set the independent variable (X) at a given value and randomly select one or more values of the dependent variable (Y) associated with the set value of X. Then the process is repeated with respect to second, third, and additional values of X. This process makes it clear that X is not a random variable. By contrast, in sample correlation analysis, a unit of observation such as a student or a consumer is selected at random and the values of X and Y that happen to be associated with that unit are recorded as the observation. In correlation sampling problems we are studying the joint variation of two variables.

As an example of a correlation analysis, suppose a study is being made of the relationship between annual family income (X) and weekly food expenditures (Y). A random sample of twenty families in a certain market areas is selected, and the data shown in Table 14.6 are recorded. Note that it is the families that are selected at random. Consequently, both X and Y are free to vary jointly at random. The scattergram for these twenty data points appears in Figure 14.5.

Table 14.6   Economic Data for a Sample of Families

Family	Income* (X)	Food expense (Y)	Family	Income* (X)	Food expense (Y)
1	49	47	11	47	54
2	46	55	12	49	57
3	48	57	13	54	58
4	38	35	14	48	65
5	63	73	15	56	66
6	58	64	16	51	53
7	40	38	17	55	69
8	55	73	18	28	18
9	66	87	19	63	76
10	54	68	20	35	49

* in hundreds of dollars

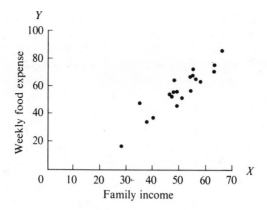

Figure 14.5   Scattergram for Consumer Expense Data.

From Table 14.6 we can find the sums needed to calculate the value of the sample correlation coefficient (r):

$$n = 20, \qquad \sum XY = 60{,}943,$$
$$\sum X = 1{,}003, \qquad \sum Y = 1{,}162,$$
$$\sum X^2 = 52{,}065, \qquad \sum Y^2 = 72{,}280,$$

where the calculation procedure is identical to that used in Table 14.2.

We could use Equation (14.7) for the sample coefficient of determination (r^2) and take the square root of it to find the value of the sample correlation coefficient (r). An algebraic equivalent to this equation is useful with a calculator; so we shall present it instead:

$$r = \frac{n \sum XY - (\sum X)(\sum Y)}{\sqrt{[n \sum X^2 - (\sum X)^2][n \sum Y^2 - (\sum Y)^2]}}. \tag{14.14}$$

For our example,

$$r = \frac{20(60{,}943) - (1{,}003)(1{,}162)}{\sqrt{[20(52{,}065) - (1{,}003)^2][20(72{,}280) - (1{,}162)^2]}} = 0.92.$$

From our work in Chapter 3, we know that 84. 6 percent $[(0.92)^2(100)]$ of the sample variability in food expenditure is explained by variations in annual income.

To make hypothesis tests about the value of the population correlation coefficient (ρ), we must note some characteristics of the sampling distribution of the sample correlation coefficient (r). When the population is distributed in accordance with the bivariate normal distribution (see Reference 3) and when there is no correlation in the population $(\rho = 0)$, the variable

$$t = \frac{r\sqrt{n-2}}{\sqrt{1-r^2}} \tag{14.15}$$

forms a t distribution with $n - 2$ degrees of freedom. For our example,

$$t = \frac{0.92\sqrt{20-2}}{\sqrt{1-0.846}}, \quad \text{or } 9.95.$$

This is well beyond any tabled value of t for 18 degrees of freedom in Appendix B. Consequently we can reject the hypothesis of no correlation in the population.

When ρ is not zero and the population is bivariate normal, the sampling distribution of r becomes increasingly skewed as ρ approaches either $+1$ or -1. To test a hypothesis about ρ having a value other than zero, we use the statistic w†, where

$$w = \frac{1}{2} \ln \left(\frac{1+r}{1-r} \right) \tag{14.16}$$

and where ln designates the natural logarithm. For samples of 10 or more observations, w is very nearly normally distributed with mean

$$\mu_w = \frac{1}{2} \ln \left(\frac{1+\rho}{1-\rho} \right) \tag{14.17}$$

and variance

$$\sigma_w^2 = \frac{1}{n-3}, \tag{14.18}$$

where n is the number of observations and ρ is the population correlation coefficient. Appendix G provides values of w for values of r or ρ.

Suppose we have reason to think that ρ should be 0.50 for the expenditure-income study. We must find

$$z = \frac{w - \mu_w}{\sigma_w} \tag{14.19}$$

from Equations (14.16), (14.17), and (14.18). Then we compare z with the normal table value for the selected level of significance. If the level of significance is 0.05, then the criterion value of z is 1.96 for a two-tailed test. From Appendix G, w is 1.58902 when $r = 0.92$, and μ_w is 0.54931 when ρ is 0.50. From Equation (14.18),

$$\sigma_w^2 = \frac{1}{20-3}, \quad \text{or } 0.0588,$$

when n is 20 observations. Then

$$z = \frac{1.589 - 0.549}{\sqrt{0.0588}}, \quad \text{or } 4.29$$

14.4   Hence, we reject the hypothesis that the population correlation coefficient (ρ) is 0.50.

Multiple Regression

The concepts and methods of regression with two variables as discussed earlier in this chapter can be extended to *multiple* regression, in which there are three, four, or more variables. In multiple regression we seek to estimate the value of a dependent

† This is often referred to as "Fisher's z statistic." Student confusion with z, the unit normal variate ($\mu = 0$, $\sigma = 1$) led us to adopt a different symbol.

variable by substituting the known value of each independent variable into a regression relationship. For example, in a study of weight increase in chickens during their first three months of life, the independent variables are birth weight and food intake. Call these variables Y, X_1, and X_2, respectively. Then the sample regression equation can be of the form

$$Y_c = a + b_1 X_1 + b_2 X_2.$$

If a third independent variable, adult weight of the mother (X_3), were included in the study, the form of the equation could be

$$Y_c = a + b_1 X_1 + b_2 X_2 + b_3 X_3.$$

Extension to more independent variables can follow the same linear pattern. Alternatively, the relationship selected for the variables can be curvilinear.

The least squares criterion discussed in the two-variable case also is applied in the majority of cases with more than two variables. Least squares procedures have been developed for both linear and curvilinear models. As before, these procedures yield values of the coefficients "a" and the "b's" that minimize the sum of squared deviations of the sample observations from the regression relationship.

In a multiple regression study of even moderate size, there is a rather formidable amount of calculation. Consequently, computer programs have been developed to reduce this task to practical levels. Many of these programs proceed in a stepwise fashion. A typical program searches among the independent variables to find the one that will maximize the explained portion of variance in the dependent variable when only a two-variable relationship is assumed. The program then prints this "best" two-variable relationship along with associated statistics. The program next goes on to find the best three-variable relationship, the best four-variable relationship, and so on, until all independent variables have been included.

Some of the associated statistics printed with each step in the program permit tests of hypotheses about the values of coefficients in the population regression relationship and about the population standard error of estimate. Analysis of variance tests that are often included for hypotheses about linear and curvilinear relationships are based on the same concepts as those discussed earlier for the two-variable case. Analysis of variance also makes it possible to test whether the inclusion of the successive independent variables in a stepwise routine improves the relationship significantly. Such a test can be used to stop adding variables when the next one fails to increase the fit of the relationship significantly.

More detailed descriptions of multiple regression are contained in Reference 4.

Summary

For a regression analysis based on a random sample from the model population as illustrated in Figure 14.3, analysis of variance can be used to test for significant regression in the population. In addition, if multiple observations of the dependent variable (Y) are made for at least three preselected values of the independent variable (X), then analysis of variance can be used to detect a departure from a straight-line regression relationship between the variables.

Both point and interval predictions of values of the dependent variable (Y) for a given value of the independent variable (X) can be based on the sample regression equation. Point prediction requires only the substitution of the value of X in the equation. Interval prediction is for either a single value of Y or for the mean value of Y at the given value of X. In either case the prediction interval for a stated level of confidence widens as X departs from the sample mean of the X observations ($\overline{X}$). The widening reflects the fact that sample regression lines fluctuate and, as a result, are less stable at their extremes than near $\overline{X}$.

Sample correlation analysis has a different basis than does sample regression analysis. Both variables in correlation analysis are random. In addition, the population of (X, Y) pairs is assumed to conform to a different model, the bivariate normal distribution. When it does, hypotheses about the value of the population correlation coefficient (ρ) can be tested from a knowledge of the sample correlation coefficient (r).

Extensions of regression to more than two variables exist. For the most part, these extensions are based on least squares techniques, also. Hypothesis tests and estimates for multiple regression are of the same general types as those discussed for the two-variable case.

See Self-Correcting Exercises 14.B

Exercises.

1. Perform an analysis of variance on the data in Exercise 1 of the previous set in Chapter 14 to test for significant linear regression. Do not find SSM nor test for nonlinearity. Compare this procedure with that used in Exercise 3 of the previous set, page 294.

2. Test the hypothesis that the linear regression model applies to the data in Exercise 1 of the previous set, page 293.

3. In Exercise 4 of the previous set in this chapter, page 294, data on length of employment and ability to assemble a component were presented. The supervisor has an employee who has been on the job 3 days and the supervisor wants a 0.95 interval estimate of this employee's assembly time for today, the fourth day on the job, before the employee begins work. Find the estimate.

4. The supervisor described in Exercise 3 wants a 0.95 interval estimate of the mean time per assembly for all workers during their fourth day on the job. Find the estimate, compare it with the answer to Exercise 3, and discuss the difference in the two estimates.

5. Is the assembly time sample in Exercise 4 of the previous set suitable for correlation analysis? Why or why not?

6. A personnel director in a large organization wants a correlation analysis of scores on a test of clerical aptitude (Y) and scores on a spelling test (X). A random sample of 10 persons who have taken both tests is selected and the partial results are:

$$\sum X = 70, \qquad \sum Y = 65, \qquad \sum XY = 472,$$
$$\sum X^2 = 624, \qquad \sum Y^2 = 533, \qquad n = 10.$$

Find the sample correlation coefficient (r) and interpret its algebraic sign. Would it have made any difference in the value of r if X and Y had been interchanged as symbols for the two variables?

7. Perform a t test of the hypothesis that no correlation is present between test scores in Exercise 6.

8. In a study of the relationship between education and income of salaried workers, the correlation coefficient was 0.70 for a sample of 103 persons. In an earlier study based on census data for the entire nation, the correlation coefficient was 0.85. Test the null hypothesis that the correlation coefficient is 0.85 in the population at present. Use $\alpha = 0.01$ and a two-tailed test.

Glossary of Equations

$Y_c = a + bX$

The sample least squares regression line has a Y intercept of a and a slope of b.

$$b = \frac{n(\sum XY) - (\sum X)(\sum Y)}{n(\sum X^2) - (\sum X)^2}$$

$$a = \frac{\sum Y - b(\sum X)}{n}$$

These usually are the most convenient equations to find the slope and intercept of the sample least squares regression line.

$$s_{Y \cdot X} = \sqrt{\frac{\sum (Y - Y_c)^2}{n - 2}}$$

The sample standard error of estimate is the sum of squared deviations of the values of the dependent variable from the sample least squares regression line divided by the number of observations in the sample less two.

$$\sum [(Y - Y_c)^2] = \sum Y^2 - a(\sum Y) - b(\sum XY)$$

The sum of squared deviations of the values of the dependent variable from the regression line can be found from the slope and intercept and certain sums from the sample observations.

$$r^2 = \left[1 - \frac{\sum (Y - Y_c)^2}{\sum (Y - \bar{Y})^2} \right]$$

The sample coefficient of determination is the portion of total variability in the observed values of the dependent variable that is attributable to the sample least squares straight line regression with the independent variable.

$$t = \frac{(b - \beta)\sqrt{\sum (X - \bar{X})^2}}{s_{Y \cdot X}}$$

The difference between the slopes of the sample regression line (b) and the population regression line (β) can be expressed as a t statistic with $n - 2$ degrees of freedom, where n is the sample size.

$$SST = SSR + SSM + SSW$$

$$SST = \sum_i \sum_j (Y_{ij}^2) - \frac{(\sum_i \sum_j Y_{ij})^2}{n}$$

$$SSR = (r^2)SST$$

$$SSW = \sum_i \sum_j (Y_{ij}^2) - \sum_j \left[\frac{(\sum_i Y_{ij})^2}{n_{\cdot j}} \right]$$

In a sample regression analysis with multiple observations of the dependent variable (Y) at more than two values of the independent variable (X), analysis of variance can be used to test for regression and nonlinearity. The total sum of squares is partitioned into a sum of squares for regression, a sum for treatment-level means and a sum within treatment levels.

$$Y_c \pm (t_{\alpha/2})(s_{Y \cdot x})\sqrt{1 + \frac{1}{n} + \frac{(X' - \overline{X})^2}{\sum (X - \overline{X})^2}}$$

The $(1 - \alpha)$ prediction interval for a single value of the dependent variable (Y) associated with a given value of the independent variable (X') is centered on the point prediction (Y_c) from the sample regression equation.

$$Y_c \pm (t_{\alpha/2})(s_{Y \cdot x})\sqrt{\frac{1}{n} + \frac{(X' - \overline{X})^2}{\sum (X - \overline{X})^2}}$$

The $(1 - \alpha)$ prediction interval for the mean value of the dependent variable $(\mu_{Y \cdot x})$ associated with a given value of the independent variable (X') is centered on the point prediction (Y_c) from the sample regression equation. The interval is narrower than the one for a single value (above).

$$r = \frac{n \sum XY - (\sum X)(\sum Y)}{\sqrt{[n \sum X^2 - (\sum X)^2]\,[n \sum Y^2 - (\sum Y)^2]}}$$

The sample correlation coefficient (r) can usually be found most conveniently with this equation.

$$t = \frac{r\sqrt{n - 2}}{\sqrt{1 - r^2}}$$

When there is no correlation in a bivariate normal population, sample correlation coefficients from samples of a given size (n) can be transformed into a t distribution with $n - 2$ degrees of freedom.

$$z = \frac{w - \mu_w}{\sigma_w}$$

Fisher's transformation can be used to convert the sampling distribution of correlation coefficients to a standard normal distribution when there are at least ten observations.

15

Nonparametric Statistics

The hypothesis tests discussed up to now usually have been *parametric* in form. In other words, the hypothesis to be tested was a statement about the value of a parameter such as the population mean or variance. The test used sample information to assess the plausibility of the hypothesized value. In this chapter we will consider *nonparametric* tests—that is, tests for which the hypotheses are not explicit statements about the value of some population parameter.

In our earlier hypothesis tests we usually made an assumption about the distribution of the population. In most cases the population was assumed to be normally distributed. By contrast, none of the tests we are about to discuss require an assumption about the distribution of the population. Such tests are said to be distribution free. Although it is not strictly correct, it is customary to refer to a test as nonparametric if it is either distribution free or nonparametric, or both.

Business and economics offer many opportunities for applying nonparametric tests. Sales and income distributions measured in money units usually exhibit strong positive skewness. Distribution-free tests often are appropriate in such cases. Marketing studies involving customers and behavioral studies involving an organization's personnel frequently permit measuring variables only on nominal or ordinal scales. Judgments of product adequacy and personnel performance ratings are cases in point. As compared with parametric tests, which are designed for use with interval and ratio scales, most nonparametric tests require no more than ordinal scaling. Nevertheless, nonparametric tests sometimes can be substituted for parametric tests.

Typically, nonparametric tests are not so powerful as their parametric counterparts, but a few of them are very nearly as powerful. We shall describe some of these. When there is reasonable doubt about the assumptions needed for a parametric test, a nonparametric test that is nearly as powerful if the assumptions apply, and broadly applicable if they do not, can be a most attractive alternative.

We have already encountered two nonparametric procedures in previous chapters. The first procedure made use of Chebyshev's inequality. We used it to perform tests about and make interval estimates of population means when we knew the value of the population variance. Although these uses were parametric, the inequality requires no assumption about the population distribution. Hence, Chebyshev's inequality is the basis for a set of distribution-free procedures. On several occasions we pointed out the loss in power that resulted when we used the inequality, as compared with a technique based on an assumption of normality for the population distribution.

The second nonparametric technique we encountered was the chi-square (χ^2) test for independence discussed in the latter portion of Chapter 12. A typical use of this test is to compare frequencies observed in the cells of a cross-classified table with those we would expect to find in the cells if the variables used to classify the data were independent. In such a test we do not ask about the value of a population parameter. Hence, the test qualifies as being nonparametric.

The Chebyshev inequality and chi-square tests, as well as those to be described in this chapter, are only a few of the many nonparametric tests developed in recent years. References 1 and 2 are devoted entirely to discussion of nonparametric tests.

Differences in Location for Two Populations

A test that is particularly sensitive to differences in location for two populations is the Wilcoxon rank sum test. The test is a distribution-free alternative to the two-sample t test we discussed in Chapter 11 for testing the hypothesis that two populations have the same mean [see Equation (11.10) and the related discussion]. The rank sum test, however, requires measurement only up to an ordinal (ranking) scale. A logical equivalent is the Mann-Whitney test, but we will discuss only the Wilcoxon test.

In the rank sum test, the null hypothesis states that the two populations from which the samples come are identical. The assumptions for the test are either independent random sampling from an infinite population or sampling with replacement between draws.

To illustrate the rank sum test, let us suppose that a new training program for salesmen has been in use for the past two years in a certain company. Composite performance ratings for the year following training are available for four salesmen trained under the new plan and six salesmen trained under the old. In the past, distributions of composite ratings have had marked positive skewness. Because of this and because ratings are subject to serious question with regard to the specifications for interval measurement, the two-sample t test is not applicable.

The rating scores for the ten salesmen, arranged by training program, are shown in Table 15.1. We want to test the null hypothesis (H_o) that both composite score

Table 15.1   Salesmen's Performance Ratings under
Two Training Programs

Program	Scores					
New	82	79	78	65		
Old	75	71	62	59	43	37

population distributions are identical against the upper one-tailed alternative (H_A) that the distribution of ratings under the new program is located in a higher range of composite scores than is the distribution under the old program. The level of significance (α) will be 0.025.

The first step in conducting the rank sum test is to rank all observations as if they belonged to a single set. By convention, ranking for this test usually is done from the smallest score to the largest. The ranks are next assigned to the applicable training program. Finally, the sum of the ranks assigned to the smaller sample is found. This sum constitutes the test statistic, W.

The steps just described have been carried out for the rating scores in our example and the results are

Program			Ranks				Sum
New	10	9	8	5			$W = 32$
Old	7	6	4	3	2	1	

The smallest score in the set of 10 is 37. Therefore, this score is assigned a rank of 1. The ranking continues through all ten numbers. The largest, 82, is given the highest rank. The rank numbers are next assigned to the applicable training program. Finally, the rank numbers for the smaller sample are summed to produce the value 32 for W, the test statistic.

We next consider how to interpret the value of the test statistic. Under the null hypothesis both programs are presumed to produce identical distributions of rating scores. Given any sample composed of four observations for the new program and six for the old, we can reason that under the null hypothesis any set of four of the ten rank numbers has the same probability of being assigned to the new program as does any other set of four. There are $_{10}C_4$, or 210, possible combinations of rank numbers, any one of which could be assigned to the smaller sample under the designated conditions. Of these, only rank combinations (10, 9, 8, 7), (10, 9, 8, 6), (10, 9, 8, 5), or (10, 9, 7, 6) would produce values of W (rank sums) equal to or greater than 32, the observed value. The probability of getting a value of 32 or greater for W under the null hypothesis is, therefore, 4/210, or 0.019. Since 0.019 is less than our preselected level of significance $(\alpha = 0.025)$, we reject the null hypothesis and accept the alternative. The new training program appears to be producing higher performance rating scores.

Table H gives critical *lower* tail values (W_α) of the rank sum for the n_1 observations in the smaller sample classified by the number of observations in the larger sample (n_2) and levels of significance (α). For instance, for a *lower* tail test with $n_1 = 4$, $n_2 = 6$, and $\alpha = 0.025$, the value of $W_{.025}$ is seen to be 12. To find critical *upper* tail values $(W_{1-\alpha})$ for one-tailed tests, we make use of the fact that the distribution of W for any values of n_1 and n_2 is symmetrical with respect to its mean. Given W_α, then $W_{1-\alpha}$ will be as far above the mean (μ_W) as W_α is below it. Algebraically, $W_{1-\alpha} = \mu_W + (\mu_W - W_\alpha)$, or

$$W_{1-\alpha} = 2\mu_W - W_\alpha. \tag{15.1}$$

For our rating-score example, n_1 is 4 and n_2 is 6. The table value of $2\mu_W$ for these values of n_1 and n_2 is 44. We have already seen that $W_{.025}$ is 12. Hence,

$$W_{.975} = 44 - 12, \quad \text{or } 32.$$

Under the null hypothesis, the probability is 0.025 *or less* that the rank sum will be as large as or larger than 32. Actually, from our earlier analysis we know the exact probability is 0.019, but it would be inconvenient to table exact probabilities for all combinations of n_1 and n_2. For representative values of α, the table shows values of W for which the probabilities are as large as possible without exceeding the selected values of α.

For two-tailed tests, the values of α shown in the table must be doubled to find the level of significance. If we had chosen to perform a two-tailed test on the rating scores, then $12 < W < 32$ would have been the acceptance region for a test at or below the 0.05 level of significance, rather than at or below the 0.025 level.

Table H presents critical values of W for sample combinations up to 12 and 25 for the smaller and larger samples, respectively. For values of n_1 and n_2 in which n_1 is 12 or greater, the rank sum is very nearly normally distributed with mean and variance

$$\mu_W = \frac{n_1(n_1 + n_2 + 1)}{2} \tag{15.2a}$$

and

$$\sigma_W^2 = \frac{n_1 n_2(n_1 + n_2 + 1)}{12}. \tag{15.2b}$$

As an illustration, let n_1 be 12, n_2 be 15, and α be 0.025. Then

$$\mu_W = \frac{12(12 + 15 + 1)}{2}, \quad \text{or } 168,$$

and

$$\sigma_W^2 = \frac{12(15)(28)}{12}, \quad \text{or } 420,$$

from which

$$\sigma_W = \sqrt{420}, \quad \text{or } 20.494.$$

If we assume that W is normally distributed, we can find the value that cuts off a portion of the lower tail equal to α from the relationship

$$W_\alpha = \mu_W + z_\alpha \sigma_W.$$

When α is 0.025, we know from our earlier work that z is -1.96. Then

$$W_{.025} = 168 + (-1.96)(20.494) = 127.832.$$

As with the normal approximation to the binomial the approximation for W_α is improved if we make a continuity correction by subtracting 0.5 from the above result. For the illustration, the corrected approximation is 127.332, or 127 to the nearest whole number. This is exactly the same as the number which appears for $n_1 = 12$, $n_2 = 15$, and $\alpha = 0.025$ in Appendix H.

Under rather broad variations in assumptions about the shape of the populations, the rank sum test performs very effectively as compared with alternatives. For instance, when all conditions are met for the two-sample t test mentioned earlier in this chapter, the rank sum test has an efficiency of 0.95 compared to 1.00 for the t test. In other words, a sample of 106 observations with the Wilcoxon test is as sensitive as a sample of 100 with the t test. Since the rank sum test does not require assumptions that are often difficult to make, it is more widely applicable than the two-sample t test.

The theory on which the distributions of rank sums is based assumes that no ties can occur. Effort should be made to measure with enough precision before ranking to prevent tied ranks. When ties do occur, however, one of two situations may arise. Suppose that in Table 15.1 the score in the top row had been 79 rather than 78. The two scores of 79 would account for ranks 8 and 9. The total value of 32 for W will remain the same no matter which of the two scores is assigned either rank. In this first situation, the tie causes no special difficulty.

To illustrate the second type of tie, we can replace the score of 78 in the top row of Table 15.1 with 75. It is now tied with the highest score in the second row, and these two scores share ranks 7 and 8. In this circumstance, we first assume that the tied score in the top row is smaller than its counterpart in the second. Rank 7 would fill in the top row, and W would be 31. We then assume that the reverse is true, so that rank 8 falls in the top row and W is 32. If both values of W fall in the rejection region or if both fall in the acceptance region for the null hypothesis, the issue is resolved. If one value of W leads to acceptance and the other to rejection, we must call the result inconclusive and repeat the experiment or reconsider our judgment about the level of significance.

The Sign Test for Matched Pairs

The *sign test* is valuable for those occasions in which two treatments are to be compared under a variety of conditions. For example, two grievance procedures for dealing with several standard types of complaint in a work group are to be rated, or two types of lawn seed are to be compared with respect to the qualities of the lawn they produce under a variety of growing conditions. When independent pairs of observations can be made, when each pair is matched with respect to relevant conditions other than the treatment, and when pairs are observed under different conditions, the sign test can be highly effective. Observing pairs under widely varying conditions can invalidate the t test for use with matched pairs, discussed in Chapter 11. The normality assumption may not be justified.

To illustrate the sign test, we shall compare two different types of house paint for wearing quality. Adjacent patches of the two types are painted on each of a number of houses in different parts of the country. Thus the patches of each pair are matched as to color, the climate and chemical pollutants in the environment, and many other possible conditions that could affect wear. The differences in environmental conditions among pairs are great. The months of satisfactory service for each patch are recorded in Table 15.2. In addition to the service lives, the table shows the sign of the differences when the service life of B is subtracted from that of A. Because of some unusually long pairs of service lives, the distribution of difference in service might reasonably be expected to have considerable positive skewness.

Table 15.2   Service Months for Two Paint Types
under Matched Conditions

Pair	Paint A	B	Sign of difference	Pair	Paint A	B	Sign of difference
1	19	22	−	9	20	21	−
2	27	26	+	10	38	41	−
3	13	14	−	11	21	23	−
4	8	18	−	12	19	20	−
5	28	30	−	13	8	10	−
6	7	6	+	14	12	16	−
7	27	28	−	15	21	26	−
8	18	19	−	16	13	8	+

As in the rank sum test, the sign test assumes that ties cannot occur within any pair. Should a tie occur, however, the difference for the pair will be zero and no sign can be attached. Exclude such ties from the analysis and reduce the number of paired observations by the number excluded.

The null hypothesis for the house paint example is that the median service lives of the two types of paint are equal under any one of the 16 sets of conditions. Under this hypothesis the probability of observing a negative difference in service lives for any pair is equal to the probability of observing a positive difference, and both are equal to 0.5. We can regard the sample as 16 independent draws from a binary population of + and − signs, wherein a + sign has a constant probability of occurrence equal to 0.5 for every draw. These are the specifications for a binomial sampling distribution, discussed in Chapter 6 and again in Chapter 12. If we let the number of pairs be the number of observations (n), then r is the number of + signs in the sample. The parameter (π) of the binary population will be 0.5 under the null hypothesis, and we can use Appendix E to find relevant probabilities for 16 observations or fewer.

Suppose that the paint manufacturer in our example wants to continue making paint A unless there is conclusive evidence that paint B is superior. A lower one-tailed test is called for, and an appropriate hypothesis statement is

$$H_O: \pi \geq 0.5,$$
$$H_A: \pi < 0.5,$$

where π is the probability that a + sign occurs. The selected level of significance (α) is 0.025.

In the experiment composed of 16 observations, three plus signs occurred. Entering Appendix E with n and π values of 16 and 0.5, respectively, we find that the probabilities for r values of three or fewer are 0.0085, 0.0018, 0.0002, and 0.0000. The sum of these probabilities is 0.0105. Hence, the probability of observing three or fewer plus signs under the null hypothesis is only 0.0105. This is less than the preselected value of $\alpha(0.025)$. Consequently, we accept H_A and conclude that paint B has demonstrated superior service life characteristics.

For values of n larger than 16, the normal approximation to the binomial can be used. This approximation and its application were discussed under the heading "Hypothesis Tests for a Single Proportion" in Chapter 12.

Sample size is important in the sign test. When the number of observations is very small, there may be no way to reject the null hypothesis for the designated level of significance. If n is 3, for example, the probabilities for possible values of r are all 0.125 or greater. Also, in comparison to the t test the information regarding the signs of differences is used, but the information regarding the amounts of differences is ignored. This results in a loss of power that must be compensated with additional observations.

Modifications and extensions of the sign test exist. Pairs need not be matched to compare two treatments, for instance. This and other variations are described in the references cited in the introduction to this chapter.

Summary

The Wilcoxon rank sum test applies to two sets of data measured on a common ordinal scale. It tests the hypothesis that both sets of data are from the same population and is highly sensitive to differences in central location of the populations. This test is distribution free and nearly as sensitive as the t test when the latter applies. The Wilcoxon test applies in many more situations than does the t test.

The sign test for matched pairs is a more broadly applicable, but less sensitive, alternative to the t test for matched pairs. The form of the sign test discussed here tests the null hypothesis that the medians of the populations from which each pair came are equal.

See Self-Correcting Exercises 15.A

Exercises

1. The impact strengths of samples of two materials being considered for milk cartons are given below. At the 0.01 level, test the hypothesis that A is stronger than B, against the alternative that B is at least as strong.

Material A	.95	.98	.99	.96	.90	.89	.92	.94
Material B	.93	.88	.91	.83	.87	.86	.82	.79

2. Measurements of the abrasive materials contained in two brands of toothpaste yielded the following results (in milligrams):

Brand A	26	23	25	24	28
Brand B	19	22	24	21	20

    At the 0.05 level, test the null hypothesis that Brand B has at least as much abrasive material per tube as does Brand A.

3. The clerical force in a large office was divided into two groups to test the effectiveness of two new makes of typewriter. Fifteen typists used typewriter Brand A and eighteen used Brand B. When the productivities of all 33 machines were ranked, with the highest productivity being given rank 1, the sum of the ranks for Brand A was 171. At the 0.01 level, make a two-tailed test of the hypothesis that there is no difference in productivity for the two brands.

4. In the "before and after" test of the reducing regimen described in Exercise 1, page 240, test the null hypothesis that adult males weigh at least as much after the regimen as before. Use a matched-pairs signed rank test and a 0.05 level of significance.

5. For Exercise 2, page 240, does the matched-pairs signed rank test support the conclusion that productivity improved? Use a one-tailed test and let α be 0.05.

6. Forty automobiles were entered in a matched-pairs signed rank test of mileages from two brands of gasoline. For each automobile, the mileage with Brand A was compared with the mileage when Brand B was used. At the 0.01 level, make a two-tailed test of the null hypothesis that there is no difference in mileages if Brand B got better mileage in 32 of the automobiles. Don't forget the continuity correction.

7. In Exercise 4, discuss the reasons why the matched-pairs sign test resulted in a conclusion different from that of the original t test in Chapter 11.

Differences in Location for Many Populations

The Kruskal-Wallis test, a direct extension of the Wilcoxon rank sum test for two populations, is sensitive to differences in location for two or more populations. Hence, it is a distribution-free counterpart of one-way analysis of variance as described in Chapter 13. If the assumptions of normality and homogeneity of variances applicable to one-way analysis of variance are met, the Kruskal-Wallis test has an efficiency of 0.955 relative to the F test. For marked departures from these assumptions, the Kruskal-Wallis test is to be preferred to the F test. Consequently, it provides a broadly applicable and highly effective test for differences in central locations for treatment levels.

When applied to just two populations, the Kruskal-Wallis test is identical to the Wilcoxon rank sum test. Therefore, to illustrate it, we shall consider a situation with three populations. A firm that grows rose bushes on a mass production basis for sale to nurseries wants to determine which of three climates produces the best bushes. Nineteen bushes of comparable ages and varieties are used in the experiment. Six bushes come from region A, five from region B, and eight from C. A panel of qualified judges scores each bush on a composite of characteristics. Then, without regard to the region from which they came, the nineteen bushes are ranked on the basis of the judges' scores, the smallest score, 1, being assigned to the poorest bush. The results appear in Table 15.3. Below the rankings, are the quantities needed to calculate the value of the test statistic, H. The first row shows the sum of the ranks for each region (T_j); the second row contains the squares of the numbers in the first row (T_j^2); the third row is the number of observations in each treatment (n_j)—the total number of observations in the experiment (n) is, of course, the sum of the values in the third row; the final row is the square of the sums in each column divided by the number of observations in that column (T_j^2/n_j). The test statistic (H) is defined as follows

$$H = \frac{12}{n(n + 1)} \left(\sum \frac{T_j^2}{n_j} \right) - 3(n + 1). \tag{15.3}$$

Table 15.3   Composite Score Ranks of Nineteen Plants

| | Growing region | | |
	A	B	C
	2	10	1
	3	11	5
	4	15	8
	6	16	9
	7	19	13
	12		14
			17
			18
Sum (T_j)	34	71	85
Sum squared (T_j^2)	1156	5041	7225
Number of observations (n_j)	6	5	8
T_j^2/n_j	192.7	1008.2	903.1

In this expression, recall that n is the total number of observations in the experiment. The rationale leading to this equation is given below. For our example,

$$H = \frac{12}{19(20)} [192.7 + 1008.2 + 903.1] - 3(20),$$

H is distributed as χ^2 with K-1 df

or

$$H = 6.44.$$

Sound procedure requires that we state the null hypothesis before we look at this or any set of observations. Then, given the hypothesis statement, we must consider how the test statistic (H) will be distributed under the null hypothesis. Our null hypothesis will be that the populations of scores for each of the three growing regions have identical distributions. They are then, in effect, all the same population. Now imagine that we repeat our experiment a very great number of times when the null hypothesis is true. We repeatedly select 19 ranked scores at random from the population and always assign six of the rank numbers at random to the first column, five to the second, and the remaining eight to the third.

Consider any one column. Each of the n_j rank numbers in that column can be viewed as a random selection from a rectangular population of the numbers 1 through 19, sampled without replacement between draws. The mean (μ) of a rectangular population composed of consecutive integers 1 through n is $(n + 1)/2$ and the variance (σ^2) is $(n - 1)(n + 1)/12$.

Now consider the sum (T_j) of the n_j rank numbers in a given column. The mean (μ_T) of the distribution of this sum will be the product $n_j\mu$, where μ is $(n + 1)/2$. The variance (σ_T^2) will be $n_j \sigma^2 (n - n_j)/(n - 1)$, where σ^2 is $(n - 1)(n + 1)/12$, and $(n - n_j)/(n - 1)$ is the correction for sampling without replacement. By the central limit theorem, as n_j grows large, the distribution of the variable T_j approaches normality. Furthermore, the distribution of

$$z = \frac{T_j - \mu_T}{\sigma_T}$$

approaches the standard normal distribution. Suppose we square z for each of the k columns in the experiment. Then it can be shown that the sum of the squared z values is distributed very nearly as chi-square (χ^2) with $k - 1$ degrees of freedom provided there are at least five observations in each of three treatments (columns). The statistic H in Equation (15.3) is an algebraic transformation of the sum just described. This transformation is a convenient form of the test statistic for calculation purposes.

What happens if the null hypothesis is not true? Specifically, consider the case in which one of the treatment means is very much smaller than the other means. The other treatments will, on the average, contain more large rank numbers than they would under the null hypothesis, and their values of T_j also will be large. Conversely, T_j will tend to be unusually small for the one treatment we mentioned. When we square the values of T_j, those that are unusually large will produce squares that more than compensate for the square of the unusually small sum. Consequently, the value of H will tend to be much larger than it would be under the null hypothesis. The clue to a departure from the null hypothesis is an unusually large value of H.

Returning to our example, we have a sample value of H equal to 6.44. There are five or more observations in each of the minimum of three treatments. We can, therefore, employ the chi-square (χ^2) approximation with two degrees of freedom. If we select the 0.05 level of significance, Table C shows that 5.99 is the critical value. Because our sample value of H exceeds the critical value, we reject the null hypothesis of equal growing effectiveness for the three regions.

When we examine the mean squared sums (T_j^2/n_j) in the last row of Table 15.3, the quantity for region A is very small relative to the other two regions, which seem to be about on a par with one another. A preponderance of low ranks in column A stems from low scores and poor rose bushes. Hence, it would appear that region A produces considerably poorer bushes than do the other two regions.

In some situations there will be fewer than five observations for at least one of the minimal three treatments needed for the chi-square approximation. Reference 3a presents tables based on exact distributions of the H statistic for small samples that can be used in such situations.

This test, like the Wilcoxon rank sum test assumes that the underlying variable is continuous, so that ties cannot occur. In practice, of course, ties do occur. When they do, they should be handled in the same manner described for the Wilcoxon test earlier in this chapter.

Rank Difference Correlation

In Chapters 3 and 14 we discussed the Pearson correlation coefficients (ρ and r) for a population and for a sample. This was seen to be an effective measure of association between pairs of values for two variables measured on interval or ratio scales. When the statistical population of paired values has a bivariate normal distribution, the sampling distribution of r is known and inferences about population values of ρ are possible.

When observations are judgments, attitudes, preferences, or aspects of behavior, only ordinal scales may apply. In other instances, we may not be able to support the assumption of bivariate normality. The Spearman rank difference correlation statistic (r_s) often is an effective alternative in these circumstances.

Let us consider an industrial training program in which employees are selected and trained at company expense. In an effort to improve the program a test designed to measure motivation was given to 12 trainees before they entered the program. At the end of the program, the comprehensive course scores were collected for these same people. If there is a strong relationship between these two measures—motivation and scores—the test may be helpful in selecting trainees. The null hypothesis is that there is no correlation between motivation upon entering the course and performance upon leaving it. The distribution of scores for several hundred trainees exhibits rather marked negative skewness. Hence, rank difference correlation is chosen to test for a relationship between motivation and performance.

After ranking each set of scores for the 12 trainees, we have recorded the results in Table 15.4. The fourth column shows the difference between the ranks of the pairs.

Table 15.4   Ranks of Trainees on Two Tests

	Ranks			
Trainee	*Motivation*	*Performance*	*Difference (d)*	*d^2*
1	3	7	−4	16
2	7	2	5	25
3	1	4	−3	9
4	11	9	2	4
5	2	1	1	1
6	9	12	−3	9
7	6	5	1	1
8	4	6	−2	4
9	5	3	2	4
10	10	8	2	4
11	12	11	1	1
12	8	10	−2	4
				$\Sigma d^2 = 82$

The fifth column shows the squared differences and the sum of the squared differences ($\sum d^2$). We can find the value of the rank difference correlation coefficient (r_s) from $\sum d^2$ and n, the number of paired observations. In Table 15.4, n is 12. The computation equation for the Spearman rank difference correlation coefficient is

$$r_s = 1 - \frac{6 \sum d^2}{n(n^2 - 1)}. \qquad (15.4)$$

For our example,

$$r_s = 1 - \frac{6(82)}{12(143)}, \quad \text{or } +0.713.$$

As with the Pearson correlation coefficient, the Spearman correlation coefficient can range from −1 through 0 to +1. A value of 0 indicates no relationship, and sample values of either +1 or −1 suggest a strong relationship in the population. In our example, if we assume that low numbers were assigned to trainees with poor motivation and poor performance, then + 0.713 suggests that poor motivation at the outset of the program is followed by poor performance, and *vice versa*.

The null hypothesis states that there is no association between motivation and performance in the population. Then, for a given sample set of n rank scores for motivation there are $n!$ equally likely arrangements of n rank scores for performance. Each arrangement will produce a value for the rank difference correlation coefficient (r_s). The number of such arrangements that produce a given value of r_s can be divided by $n!$ to obtain the probability for that value under the null hypothesis. When this is done for all possible values of r_s, the result is the sampling distribution of the test statistic r_s. The distribution of r_s ranges from -1 to $+1$ and is symmetrical for any sample size. As sample size increases, values of r_s tend to become more concentrated around zero and the probabilities of values in the vicinity of $+1$ and -1 become very small.

For small numbers of paired observations, References 1b and 3b present applicable tables. When there are 10 or more pairs, however, and when we want to test the hypothesis of no correlation in the population, the statistic

$$t = r_s \sqrt{\frac{n - 2}{1 - r_s^2}} \tag{15.5}$$

can be assumed to come from a t distribution with $n - 2$ degrees of freedom.

For our example, we obtained a sample rank difference correlation coefficient (r_s) of $+0.713$ from 12 pairs of ranks. We want to test the null hypothesis that the population rank difference correlation coefficient (ρ_s) is 0 or negative against the one-tailed alternative that ρ_s is positive. The level of significance (α) will be 0.01. In Appendix B we read that $t_{.99}$ is 2.764 for 10 degrees of freedom ($n - 2$). This is the critical value for the test statistic. From Equation (15.5) we find that

$$t = r_s \sqrt{\frac{n - 2}{1 - r_s^2}},$$

$$= 0.713 \sqrt{\frac{10}{1 - 0.713^2}}, \quad \text{or } +3.22.$$

We can reject the null hypothesis and accept this result as evidence of positive association between motivation and performance in the population.

It can be shown that the Pearson correlation coefficient and the Spearman rank difference correlation coefficient are exact algebraic equivalents for ranked data. When the bivariate normality assumptions described in Chapter 14 are satisfied, the efficiency of the Spearman coefficient applied to ranked data is 0.91 relative to the Pearson coefficient applied to the data as measured on the original interval or ratio scale.

As with the other ranking techniques in this chapter, the assumptions underlying rank difference correlation prevent ties in ranks. Since ties do occur in practice, however, the procedure here is the same as that discussed in conjunction with the Wilcoxon rank sum test.

Summary

The Kruskal-Wallis test applies to two or more sets of data measured on a common ordinal scale. It tests the hypothesis that all of the sets come from populations with the same central location and can substitute for one-way analysis of variance when the necessary conditions for the latter do not hold. Even when the conditions hold, the Kruskal-Wallis test is almost as sensitive as the F test. When there are only two sets of data, the Kruskal-Wallis test is equivalent to the Wilcoxon rank sum test discussed earlier in this chapter.

Rank difference correlation can be used to test the hypothesis that a sample of paired rank numbers came from a bivariate population in which there is no correlation. The test is more broadly applicable than the one based on the Pearson coefficient and is nearly as sensitive as the latter when the necessary conditions are met.

It should be reemphasized that this chapter considers only four from a very large number of nonparametric tests. References to others have been cited throughout the chapter.

See Self-Correcting Exercises 15.B

Exercises

1. Transparent sacks, paper bags, and bulk display were the three methods used to market oranges in 20 supermarkets, all belonging to the same chain. The same grade of oranges was sold at the same price in all markets during the test period. Sales in pounds per day for each type of display were:

Transparent sacks	Paper bags	Bulk display
36	29	16
40	30	26
47	45	17
49	39	22
23	13	28
41	32	
38	18	
	31	

Test the null hypothesis that all 3 displays result in the same mean sales per day by using the Kruskal-Wallis procedure. What display would you select as a result of this test?

2. Suppose that the first observation in the left column in Exercise 1 is 39 instead of 36. What change in procedure is indicated to conduct the same test? Discuss fully.

3. A panel of housewives and a panel of college students rated 12 television programs. The mean ratings for the programs were:

Program	Housewives' rating	Students' rating
1	61	86
2	55	70
3	83	58
4	42	87
5	31	52
6	64	48
7	69	78
8	78	92
9	92	67
10	94	82
11	70	75
12	28	59

Calculate the Spearman rank difference correlation coefficient as a measure of the consistency of the two rating efforts.

4. Test the null hypothesis that there is no consistency in the ratings by the two groups in Exercise 3.

5. A group of party workers and a group of independent voters were asked to rank four candidates for office. The two sets of ranks are:

Candidate	A	B	C	D
Party	2	1	4	3
Independents	1	3	4	2

Calculate the Spearman correlation coefficient.

6. Apply Equation (14.14) to the two sets of ranks in Exercise 5 to find the Pearson correlation coefficient.

7. Compare the results of Exercise 5 and 6. What general conclusion is suggested?

Glossary of Equations

$$\mu_W = \frac{n_1(n_1 + n_2 + 1)}{2}$$

$$\sigma_W^2 = \frac{n_2 \mu_W}{6}$$

The mean (μ_W) of the Wilcoxon rank sum statistic for two identical populations is a function of the number of observations in the smaller (n_1) and larger (n_2) samples. The variance (σ_W^2) can be found from the mean. The distribution is essentially normally distributed when n_1 is 12 or greater.

$$H = \frac{12}{n(n+1)}\left(\sum \frac{T_j^2}{n_j}\right) - 3(n+1)$$

The test statistic (H) for the Kruskal-Wallis test for differences in location of three or more populations is a function of the total number of observations in the experiment (n), the sum of the ranks (T_j) in each treatment and the numbers of observations in each treatment (n_j).

$$r_s = 1 - \frac{6\sum d^2}{n(n^2 - 1)}$$

The Spearman rank difference correlation coefficient (r_s) is found from the sum of the squared rank differences ($\sum d^2$) for the paired observations and the number of pairs (n).

$$t = r_s \sqrt{\frac{n-2}{1-r_s^2}}$$

When there is no association between the two variables in the population, the sample rank difference correlation coefficient (r_s) is distributed approximately as t with $n-2$ degrees of freedom for 10 or more paired observations.

16

Sample Survey Methods

Up to now, our methods for statistical inference have all assumed that sample statistics were generated by probability processes in which the successive draws from the underlying population were independent of one another. Confidence intervals, tests of hypotheses, and so forth, have assumed random sampling. In the designed experiments of Chapter 13, this meant that experimental materials were assigned to treatment groups at random. In other contexts, there was an identifiable universe of units of observation, from which units were randomly selected for inclusion in a sample study. Now we shall look in more detail at selection procedures. When a universe of units of observation is to be surveyed, we need a selection method which insures that the survey statistics are the result of a genuine probability process. Such methods are called *probability sampling* techniques. We will examine a number of probability sampling techniques besides simple random sampling. We will look first at the theory for each method along with an abbreviated example. In the second part of the chapter we consider the design of sample surveys and how to carry them out.

The distinctive feature of probability sampling methods is that they permit a quantitative statement of sampling error, which may be the standard error of the mean or the standard error of a proportion, depending on the kind of data collected in the survey.

The map in Figure 16.1 shows all of the occupied dwelling units in a hypothetical village called Georges Mills. Table 16.1 indicates whether each dwelling unit is owner- or renter-occupied, and the most recent week's grocery bill for the occupant family. The list, which covers completely the universe of units of observation, will be used to illustrate the principles of several kinds of probability sampling methods.

Table 16.1   Listing of Dwelling Units with Ownership Status*
and Weekly Grocery Bill

1		$49.6	31		$67.1
2		45.2	32		58.2
3	R	41.5	33	R	35.6
4	R	28.0	34		52.6
5	R	41.2	35		56.4
6	R	29.8	36		52.9
7	R	38.4	37	R	41.3
8		52.8	38	R	35.1
9	R	59.1	39	R	61.1
10	R	31.3	40	R	48.8
11		36.3	41		66.0
12	R	45.3	42		52.6
13		57.0	43	R	44.5
14		60.8	44	R	48.2
15	R	30.4	45		65.4
16		59.8	46		55.2
17	R	45.1	47	R	52.2
18		43.2	48		72.5
19		48.2	49		60.7
20	R	31.6	50		71.2
21	R	35.8	51		57.3
22	R	26.0	52		60.2
23		51.9	53	R	53.6
24		33.1	54		71.6
25	R	30.4	55		55.0
26		55.4	56		58.8
27		57.0	57	R	47.5
28		43.5	58	R	49.0
29		50.0	59		63.1
30		53.9	60		62.1

* R indicates renter-occupied, all others are owner-occupied.

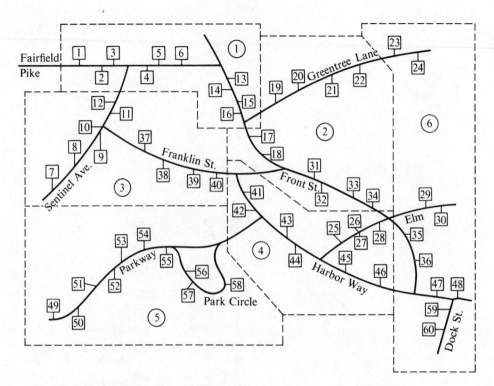

Figure 16.1 Dwelling Unit Map for Village of Georges Mills

Simple Random Sampling

Definition | A **simple**, or **unrestricted**, **random sample** of n units from a universe of N units of observation is defined as a sample resulting from a process in which each possible combination of n out of N units has an equal probability of selection.

The definition above is phrased in terms of sampling without replacement. In the village of Georges Mills we have $N = 60$ units of observation, or households. If we were contemplating a sample of $n = 5$ households, a simple random sample is one in which each of the $_{60}C_5 = 5,461,512$ combinations of 60 households taken 5 at a time has an equal probability of comprising the ultimate sample. The formulas we have dealt with so far have applied to sampling with replacement. For example the basic formula we have used for the standard error of a sample mean is

$$\sigma_{\bar{X}} = \frac{\sigma_X}{\sqrt{n}}.$$

When sampling is conducted without replacement, a factor called the *finite popula-tion modifier* is applied in finding the standard errors of sample statistics. In sampling without replacement the standard error of the mean is

$$\blacksquare \qquad \sigma_{\bar{X}} = \frac{\sigma_X}{\sqrt{n}} \sqrt{\frac{N-n}{N-1}}. \qquad \blacksquare \qquad (16.1)$$

The fraction under the radical is the finite population modifier. The effect of sampling without replacement is to reduce the standard error. For example, the variance of the 60 expenditure figures for the households in Georges Mills is 137.3. If we were to take a sample of size $n = 12$ *with replacement*, the standard error of the sample mean is

$$\sigma_{\bar{X}} = \frac{\sigma_X}{\sqrt{n}} = \frac{\sqrt{137.3}}{\sqrt{12}} = \sqrt{11.4} = 3.38.$$

If the sample is taken *without replacement*, we have

$$\sigma_{\bar{X}} = \frac{\sigma_X}{\sqrt{n}} \sqrt{\frac{N-n}{N-1}} = 3.38 \sqrt{\frac{60-12}{60-1}} = 3.38\sqrt{0.814} = 3.05.$$

When a sample drawn without replacement is a small fraction of the units in the universe, the finite population modifier, while still correct, is often omitted. For a sample of 1000 households in a city of 50,000 households, the modifier would be

$$\sqrt{\frac{50,000 - 1,000}{50,000 - 1}} = \sqrt{\frac{49,000}{49,999}} = \sqrt{0.980} = 0.990.$$

For the samples drawn from our small illustrative population, we will include the finite population modifier. It can be omitted with but small effects when the sample size is less than 5 percent of the population size.

Drawing A Simple Random Sample

To draw a *simple random sample* we must number all the units of observation in the universe. On our map the dwelling units are numbered. A list of addresses numbered from 01 to 60 would do as well. We then enter the random number table (Appendix F) and read successive two-digit numbers to select our sample. We will select a sample of $n = 12$ dwelling units. Only the numbers 01–60 will draw a dwelling unit. Should we encounter 00 or any two-digit number in the range 61–99, we will simply pass along to the next two-digit number. Table 16.2 shows the sequence of random two-digit numbers found in the first row of Appendix F. The third two-digit number, 73, does not draw one of the 60 dwelling units. Notice that the thirteenth number, 35, has already been drawn (the ninth number). It is ignored and we continue to read two-digit numbers until twelve nonrepeating numbers between 01 and 60 have been drawn.

The mean and variance of the expenditure data for the sample are shown in Table 16.2. We must now use the sample data to estimate the standard error of the mean.

$$s_{\bar{X}} = \frac{s_X}{\sqrt{n}} \sqrt{\frac{N-n}{N-1}}, \qquad (16.2)$$

$$s_{\bar{X}} = \frac{\sqrt{166.2}}{\sqrt{12}} \sqrt{\frac{60-12}{60-1}} = 3.36.$$

Table 16.2   Selection of a Random Sample of Twelve
Dwelling Units and Calculation of Mean and Variance
of Grocery Expenditures for the Sample

Random number	X
10	31.3
09	59.1
73	—
25	30.4
33	35.6
76	—
52	60.2
01	49.6
35	56.4
86	—
34	52.6
67	—
35	repeat
48	72.5
76	—
80	—
95	—
90	—
91	—
17	45.1
39	61.1
29	50.0

$\Sigma X = 603.9$

$\Sigma X^2 = 32219.41$

$(\Sigma X)^2/n = 30391.27$

$\Sigma x^2 = 32219.41 - 30391.27 = 1828.14$

$s^2 = 1828.14/11 = 166.2$

$\bar{X} = 603.9/12 = \$50.32$

To form a confidence interval for the population mean based on the sample data, we use the Student t distribution. This involves an assumption that the X population of expenditures is normally distributed. For a 95-percent confidence interval we find the proper t value for $m = n - 1 = 11$ degrees of freedom to be 2.201. The interval then is

$$\$50.32 \pm 2.201 \ (\$3.36), \quad \text{or} \quad \$50.32 \pm \$7.39.$$

The actual mean of the 60 expenditure figures is \$49.79, and our interval estimate is, in fact, correct. However, in taking a sample of size $n = 12$ and basing our estimate wholly on sample information, we have put ourselves in a role where we do not know any parameter values. If the assumption of normality seems unduly restrictive, recall from Chapter 9 that when n is large, that is, $n \geq 50$, one may calculate $\bar{X} \pm z_{\alpha/2} s_{\bar{X}}$ without having to be concerned about normality of the X population. This would be the more commonly encountered practical situation.

Stratified Random Sampling

In *stratified random sampling*, the universe of units of observation is divided into groups called strata and simple random samples are drawn separately from each stratum. A complete listing is made as in unrestricted (nonstratified) random sampling. In addition, the characteristic on which the stratification is based must be known for each unit in the universe. The basis for stratification will depend on the nature of the survey; common bases are sex, geographic area, occupational class, political party, and so forth. If the ratio of units drawn to total units is the same for each stratum, the term *proportionate stratified random sampling* is applied. This ratio is called the *sampling fraction*, or *sampling ratio*.

To illustrate proportionate stratified random sampling let us return to Georges Mills. Before selecting a sample (again of 12 dwelling units) we have determined for each dwelling unit whether it is owner-occupied or rented. It occurs to us that there may well be an association between home ownership and expenditures on groceries. Specifically we think that home owners may be more affluent and spend more for groceries on the average than do renters. Therefore it would seem that we should take care to have owners and renters included in the sample in the same respective proportions that they represent in the population.

Table 16.1 shows that there are 35 owner-occupied dwelling units and 25 renter-occupied units. A proportionate stratified random sample will require that $35/60 = 7/12$ of the units drawn be owner-occupied and $25/60 = 5/12$ be renter-occupied. For a total sample of 12 units this means 7 owner-occupied and 5 renter-occupied units. We now divide the universe into the two strata and select simple random samples of the required size from each stratum.

Definition	A **proportionate stratified random sample** is a sample in which the numbers of observation units drawn from each stratum are proportional to their respective numbers in the universe. The resulting subsamples are drawn on a simple random basis from the respective strata.

In Table 16.3(a) the original listing of observational units has been segregated by

Table 16.3(a)   Renumbering of Dwelling Units by Strata

Owner-occupied				Renter-occupied			
New number	Old number	New number	Old number	New number	Old number	New number	Old number
1	1	19	34	1	3	19	40
2	2	20	35	2	4	20	43
3	8	21	36	3	5	21	44
4	11	22	41	4	6	22	47
5	13	23	42	5	7	23	53
6	14	24	45	6	9	24	57
7	16	25	46	7	10	25	58
8	18	26	48	8	12		
9	19	27	49	9	15		
10	23	28	50	10	17		
11	24	29	51	11	20		
12	26	30	52	12	21		
13	27	31	54	13	22		
14	28	32	55	14	25		
15	29	33	56	15	33		
16	30	34	59	16	37		
17	31	35	60	17	38		
18	32			18	39		

Table 16.3(b)   Selection of a Proportionate Stratified Random Sample of Twelve Dwelling Units and Calculation of Mean and Variance of Grocery Expenditures for Each Stratum Subsample

Random number	Owners X_1	Random number	Renters X_2
27	60.7	05	38.4
49 (14)	43.5	64 (14)	30.4
45 (10)	51.9	89 (14)	Repeat
37 (2)	45.2	47 (22)	52.2
54 (19)	52.6	42 (17)	35.1
20	56.4	96 (21)	48.2
48 (13)	57.0		

ΣX_1	$= 367.3$	ΣX_2	$= 204.3$
$\bar{X}_1$	$= 52.47$	$\bar{X}_2$	$= 40.86$
ΣX_1^2	$= 19510.11$	ΣX_2^2	$= 8678.81$
$(\Sigma X_1)^2/n_1$	$= 19272.76$	$(\Sigma X_2)^2/n_2$	$= 8347.70$
Σx_1^2	$= 237.35$	Σx_2^2	$= 331.11$
s_1^2	$= \dfrac{237.35}{6} = 39.56$	s_2^2	$= \dfrac{331.11}{4} = 82.78$

owner and renter categories and the dwelling units renumbered. We now select 7 owner-occupied units and 5 renter-occupied units, and the results are presented in Table 16.3(b). The random numbers there begin with the 23rd two-digit number in the first row of Appendix F (our previous sample selection ended with the 22nd number). To avoid having to pass over a large number of nonapplicable two-digit random numbers, let random numbers in the interval 36–70 as well as 1–35 select owner-occupied units. For rented units numbered 1–25 let the random numbers 26–50, 51–75, and 76–00 select a numbered dwelling unit as well as the random numbers 01–25. For example, the first random number selected for owners is 27, and the 27th owner-occupied unit is number 49 from the original list in Table 16.1. The second random number drawn is 49. This selects the 14th owner-occupied unit in the list, since 49 is the 14th number in the interval 36–70. The 14th owner-occupied unit is dwelling unit number 28 in the original listing, with a weekly grocery bill of $43.5.

Theory for Stratified Sampling

Having illustrated the process of drawing a (proportionate) stratified random sample and secured our sample data, we now consider the *why* of stratified sampling. What do we hope to accomplish? By sampling from individual strata we hope to accomplish a reduction in the standard error of the mean as compared with an equal size simple random sample. The expected reduction in sampling error is brought about by eliminating the variation among stratum means as a source of sampling error. The greater the differences among the stratum means, the greater will be the expected reduction in sampling error brought about by stratification. This can be illustrated most clearly by analyzing our population data for the 60 households.

The population stratum means are \$55.90 for the 35 homeowners and \$41.23 for the 25 renters. Thus our intuition that homeowners spend more for groceries than renters is confirmed. Since there are 35 homeowners and 25 renters, the mean of the 60 expenditure values can be obtained by

$$(35/60)(\$55.90) + (25/60)(\$41.23) = \$49.79.$$

After partitioning each observation into a deviation from its stratum mean and a deviation of its stratum mean from the grand mean, the sum of squared deviations among strata is

$$35(55.90 - 49.79)^2 + 25(41.23 - 49.79)^2 = 3139.$$

The sum of squared deviations and resulting variances within population strata are, for home-owners (X_1),

$$\sigma_1^2 = \frac{\sum (X_1 - \mu_1)^2}{N_1} = \frac{\sum (X_1 - 55.90)^2}{35} = \frac{3933}{35} = 112;$$

and for renters (X_2),

$$\sigma_2^2 = \frac{\sum (X_2 - \mu_2)^2}{N_2} = \frac{\sum (X_2 - 41.23)^2}{25} = \frac{1165}{25} = 46.6.$$

When a stratified sample is taken, the best point estimate of the population mean will be

$$\blacksquare \qquad \overline{X} = \sum_j (w_j \overline{X}_j), \qquad \blacksquare \tag{16.3}$$

where $\overline{X}_j$ is the sample mean for the jth stratum, and w_j is the universe weight for the jth stratum, that is, $w_j = N_j/N$. In our example this estimate will be

$$\overline{X} = (35/60)(\overline{X}_1) + (25/60)(\overline{X}_2).$$

The sample mean from a given stratum has a variance that depends only on the variance of X within that stratum. Since we have computed σ_1^2 and σ_2^2 from the population data we know what these variances are. They are, using appropriate finite population modifiers,

$$\sigma_{\overline{X}_1}^2 = \frac{\sigma_1^2}{n_1} \left(\frac{N_1 - n_1}{N_1 - 1} \right) = \frac{112.4}{7} \left(\frac{35 - 7}{35 - 1} \right) = 13.2,$$

$$\sigma_{\overline{X}_2}^2 = \frac{\sigma_2^2}{n_2} \left(\frac{N_2 - n_2}{N_2 - 1} \right) = \frac{46.6}{5} \left(\frac{25 - 5}{25 - 1} \right) = 7.77.$$

Now, $\overline{X}$ in Equation (16.3) is a weighted mean of the several stratum means, where the weights are constants determined by the sizes of the strata. Further, the samples from the several strata are independent of one another. Under these conditions the variance of the weighted mean, or the variance of the mean estimated from a stratified random sample, is

$$\blacksquare \qquad \sigma_{\overline{X}}^2 = \sum_j (w_j^2 \sigma_{\overline{X}j}^2). \qquad \blacksquare \tag{16.4}$$

In our case this variance is

$$\sigma_{\bar{X}}^2 = (7/12)^2(13.2) + (5/12)^2(7.77) = 5.84,$$

from which

$$\sigma_{\bar{X}} = \sqrt{5.84} = 2.42.$$

This standard error is materially smaller than the standard error for an unrestricted random sample of size 12 from the Georges Mills expenditure data. That figure was calculated (page 325) to be 3.05. The reason for this improvement through stratification is that the among-strata sum of squares is not a part of Equation (16.4). This is a consequence of drawing separately from the strata and combining the stratum means by known population weights. A summary of the population sums of squares and variances is given in the table below.

	Sum of squared deviations	N	Variance
Within strata			
Homeowners	3933	35	112.4
Renters	1165	25	46.6
Among strata	3139	60	52.3
Total	8237	60	137.3

Stratified random samples permit more reliable estimates of parameters than unrestricted random samples of the same size to the extent that the basis of stratification is associated with the variable(s) of interest in the survey. When such correlation is present, the stratum means differ; and stratification allows the among-strata variance to "escape" from the sampling error. By contrast, an unrestricted random sample is drawn over all the units of observation at once, and what would otherwise be an among-strata variance is included in the sampling error. Cutting down on sampling error is the prime motivation for stratification.

Interval Estimation from the Sample Data

The data from our stratified random sample were given in Table 16.3(b). Using Equation (16.3) we calculate our point estimate of the population mean:

$$\bar{X} = \sum_j (w_j \bar{X}_j) = (35/60)(\$52.47) + (25/60)(\$40.86) = \$47.63.$$

Then, using the sample data we estimate the variances of the individual stratum means

$$s_{\bar{X}_1}^2 = \frac{s_1^2}{n_1}\left(\frac{N_1 - n_1}{N_1 - 1}\right) = \frac{39.56}{7}\left(\frac{35 - 7}{35 - 1}\right) = 4.65,$$

$$s_{\bar{X}_2}^2 = \frac{s_2^2}{n_2}\left(\frac{N_2 - n_2}{N_2 - 1}\right) = \frac{82.78}{5}\left(\frac{25 - 5}{25 - 1}\right) = 13.8.$$

Now we apply Equation (16.4) using the estimated variances of the stratum means.

$$s_{\bar{X}}^2 = \sum_j (w_j^2 s_{\bar{X}_j}^2) \tag{16.5}$$

$$= (7/12)^2(4.65) + (5/12)^2(13.8),$$

$$s_{\bar{X}}^2 = 3.98 \quad \text{and} \quad s_{\bar{x}} = \sqrt{3.98} = \$1.99.$$

The confidence interval is $\bar{X} \pm t_{\alpha/2}\, s_{\bar{x}}$. The degrees of freedom for t are $\sum_j (n_j - 1)$, in our case $(7 - 1) + (5 - 1) = 10$. For our 95-percent confidence interval, the t multiple is then 2.228, and we have

$$\bar{X} \pm 2.228\, s_{\bar{x}},$$

$$\$47.63 \pm 2.228(\$1.99),$$

$$\$47.63 \pm \$4.43.$$

The half-width of our 95-percent confidence interval for the stratified random sample ($4.43) is much narrower than the corresponding half-width we obtained for the simple random sample ($7.39). This is a greater reduction than our population analysis would lead us to expect. We happened to get a sample that yielded an estimated stratum variance for homeowners considerably smaller than the actual stratum variance.

Cluster Sampling

The units of observation in a universe often come in convenient groupings, or clusters. Examples of clusters are pages in a listing of phone subscribers, blocks of dwelling units in a city, places of employment of individuals, and so on.

Definition	**Cluster sampling** is a procedure whereby units of observation in the universe are divided into clusters, or *primary sampling units*. Then sampling proceeds by stages. First a sample of clusters is selected, and then units of observation are drawn from the clusters that have been selected. Units drawn at the final stage are called *elementary units*.

Thus, we might draw 50 pages at random from a telephone directory, and from each page drawn select 4 subscribers at random. We would have a cluster sample of $n = 200$ with 50 primary sampling units and 4 elementary units per primary sampling unit. Many other combinations could produce the sample of 200 subscribers. One attractive feature of cluster sampling is the saving in listing and numbering. Instead of having to list and number the entire universe of elementary units, we need list and number only the primary sampling units and then the elementary units within the primary units that are drawn.

On the map of Georges Mills (Figure 16.1) the village has been divided into six geographic areas, with ten dwelling units in each area. We will select a sample of twelve dwelling units by selecting three clusters and then selecting four dwelling units from each of the selected clusters. Both stages of selection will be conducted without replacement.

The areas are already numbered from 1 to 6. Our drawing of random numbers to select the stratified sample ended with the 10th two-digit number of the second row in Appendix F. We now draw one-digit numbers ignoring 0, 7, 8, and 9. The sequence is 2-4-8-0-5. The clusters selected are, thus, areas 2, 4, and 5. In Table 16.4(a) we show

Table 16.4(a)   Cluster Sample—Selection of Dwelling Units from Geographic Clusters of Dwelling Units

Area 2 (cluster 1)			Area 4 (cluster 2)			Area 5 (cluster 3)		
Random number	Observation number	X_1	Random number	Observation number	X_2	Random number	Observation number	X_3
2	18	43.2	7	43	44.5	3	51	57.3
4	20	31.6	2	26	55.4	6	54	71.6
0	34	52.6	0	46	55.2	1	49	60.7
3	19	48.2	6	42	52.6	0	58	49.0
		175.6			207.7			238.6

the selection of four dwelling units from each of these areas and the resulting expenditure data. We continue reading one-digit random numbers and regard the dwelling units in each area as renumbered from 1 to 10 with the lowest numbered unit reassigned the number 1 and the highest numbered unit in an area reassigned the number 10. The original numbers are on Figure 16.1 and the associated X data in Table 16.1. We associate the random digit "0" with the tenth, or highest-numbered dwelling unit in an area. The foregoing is just one of a number of ways of achieving a one-to-one relation between each random number and a specific dwelling unit within a cluster. We establish our rules for this before we draw the random numbers.

Theory for Cluster Sampling

Cluster sampling involves a two-stage selection process whereby we first select clusters and then select elementary units from each cluster selected. In cluster sampling, each elementary unit is given an equal probability of selection. This can be accomplished in one of two ways. In the first, each primary sampling unit (cluster) is given an equal probability of selection and then a constant fraction of elementary units within each selected primary unit is drawn. In the second method, primary sampling units are selected with probabilities proportional to the numbers of elementary units they contain, and a constant number of elementary units are drawn from each primary unit. The first method is more common, because often the count of elementary units in the primary units is not known, and the constant sampling ratio is often easier to carry out in the survey. In our example the clusters are of equal size, so that the two methods amount to the same thing.

As long as each elementary unit has an equal probability of selection in a cluster sample, the best point estimate of the population mean is the mean of the sample observations:

$$■ \quad \bar{X} = \frac{\sum_k \sum_i X_{ik}}{n}, \quad ■ \tag{16.6}$$

where X_{ik} is the ith observation in the kth cluster selected in the sample.

The variance of the cluster sample mean depends on two sources of variation. The first is the variance among cluster means and the second is the variance within clusters. The variance among cluster means reflects sampling variability introduced in the selection of clusters for the sample, and the variance within clusters reflects sampling variability introduced by selecting elementary units from clusters. Since we have access in our example to the population data, we can figure these variances. We use the following symbols:

N = the number of elementary units in the population,
N_j = the number of elementary units in the jth cluster,
X_{ij} = the value of the ith observation in the jth cluster,
n = sample size of the cluster sample,
J = the number of clusters in the population,
K = the number of clusters drawn into the sample,
n_k = number of sample observations drawn from the kth cluster,
X_{ik} = the value of the ith sample observation in the kth cluster drawn,
μ = population mean,
μ_j = cluster mean (population).

In our example $N = 60$, $N_j = 10$, $n = 12$, $J = 6$, $K = 3$, and $n_k = 4$. The among-cluster variance is

$$\sigma_a^2 = \frac{\sum_j N_j[(\mu_j - \mu)^2]}{N} \left(\frac{J - K}{J - 1}\right),$$

$$\sigma_a^2 = \frac{1880}{60} \left(\frac{6 - 3}{6 - 1}\right) = 18.8.$$

The within-cluster variance is

$$\sigma_w^2 = \frac{\sum_j \sum_i (X_{ij} - \mu_j)^2}{N} \left(\frac{N - n}{N - 1}\right),$$

$$\sigma_w^2 = \frac{6357}{60} \left(\frac{60 - 12}{60 - 1}\right) = 86.2.$$

Here, notice that the sums of squares, 1880 and 6357, add to $8237 = \sum_{ij} (X_{ij} - \mu)^2$. The deviation of any X_{ij} from the population mean can be expressed as

$$(X_{ij} - \mu) = (\mu_j - \mu) + (X_{ij} - \mu_j).$$

The two-stage sampling process can be regarded as drawing a value of $(\mu_j - \mu)$ at the primary stage and drawing a value of $(X_{ij} - \mu_j)$ at the ultimate stage. Note that before the sample is drawn we do not know the μ_j for the ultimate stage. In effect we draw n_k identical values of $(\mu_j - \mu)$ at a time. The sampling variance of $n_k(\mu_j - \mu)$

is $n_k^2\sigma_a^2$. For the K primary draws the variance of $\sum_k n_k(\mu_j - \mu)$ is $\sum_k (n_k^2\sigma_a^2)$, because the draws are random. The sampling variance of $(X_{ij} - \mu_j)$ is σ_w^2. The sampling variance of the sum of these elements for the sample is $n\sigma_w^2$. The variance of $\sum X_{ij}$ obtained in the cluster sample is then

$$\sum_k (n_k^2\sigma_a^2) + n\sigma_w^2,$$

and the variance of $\overline{X}$ is

$$\sigma_{\overline{X}}^2 = \frac{\sum_k (n_k^2\sigma_a^2) + n\sigma_w^2}{n^2}.$$

If n_k, the number of elementary sampling units per cluster, is constant, then $n = Kn_k$ and the above expression reduces to

$$\blacksquare \qquad \sigma_{\overline{X}}^2 = \frac{n_k\sigma_a^2 + \sigma_w^2}{n}. \qquad \blacksquare \tag{16.7}$$

For our sample mean the variance is

$$\sigma_{\overline{X}}^2 = \frac{4(18.8) + 86.2}{12} = 13.4;$$

thus

$$\sigma_{\overline{X}} = \sqrt{13.4} = 3.66.$$

To see most clearly the effect of clustering on sampling error, consider the case of sampling with replacement. Here the formulas are simpler because we have no finite population modifiers to be concerned with. In this situation

$$\frac{\sum_{ij}(X_{ij} - \mu)^2}{N} = \frac{\sum_j N_j(\mu_j - \mu)^2}{N} + \frac{\sum_i \sum_j (X_{ij} - \mu_j)^2}{N},$$

or

$$\sigma^2 = \sigma_a^2 + \sigma_w^2.$$

We can rearrange Equation (16.7) as follows:

$$\sigma_{\overline{X}}^2 = \frac{n_k\sigma_a^2 + \sigma_w^2}{n} = \frac{n_k\sigma_a^2 + (\sigma^2 - \sigma_a^2)}{n} = \frac{\sigma^2 + \sigma_a^2(n_k - 1)}{n},$$

$$\sigma_{\overline{X}}^2 = \frac{\sigma^2}{n} + \frac{\sigma_a^2(n_k - 1)}{n}.$$

Now we can see what happens. The variance of the mean of an unrestricted random sample is σ^2/n. If one elementary unit is drawn per cluster selected, $(n_k - 1)$ in the expression above is zero; and the cluster sample will have the same error as a random sample. But if $n_k \geq 2$, the cluster sampling method will yield greater sampling error than a random sample of the same size. How much greater will depend on both the number of units drawn per cluster and the size of the variance among clusters. Application of this principle will be left to our discussion of survey designs. It should be clear that the motivation for cluster sampling must be lower cost, since it is never a more reliable method than random sampling.

Interval Estimation from the Sample Data

To establish a confidence interval from the data of a cluster sample we must first estimate σ_a^2 and σ_w^2. These estimates are furnished by

$$s_a^2 = \frac{\sum_k (\bar{X}_k - \bar{X})^2}{K - 1} \left(\frac{J - K}{J - 1}\right) \tag{16.8}$$

and

$$s_w^2 = \frac{\sum_k \sum_i (X_{ik} - \bar{X}_k)^2}{\sum_k (n_k - 1)} \left(\frac{N - n}{N - 1}\right). \tag{16.9}$$

In Table 16.4(b) we continue from the data of Table 16.4(a) to obtain the sums of squares required by the equations.

Table 16.4(b)   Calculations from Cluster Sample Data to Establish Confidence Interval for Population Mean

	Area 2 (cluster 1)	Area 4 (cluster 2)	Area 5 (cluster 3)	Total
$\sum_j X_{ik}$	175.6	207.7	238.6	$621.9 = \sum_k \sum_i X_{ik}$
$\sum_i X_{ik}^2$	7954.80	10863.21	14495.34	
$(\sum_i X_i)^2 / n_k$	7708.84	10784.82	14232.49	
$\sum_i (X_{ik} - \bar{X}_k)^2$	245.96	78.39	262.85	$587.20 = \sum_k \sum_i (X_{ik} - \bar{X})^2$
$\bar{X}_k$	43.900	51.925	59.650	
$\bar{X}$	51.825	51.825	51.825	
$(\bar{X}_k - \bar{X})^2$	62.806	0.010	61.231	$124.047 = \sum_k (\bar{X}_k - \bar{X})^2$

Following Equation (16.6), our point estimate of the population mean is

$$\bar{X} = \frac{\sum_k \sum_i X_{ik}}{n} = \frac{621.9}{12} = \$51.825.$$

In our example, $J = 6$, $K = 3$, $N = 60$, and $n = 12$ and our estimates of among and within cluster variances are

$$s_a^2 = \frac{\sum_k (\bar{X}_k - \bar{X})^2}{K - 1} \left(\frac{J - K}{J - 1}\right) = \frac{124.047}{3 - 1} \left(\frac{6 - 3}{6 - 1}\right) = 37.21,$$

$$s_w^2 = \frac{\sum_k \sum_i (X_{ik} - \bar{X}_k)^2}{\sum_k (n_k - 1)} \left(\frac{N - n}{N - 1}\right) = \frac{587.20}{9} \left(\frac{60 - 12}{60 - 1}\right) = 53.08.$$

Then we estimate the variance of the sample mean, using the sample equivalent of Equation (16.7), as

$$s_{\bar{X}}^2 = \frac{n_k s_a^2 + s_w^2}{n} = \frac{4(37.21) + 53.08}{12} = 16.83,$$

and

$$s_{\bar{X}} = \sqrt{16.83} = \$4.10.$$

The degrees of freedom are $(K - 1)$ for s_a^2 and $\sum_k (n_k - 1)$ for s_w^2. These are added to obtain the degrees of freedom for $s_{\bar{X}}^2$. In our case the degrees of freedom are $2 + 9 = 11$. The t distribution is used, and for the 95 percent confidence interval we have

$$\bar{X} \pm 2.201 \, s_{\bar{X}},$$
$$\$51.825 \pm 2.201(\$4.10),$$
$$\$51.825 \pm \$9.024.$$

Replicate Sampling

The principle of replicate sampling, a procedure often useful in survey designs, is to divide the total sample survey into K independent subsurveys. Each survey is carried out as if it were the only survey—that is, independently of the other subsurveys—and each provides an independent estimate of the parameter of interest. The average of these estimates provides the single best estimate of the parameter, and the variance among the estimates is used to provide the measure of sampling errors. While the method is used in more complex applied designs, it can be illustrated in the context of replicated random samples. Suppose we were to regard a random sample of $n = 400$ as ten independent subsamples of $n_k = 40$ each. We could use $\bar{X}_1, \bar{X}_2, \bar{X}_3, \ldots, \bar{X}_K$ in the following way:

$$\bar{X} = \frac{\sum_k \bar{X}_k}{K}, \tag{16.10}$$

$$s_{\bar{X}}^2 = \frac{\sum_k (\bar{X}_k - \bar{X})^2}{K - 1} \left(\frac{1}{K}\right). \tag{16.11}$$

The confidence interval for the population mean would then be figured from $\bar{X} \pm ts_{\bar{X}}$, where t is the appropriate level of Student's t statistic for $m = K - 1$ degrees of freedom. Note that the first term on the right-hand side of Equation (16.11) estimates the variance among sample means of size n_k (40 in the example suggested). But the variance of the mean of ten subsample means is one-tenth of the variance of the subsample means. Therefore, $s_{\bar{X}}^2 = s_{\bar{X}_k}^2(1/K)$, and our estimate of $s_{\bar{X}}$ has only $K - 1$ degrees of freedom. In the example suggested, we had ten subsamples. The degrees of freedom are 9, and the multiple employed for a 95-percent confidence interval will be 2.262.

The only advantage of replicate sampling apparent in the foregoing discussion is some saving in computational effort, which would be substantial in a large multi-purpose survey (with many variables of interest), even in the age of high-speed computers. A more fundamental advantage is that the replicate approach permits computation of sampling errors for surveys where no direct conventional sampling error formula is applicable. The design may not conform to any standard model for which the sampling error formula is known or easy to determine. Yet if the design is consistently applied in K independent subsurveys, confidence intervals can be constructed just as we have indicated above.

Systematic Sampling

A selection procedure that is easier to carry out than using random number tables is to select every kth unit in a listing of units of observation. Suppose there are about 40,000 records in a credit file, but they are not numbered serially. It may be easier to draw every 100th record from the file to obtain a sample of about 400 records than to serially number all the records so that 400 random numbers can be drawn. A random start would be made by selecting a single random number between 00 and 99. If the record selected by this draw were the 57th, then records 57, 157, 257, ..., 39957 would constitute the sample. These records could then be separated out by a hand or machine count.

Systematic sampling is not equivalent to random sampling. If there is a periodic order in the file related to the variable(s) under study, a systematic sample may have larger sampling error than a random sample. An example would be a file of military personnel by battalion, company, and platoon in which the order of listing within each battalion was by company with company officers listed first, and within companies was by platoon with platoon leaders listed at the head of their platoons. Various systematic sampling intervals (values of k) with certain starting points would draw chiefly platoon personnel from a list ordered this way. Others could draw samples with a greatly disproportionate number of company officers and platoon leaders. This would be a poor way to draw a sample for a survey of attitudes toward military life.

Many lists or natural orderings of units of observation have a built-in stratification rather than periodic ordering. When this is so, a systematic sample takes on the features of a stratified sample and can have a smaller sampling error than a random sample of the same size. The list of military personnel could easily be organized with all battalion-level officers first, followed by all company officers, all platoon leaders, and all platoon personnel. Then, a systematic sample with a random start will produce a nearly proportionate representation of officers versus enlisted personnel.

Use of error formulas for random sampling is not really appropriate when the sample is drawn systematically. While it is often difficult to conceive of any way in which certain lists might have a periodic ordering, one cannot be sure about it. Listings and natural orderings more often than not have unknown elements of stratification built into them. A solution can be found in the replicate subsample approach. In the example of 40,000 records and a desired sample of 400, we could take ten systematic subsamples. Each subsample would have a random starting point determined by drawing a random number between 000 and 999. Every thousandth card would then be drawn. If the starting point for the first subsample were 289, then cards 289, 1289, 2289 ..., 39289 would constitute the records to be selected for the first subsample. Each subsample of 40 observations will provide an estimate of any population mean (for example, average balance outstanding, proportion of accounts with zero balance). Confidence limits can be set as previously shown. The systematic selection procedure will cause the sample to pick up any gains in reliability from stratification related to the file order. Whether these gains are in fact small or large, they will be properly reflected in the variance among the subsample means.

Summary

In simple random sampling, selection of the units of observations is made from the entire universe without restriction. In stratified sampling, the units of observation are first separated into strata and then separate samples of predetermined size are drawn from each stratum. In cluster sampling, selection is done in two stages. Clusters are identifiable collections of elementary units of observation. First, a sample of clusters is selected, and then elementary units are drawn from the selected clusters. Stratification tends to reduce sampling error to the extent that the basis of stratification is associated with the variables measured in a survey. Clustering tends to increase sampling error, a disadvantage that may be offset by lower costs per unit sampled.

Replicate sampling provides an alternative method of calculating the sampling error for a survey. It is particularly useful when a more direct method is complex or unknown. A case in point is systematic sampling, in which units of observation are selected by sampling at a constant interval through an ordered listing of the units in the universe. There is no direct formula for the sampling error of a systematic sample. If replicate systematic samples are taken, however, the sampling error can be determined from the variance among the means of the several replicate samples.

See Self-Correcting Exercises 16.A

Exercises

1. An environmental protection agency is planning a survey in which it will sample twenty of fifty firms of a certain kind in its jurisdiction. If the finite population modifier is not employed in calculating sampling errors, by what factor will these errors be overstated?

2. Select an unrestricted random sample of five dwelling units from Georges Mills (Figure 16.1 and Table 16.1), using the dwelling-unit identification numbers. From your sample data, establish the 95 percent confidence interval for the mean weekly grocery bill.

3. Select a stratified random sample of two dwelling units from each of the six neighborhoods in Georges Mills. From Equations (16.3) and (16.5), establish the 95 percent confidence interval for mean weekly grocery expenditures from your sample.

4. The sums of squared deviations *within* each of the six neighborhoods are given below for weekly grocery expenditures for *all* dwelling units in Georges Mills.

Neighborhood	1	2	3	4	5	6	Total
$\sum_{j}(X_{ij} - \bar{X}_j)^2$	1389	1443	952	1032	590	951	6357

Given this information *and* your sample mean in Exercise 3, what is the 95 percent confidence interval for the population mean?

5. The sum of squared deviations within strata, using homeowners and renters as strata, is 5098 (see page 330). Which is the more efficient basis of stratification—home ownership or neighborhood? Explain.

6. Given that the sum of squared deviations within neighborhoods is 6537 and the sum of squared deviations among neighborhoods is 1880 for the expenditure data in Georges Mills, find the variance of the sample mean for a cluster sample of three randomly selected families from each of four randomly selected neighborhoods. Compare this with the result for three clusters of four observations each, given immediately following equation (16.7) in the text.

7. Divide the list of 60 dwelling units in Georges Mills into five segments of 12 successive dwelling units each (Table 16.1). Then select two systematic replicate subsamples of five units each by choosing two random starting numbers between 1 and 12. Using Equations (16.10) and (16.11), establish the 95 percent confidence interval for mean weekly grocery expenditures from your sample.

8. Examine the pattern of numbering for the dwelling units in Figure 16.1. What kind of implicit stratification is built into the systematic sampling procedure of Exercise 7? Compare the half-width of your confidence interval in Exercise 7 with that in Exercise 3.

9. Note the situations under Populations I and II below and the two proposed sample sizes, (a) and (b), under each.

	Population I		Population II	
	Stratum 1	Stratum 2	Stratum 1	Stratum 2
N = size of stratum	1000	2000	1000	2000
σ^2 = variance within stratum	45⅓	45⅓	64	36
n = sample size:				
Plan (a)	20	40	20	40
Plan (b)	30	30	24	36

Use Equation (16.4) to find the variance of the mean of the stratified random sample under Plan (a) and Plan (b) for each population situation. Ignore the finite population modifier in calculating variances of the mean for individual strata.

10. The situations represented by Populations I and II in Exercise 9 are comparable in that the total sum of squared deviations within strata are the same, that is, 136,000. Does it appear from your answers that there is a way to allocate a total sample size among strata that will achieve a smaller standard error of the mean than will an allocation in proportion to size of strata?

Further Elements in Sample Surveys

So far we have considered some of the basic sampling methods for surveys. A given survey may combine several of these methods. For example, a survey of characteristics of hourly wage employees in a community might employ clustering by using the firm as a primary sampling unit. Stratification might be introduced by placing the primary sampling units into standard industrial classification groupings.† Within each stratum, the ordering of firms might be made in accordance with size categories and a systematic selection of firms made. Within the firms selected, it would be convenient to include every kth employee from a prepared list of hourly wage employees. The entire

† Lists of firms by S. I. C. (Standard Industrial Classification) code are frequently available from State Employment Security Commissions.

procedure could be replicated by carrying the survey out as five independent subsurveys and confidence limits set from the five independent estimates of various parameters. In evaluating alternatives for a survey design, certain considerations should be kept in mind.

Design of Clusters and Strata

When there are alternatives, always design strata that promise to be as different as possible and clusters that promise to be as similar as possible with respect to the variable(s) of interest. Variation among strata decreases sampling error, while variation among clusters increases sampling error. Let us restate the principle as a rule of procedure.

Rule

> Select factors for stratification that will maximize differences among stratum means. This maximizes homogeneity *within* strata and thus minimizes within-strata variances.
>
> Select a basis for clustering that minimizes differences among cluster means. This means that the observations *within* a cluster should be as heterogeneous as possible.

In the employee example, we hope for some decrease in sampling error by virtue of differences in characteristics of employees among industry groups. We hope that difference among firms (within an industry group) in characteristics of employees are not so marked as to greatly increase sampling error by using firms as clusters. General knowledge of a field and results from previous surveys are, of course, useful in validating these judgments. There must be some payoff from the use of clusters. In the example cited, costs are much reduced by restricting the listing and field procedures to a limited number of firms.

A decision between a large number of small clusters and a small number of larger clusters hinges on the balance between greater heterogeneity within large clusters and increased cost of listing and interview work in the larger clusters. For example, a survey of dwelling units in a city might contemplate small clusters such as city blocks or large clusters such as census tracts or neighborhoods. Assume there is no up-to-date directory relating dwelling units to locations and field workers have to prepare a listing of dwelling units in each cluster selected in the primary sampling stage. With a smaller number of large clusters (neighborhoods) a larger proportion of clusters would probably be needed in the sample. Hence, a greater amount of listing work would be needed. This expense is not likely to be offset by the increased heterogeneity within neighborhoods as opposed to blocks. There are great differences in dwelling-unit characteristics among blocks in cities, but economic stratification often occurs on a broad grographic scale so that it is not much reduced by going to neighborhoods as a primary unit. Blocks are not ideal clusters, but they would probably be used. Their bad effects on sampling error would be limited by restricting the number of dwelling units selected per block to a small number, say two or three, and including many blocks. The number of blocks would still be a small fraction of the total in the city, and the cost saving in the listing work would be preserved.

Replicated Systematic Samples of Clusters

The listing work in the previous example assumed that the blocks (or neighborhoods) were to be selected with equal probability and that every kth dwelling unit in each cluster drawn would be included in the sample. If an up-to-date count of dwelling units by blocks and neighborhoods were not available, this would have to be the case. A specific example of the alternatives described there might be as follows:

	Primary sampling unit (p.s.u.)	
	Blocks	*Neighborhoods*
(1) Total dwelling units (d.u.'s) in city	40,000	40,000
(2) Desired sample (d.u.'s)	800	800
(3) Total p.s.u.'s in city	4,000	100
(4) Number of p.s.u.'s in sample	400	40
(5) Average number of d.u.'s per p.s.u. (1) ÷ (3)	10	400
(6) D.u.'s in sample per p.s.u. (2) ÷ (4)	2	20
(7) Primary sampling rate (4) ÷ (3)	1/10	4/10
(8) Second-stage sampling rate (6) ÷ (5)	2/10	1/20
(9) Overall sampling rate (2) ÷ (1) = (7) × (8)	2/100	2/100

The desired sample is 800 dwelling units from a universe of 40,000 dwelling units. If blocks are used as clusters, we want to sample only two dwelling units per block for reasons already explained. Given that there are 4,000 blocks in the city, the remaining figures follow. There are 100 neighborhoods, and in order to get representation of their variety, we feel we must include 40 in the sample. As a consequence, we would have to list the dwelling units in 40 percent of the city rather than in 10 percent as in the block sample. This was the unacceptable alternative in the previous section. The preferred 10-percent block sample and 20-percent sample of dwelling units per block could be carried out in a replicated systematic sample framework as follows. The blocks would be listed in a geographic ordering, with adjacent blocks close to each other in the list. Then the list is divided into 40 zones of 100 blocks each. Ten subsamples each, with one block drawn systematically from each zone by using a random start and a systematic interval of 100 blocks, would produce the 400 blocks desired. The ordering of blocks and the systematic selection provides automatic geographic stratification of the clusters.

At the second stage, we intend a 20 percent sample of dwelling units within blocks. This will yield varying numbers of dwelling units depending on how many dwelling units are in each block selected. A convenient device to use at this stage is a form which defines ten elements for each block. Exactly two of these will be brought into the sample in every case. This is done by having a prelisted form with ten lines, as follows.

Line	d.u. No.	Address	d.u. No.	Address	d.u. No.	Address
1	1		11		21	
2	2		12		22	
3	3		13		23	
4	4		14		24	
5	5		15		25	
6	6		16		26	
7	7		17		27	
8	8		18		28	
9	9		19		29	
10	10		20		30	

The interviewer enters a block in a predesignated way and lists dwelling units in a prescribed order. For example, interviewers may be told to begin at the northwest corner of a block and proceed in a clockwise fashion. Directions for listing dwelling units within multiple-family dwellings can be made specific also. The element, however, is the line number. When the listing is finished, the interviewer opens an envelope containing two previously drawn random numbers between 1 and 10. He then secures information on any and all dwelling units on that line. If the block has but six dwelling units and the random numbers are 6 and 9, then one dwelling unit will be included. If there are 23 dwelling units in the block and line 2 is selected, three dwelling units will be drawn into this element in the frame. Each subsample contains 80 elements from the frame (40×2). It may contain a few more or less than 80 dwelling units, although that is the expected number per subsample. Sampling error would be calculated by the replicate samples method, using $n_k = 80$.

Error Formulas for Proportions

Earlier in this chapter we developed sampling error formulas for means for simple random, stratified random, cluster, and replicate sampling. Many survey questions are of the yes-no and multiple-choice types, in which the proportion selecting a given alternative answer is the statistic of interest. When dealing with proportions, the product of $p(1 - p)$, or pq, takes the place of s_X^2 in the equations presented previously. For example, an unrestricted random sample without replacement has an estimated standard error of the mean given by Equation (16.2). That formula is repeated below along with the corresponding formula for the estimated standard error of a proportion (s_p).

$$s_{\bar{X}} = \frac{s_X}{\sqrt{n}}\sqrt{\frac{N-n}{N-1}}, \qquad s_p = \sqrt{\frac{pq}{n}\frac{N-n}{N-1}}.$$

A confidence interval for the proportion can be established by $p \pm z_{\alpha/2}\,s_p$ if the sample size, n, meets the minimum level given in Chapter 12 (page 249).

For a stratified random sample we need p_j, the proportions found from the sample observations in each stratum. The point estimate of the population proportion parallels the point estimate for the population mean [Equation (16.3)] as follows:

$$\overline{X} = \sum_j (w_j \overline{X}_j), \qquad p = \sum_j (w_j p_j).$$

To estimate the variance of the proportion, we need the $s_{p_j}^2$ for each stratum. These are

$$s_{p_j}^2 = \frac{p_j q_j}{n_j} \left(\frac{N_j - n_j}{N_j - 1}\right).$$

Then we apply the equivalent of Equation (16.5).

$$s_{\overline{X}}^2 = \sum_j (w_j^2 s_{\overline{X}_j}^2), \qquad s_p^2 = \sum_j (w_j^2 s_{p_j}^2).$$

For a cluster sample (with equal probability of selection of each elementary unit), the parallel formulas are

$$\overline{X} = \frac{\sum_k \sum_i X_{ik}}{n}, \qquad p = \frac{\sum_k r_k}{n},$$

where r_k is the number of successes observed in the kth cluster selected in the sample and

$$s_a^2 = \frac{\sum_k (\overline{X}_k - \overline{X})^2}{K - 1} \frac{J - K}{J - 1}, \qquad s_a^2 = \frac{\sum_k (p_k - p)^2}{K - 1} \frac{J - K}{J - 1},$$

where p_k is the sample proportion observed in the kth cluster selected, that is, $p_k = r_k / n_k$.

$$s_w^2 = \frac{\sum_k \sum_i (X_{ik} - \overline{X}_k)^2}{\sum_k (n_k - 1)} \left(\frac{N - n}{N - 1}\right), \qquad s_w^2 = \frac{\sum_k (n_k p_k q_k)}{\sum_k (n_k - 1)} \left(\frac{N - n}{N - 1}\right).$$

Then we find s_p^2, if n_k is constant,† just as we did $s_{\overline{X}}^2$ previously.

$$s_p^2 = \frac{n_k s_a^2 + s_w^2}{n}.$$

For replicate sampling with K subsamples simply substitute p_k and p for $\overline{X}_k$ and $\overline{X}$ in Equation (16.10) and (16.11).

Determination of Sample Size

Most surveys deal with a number of characteristics of the elementary units of observation. Frequently estimates are desired for each of these characteristics within various subgroups of the universe of units. A market survey might require estimates of dollar purchases and brand preference for a number of products by income level and

† If n_k is variable, then we have

$$s_p^2 = \frac{\sum (n_k^2 s_a^2 + n_k s_w^2)}{n^2}.$$

age and race of major purchaser in households in a market area. The clients for the survey want estimates for each of 24 groups formed by cross-classifying four age groups, three income groups, and two racial groups. Suppose the latest Census of Population and Housing figures are consulted to form four age groups and three income groups of equal numbers in the city and that the racial distribution is 2/3 white and 1/3 nonwhite. The following table shows the proportions of total households within each income × age × race class.

Age group \\ Race Income	White			Nonwhite		
	1	2	3	1	2	3
I	2/36	2/36	2/36	1/36	1/36	1/36
II	2/36	2/36	2/36	1/36	1/36	1/36
III	2/36	2/36	2/36	1/36	1/36	1/36
IV	2/36	2/36	2/36	1/36	1/36	1/36

If a total sample size of 720 is contemplated, we would expect only 20 observations to be available for estimating the parameters in the cells of the nonwhite sector. Put another way, the accuracy required in the smallest subgroup for which a separate estimate is desired will determine the needed total size of sample for any method with equal selection probabilities of each elementary unit. If this required a larger sample than the client could afford, he might have to settle for estimates by age group only (3/36 each) or by income group only (4/36 each) within the nonwhite sector. We can let the total sample size dictate the level to which useful estimates can be carried.

To determine how large a sample is needed, the survey statistician must get an answer to the question of how accurate the sponsors of a study wish the results to be. For a simple random sample (ignoring the finite population modifier) we have the basic relationship

$$\sigma_{\bar{X}}^2 = \frac{\sigma_X^2}{n}.$$

This can be rearranged to form an equation for n, the sample size:

$$n = \frac{\sigma_X^2}{\sigma_{\bar{X}}^2}. \tag{16.12}$$

If a sponsor says he wants to be virtually certain that the sample mean monthly rental level in a housing survey is within a dollar of the true mean level, we can secure a figure for $\sigma_{\bar{X}}^2$. "Virtually certain" can be taken to mean 99.7 percent sure, which corresponds to a z multiple of ± 3.0. The sponsor has said, in effect,

$$3.0\sigma_{\bar{X}} = 1.00,$$

$$\sigma_{\bar{X}} = 1/3,$$

$$\sigma_{\bar{X}}^2 = 1/9.$$

We now need a provisional, or planning, figure for σ_X^2. Suppose the last Census of Housing for the area showed a standard deviation of monthly rentals of $5. Using this we can calculate

$$n = \frac{\sigma_X^2}{\sigma_{\bar{X}}^2} = \frac{(5)^2}{1/9} = 225.$$

A sample of 225 housing units would be required. We assume this number of units is a small fraction of the total number of housing units in the area. If it were not, we would have to make adjustments for the finite population modifier.

In detailed work on sample size requirements, many survey statisticians work with relative variability. This is natural because accuracy requirements generally come in that form. A client is likely to say that he wants all parameters estimated within 3 percent (or some other percent) of the true value with a certain confidence. Suppose we want estimates within 3 percent with 95 percent confidence. This means we want

$$\frac{1.96 \; \sigma_{\bar{X}}}{\mu} = 0.03.$$

Let $V_{\bar{X}}$ stand for *relative error*, That is, let

$$V_{\bar{X}} = \frac{\sigma_{\bar{X}}}{\mu}.$$

Then, for our example,

$$1.96 \; V_{\bar{X}} = 0.03,$$
$$V_{\bar{X}} = 0.015 \text{ (approx.)}.$$

Designating σ_X/μ as V and remembering that $\sigma_{\bar{X}}/\mu = V_{\bar{X}}$, we can develop the following equations.

$$\sigma_{\bar{X}}^2 = \frac{\sigma_X^2}{n},$$

$$\frac{\sigma_{\bar{X}}^2}{\mu^2} = \frac{\sigma_X^2}{\mu^2}\left(\frac{1}{n}\right), \quad \text{or} \quad V_{\bar{X}}^2 = \frac{V^2}{n},$$

$$n = \frac{V^2 \dagger}{V_{\bar{X}}^2}. \tag{16.13}$$

The relative accuracy requirement has given us a value for $V_{\bar{X}}$. We must have a provisional, or planning, figure for $V = \sigma_X/\mu$. This ratio of the standard deviation to the mean is called the *coefficient of variation*. Suppose, using the housing rental example, that past surveys had produced coefficients of variation around 0.20. Our required sample size to meet the relative accuracy requirement is

$$n = \frac{V^2}{V_{\bar{X}}^2} = \frac{(0.20)^2}{(0.015)^2} = 178 \text{ (approx.)}.$$

† If the finite population modifier is included, the corresponding formula when the finite modifier is taken as $(N - n)/N$ is

$$n = \frac{NV^2}{V^2 + NV_{\bar{X}}^2}.$$

As suggested earlier, one may as often be applying this formula to find the sample size required from subgroups as from the total units in the universe of study.

In addition to subgroups, a further complication arises from the different parameters required to be estimated. These will generally be different means and different proportions. We suggested mean expenditures on various products and proportions of families preferring different brands of each product as an example in a market survey. For each brand, preference is a binary variable (prefer, not prefer). The square of the coefficient of variation for a binary variable is $(1 - \pi)/\pi$. Where π is low, the coefficient of variation is large, and a correspondingly large n will be required to keep sampling error within a specified percentage bound. Suppose one wants to estimate various proportions within 20 percent with 0.95 confidence ($z \approx 2.0$). If the true proportion is around 0.50, this requires (using the basic formula and setting 2.0 $V_{\bar{x}} = 0.20$)

$$n = \frac{V^2}{V_{\bar{x}}^2} = \frac{(1 - \pi)/\pi}{(0.10)^2} = \frac{0.50/0.50}{0.01} = 100.$$

But if π is around 0.30, we need

$$n = \frac{(1 - \pi)/\pi}{(0.10)^2} = \frac{0.70/0.30}{0.01} = 233,$$

and if π is around 0.10, we need

$$n = \frac{(1 - \pi)/\pi}{(0.10)^2} = \frac{0.90/0.10}{0.01} = 900.$$

Clearly, the total size of the sample required will depend on the relative accuracy required for the estimates of the smaller proportions. Again, we may have to "give up" at some point. We may have to settle for reasonable estimates of brand preference for the three leading brands and let the rest go to an "all other" category.

Nonsampling Errors

We complete our discussion of practical problems in survey design by considering sources of error not embodied in the various sampling error formulas. Nonsampling errors are those that would remain in a survey even if the survey method were carried out over and over again. They are errors that are not lessened by increasing the size of the sample. The terms *persistent errors* or *bias* are often used to refer to them. One writer likens sampling and nonsampling error to the two sides of a right triangle. (See Reference 1.) Total survey error, represented by the hypotenuse, is not reduced much by shortening the height (sampling error) of the triangle if the base (nonsampling error) is large. By the same token, if nonsampling errors can be made small, reduction of sampling errors through good design and/or increased sample size will be effective in reducing total error. Either way, there is motivation for holding nonsampling error to a low level. Some of the sources of these persistent errors and measures that can be taken to control them are discussed below.

Faulty sampling frame   It sometimes happens that the correspondence between the sampling frame and the elementary units of observation is shaky. The classic situation of this kind was the Literary Digest poll of 1936. Based on a very large sample, this poll predicted an election victory for Alfred Landon over Franklin Roosevelt. The sampling frame used was comprised of telephone subscribers and automobile registration lists. Unfortunately for the Literary Digest (which went out of business soon after), many voters were not included in this frame; and these members of non-auto and nontelephone families voted heavily for Roosevelt.

The only solution to faulty sampling frames is care and persistence in developing the frame. It is dangerous practice to settle for incomplete frames on the assumption that units not in the frame are like units in the frame. If surveys are worth doing at all, they should be done with care.

Incomplete response   The two major sources of incomplete returns in sampling human populations are respondents not located and refusals. Here, the frame may be good but incomplete returns can cause bias in the results. Mail-back questionnaires are particularly a problem. A thirty percent response is high even for a well-prepared mail survey that provides some motivation for a completed response. Can we afford to assume that those not responding have characteristics similar to the responding group? The usual solution is to sample the nonrespondents to find out. Follow-up reminders can be sent out and late returns can be compared with early returns to find the differences for the slower group. If there is much at stake, phone calls or personal interviews can be attempted with a sample of nonrespondents. But if the survey is important, it might be better to have conducted personal interviews from the beginning.

Refusals are much lower in well-conducted personal interviews. Even in general household interviewing, they can be kept well below ten percent. Respondents who are not at home are another problem, and the solution is a series of call-backs at different hours to bring them into the returns column. Substitution of next-door families for "not-at-homes" is bad practice. In most surveys, the not-at-homes are different from the at-homes, in that they tend to be families with multiple wage earners and younger and older families without children at home. Substitution can be worse than doing nothing, because it just compounds the bias of under-representation of certain types of families.

Performance of survey personnel   Provision should be included in any survey for checking to see that gross errors or persistent variation from planned procedures are corrected in the normal conduct of the survey. This means adequate training of staff followed by routine checks on work. Falsifying interviews, making unauthorized substitutions, and so forth, can be easily detected (and discouraged initially) by adequate field reporting forms and checks on a sample of the work reported. Accuracy of interview work can be checked by repeat interviews by highly skilled personnel with a small sample of respondents. The Bureau of the Census checks the quality of the U.S. Census of Population and Housing through just such a procedure called the Post-Enumeration Survey. When replicated sample procedures are employed, there are opportunities for building designed experiments into the replicated framework to compare the performance of different interviewers or interviewer teams. These experiments can spot significant differences among interviewers or teams. This knowledge can be helpful in achieving more uniform survey work in subsequent surveys.

Processes of coding and tabulating completed interviews can be checked through quality control procedures. Usually this involves checking periodic small samples of work and increased inspection when quality levels are suspect. Errors found are of course rectified. Use of control totals in preparing detailed breakdowns, whether by hand or machine, is standard operating procedure. Mechanical card sorters and counters, mark-sense systems, and electronic computers are not error free.

Summary

Stratification and clustering have opposite effects on sampling error. To maximize the reduction of sampling error through stratification, one seeks to maximize differences among strata. To minimize the increase in sampling error associated with clustering, one seeks to minimize differences among clusters. Systematic sampling of cluster units such as city blocks provides some of the advantages of stratification. Error formulas for proportions paralleling the basic formulas for means are available. To determine how large a sample survey is appropriate, one must first obtain a statement of the accuracy required for estimates produced by the survey. When the purpose of a survey is to estimate a variety of parameters (means and proportions) for specified subgroups as well as for the total population, determination of required sample size becomes correspondingly complicated. Control of nonsampling errors can be critical to surveys because these errors are not reflected in the various sampling-error formulas. The chief sources of nonsampling errors are faulty sampling frames, incomplete response, and inadequate supervision of and performance by survey personnel.

See Self-Correcting Exercises 16.B

Exercises

1. A sample of 1,002 persons 60 years of age and over in Chicago was obtained by selecting two individuals 70 years of age and over, and one individual 60–69 years of age in each of 334 city blocks. The blocks were selected in a systematic manner from a geographic listing of all city blocks with the probability of inclusion of any block being in direct proportion to the number of elderly in the block according to the Census of Population. The percentage of elderly who were 70 years of age and over was 44 percent in the census and females constituted 57.4 percent of all the elderly in the city. Interviewers were instructed to alternate interviews between male and female. The plan for calculating sampling errors called for stratum weights for 60–69 and 70 and over based on the census, but not for male and female. The basic design was replicated in five subsamples, and the confidence intervals for various statistics averaged 50 percent greater than an unrestricted random sample would have produced. Identify the elements in this design that tend to (a) increase and (b) decrease sampling error compared to an unrestricted random sample of the same size.

2. Are there any sources of bias in the sample design as described in Exercise 1? What additional information would be important in evaluating the sampling design?

3. A proportionate stratified sample of 1000 registered voters was conducted in a large city with 60-percent Democratic and 40-percent Republican registration. On a particular issue, 20 percent of the Democrats and 40 percent of the Republicans expressed agreement. Estimate the proportion agreeing with the issue among all registered voters and estimate the standard error of the proportion.

4. A survey with five replicate subsamples of 100 each was conducted among dwelling units in a large area to determine the proportion of units in need of repair. The numbers needing repair in the subsamples were 40, 45, 41, 53, and 51. Calculate the 95-percent confidence interval for the proportion.

5. A firm wishes to estimate within $200 the average mortgage balance on homes in an area with 95 percent confidence. It is supposed that the standard deviation of balances is around $3,000. Assuming that the sample size will be a negligible fraction of the population size, how large a sample should be taken?

6. How large a sample should be taken in the situation in Exercise 5 if there are only 2000 homes in the area? In advance of taking the sample, the average balance is thought to be on the order of $10,000.

7. The firm of Exercise 5 is also concerned to estimate the proportion of homes with zero mortgage balances within 0.03 of the true value with virtual certainty. In the absence of any idea about the proportion with zero balances, how large a sample would be required?

8. If the firm does not believe the proportion of homes with zero mortgage balance will exceed 0.20, how large a sample would be required?

9. In a brand-preference study in a large metropolitan area, a market research firm will study the standings of three competing brands in four equal market segments defined by age and ethnic background. The client wants to maintain 5 percent relative accuracy ($V_{\bar{x}}$) in estimating market share in any market segment for any brand having as low as 20 percent true preference share. How large a total sample will be required for the study?

Glossary of Equations

$$\sigma_{\bar{X}} = \frac{\sigma_X}{\sqrt{n}}\sqrt{\frac{N-n}{N-1}}$$

The fundamental formula for the standard error of a sample mean when an unrestricted random sample is drawn without replacement.

$$s_{\bar{X}} = \frac{s_X}{\sqrt{n}}\sqrt{\frac{N-n}{N-1}}$$

The formula for estimating the standard error of a sample mean from the sample data when an unrestricted random sample is drawn without replacement.

$$\bar{X} = \sum_j (w_j \bar{X}_j)$$

The point estimate of the population mean from a stratified random sample is a weighted mean of the stratum means, where the weights are the fractions of the units of observation belonging in each stratum, that is, $w_j = N_j/N$.

$$\sigma_{\bar{X}}^2 = \sum_j (w_j^2 \sigma_{\bar{X}_j}^2)$$

The fundamental formula for determining the variance of the mean of a stratified random sample from the variances of the individual stratum means.

$$s_{\bar{X}}^2 = \sum_j (w_j^2 s_{\bar{X}j}^2)$$

The formula for estimating the variance of the mean of a stratified random sample from the estimated variances of the individual stratum means.

$$\bar{X} = \frac{\sum_k \sum_i X_{ik}}{n}$$

The point estimate of the population mean from a cluster sample of K clusters is the sum of all the sample observations divided by the sample size.

$$\sigma_{\bar{X}}^2 = \frac{n_k \sigma_a^2 + \sigma_w^2}{n}$$

The fundamental formula for the variance of the mean of a cluster sample in which the number of elementary units sampled is the same within each cluster selected, that is, n_k is a constant.

$$s_a^2 = \frac{\sum_k (\bar{X}_k - \bar{X})^2}{K-1} \left(\frac{J-K}{K-1}\right)$$

$$s_w^2 = \frac{\sum_k \sum_i (X_{ik} - \bar{X}_k)^2}{\sum_k (n_k - 1)} \left(\frac{N-n}{N-1}\right)$$

Formulas for estimating σ_a^2 and σ_w^2 for use in the previous formula to obtain $s_{\bar{X}}^2$ entirely from the sample data of a cluster sample. The formulas assume selection without replacement at both sampling stages.

$$\bar{X} = \frac{\sum_k \bar{X}_k}{K}$$

The formula for the point estimate of a mean based on K equal sized replicated subsamples.

$$s_{\bar{X}}^2 = \frac{\sum_k (\bar{X}_k - \bar{X})^2}{K-1} \left(\frac{1}{K}\right)$$

The formula for estimating the variance of a mean when the mean is based on K equal-size replicated subsamples.

$$n = \frac{\sigma_X^2}{\sigma_{\bar{X}}^2}$$

The basic (with replacement) relationship between σ_X^2 and n restated as an equation for required sample size.

$$n = \frac{V^2}{V_{\bar{X}}^2}$$

The same relation as immediately above; however, here the population standard deviation and standard error of the mean are stated relative to the population mean —that is, $V = \sigma_X/\mu$ and $V_{\bar{X}} = \sigma_{\bar{X}}/\mu$.

17

Time Series and Index Numbers

Time series analysis is the study of how values of a variable change over time. In some situations detection and estimation of time-related changes may be made by methods we have already studied. For example, the sequence of observations in a typical quality control situation (for example, mean breaking strength of samples of ten products taken each half hour from a production line) constitute a time series. A decision on process control made by observing whether a sample mean lies outside established control limits is a decision about a change in the mean level (μ) of a series. Observing a sample mean outside the control limits signals a search for some element in the production process that is responsible for the presumed change in the process mean.

The changes associated with time in many series can be quite complex. Consider annual sales of a firm or yearly values of a nationwide series such as U.S. copper production. Changes in such series are the outcome of a variety of forces that make themselves felt in different ways over time. First, there may be a long-term underlying growth associated with basic factors, such as population growth and market expansion. Other changes in the series might reflect particular causes, such as gaining and losing major customers or sudden changes in access to overseas markets resulting from international political changes. Still other changes may reflect general business recessions and recoveries. For a variety of financial and technical reasons, cyclical effects of varying intensity and duration are common in many business and economic series.

If we consider monthly, weekly, or daily series, we encounter still another set of forces producing variation in the series over time. In many areas of business, activity is brisk at certain times of the year (month, or week) and slack at other times. The "causes" of this variation are usually found in natural events or customs connected with the calendar. The Christmas selling season in retail sales and bi-monthly paydays are cases in point.

A major objective of time series analysis is forecasting. Simply describing past history serves this objective if it aids our understanding of the processes at work. The attempt to anticipate the future is much more explicit when quantitative forecasts of future values are made. These forecasts are based on the assumption that certain elements of systematic variation found in past data will persist into the future.

Components of Time Series Variation

The foregoing description of forces bearing on typical time series was phrased in terms of sets of forces that are viewed as producing different patterns of variation in a time series. This may be said to constitute a common "model" of time series variation. The model is made more specific in terms of operating on the data (but more abstract in terms of underlying causal factors) by defining the elements of variation as follows.

Definitions

Trend in a time series is a long-term, smooth, irreversible pattern of change from period to period in the series.

Cyclical variation in a time series is a wave-like or oscillating pattern of variation in the series. The duration and amplitude need not be repetitive, but can vary from cycle to cycle. Variation associated with short repetitive calendar periods is excluded (see seasonal variation).

Seasonal variation is the repetitive pattern of variation occurring within a year (or shorter repetitive calendar period) in a time series.

Irregular variation in a time series is comprised of changes which cannot be described as trend, cyclical variation, or seasonal variation.

We use the symbols T, C, S, and I for these elements of time series variation. The actual time series, Y, is viewed as the result of these elements acting in combination. In the most common model for time series analysis the elements are viewed as combining multiplicatively. In a time series of annual data there are only three elements of variation. The model is

$$Y = T \times C \times I.$$

If a series contains data by months, quarters, or other calendar periods less than one year, all four elements can be present and the model is

$$Y = T \times C \times S \times I.$$

The attempt to break down the total variability of a series into these component elements of variation is called time series decomposition. The importance of time series decomposition for forecasting is that we hope to find repetitive patterns in these elements that can be projected into the future as intelligent forecasts. In the first half of this chapter we introduce the methods for time series decomposition. The second half is devoted to application of the analysis and to the construction of index numbers of related time series.

Determination of Trend

The fitting of a simple equation to the data of a time series is the most common way of describing the "smooth, irreversible pattern of change" that we call trend. A linear equation with a constant absolute change in level per time period is the simplest trend model. Many equations besides straight lines can be employed as models for trend in time series as long as the result is a description or forecast that is always increasing or always decreasing. The type of equation chosen to represent trend is based on an examination of the graph of the time series. Growth patterns that are obviously curvilinear should not be forced into straight lines.

To keep our examples simple we will deal only with straight-line trends. Methods for fitting other equations are presented in Reference 1. A variety of equations are commonly incorporated into "canned" computer programs for trend description.

Figure 17.1 shows two time series. They could represent sales for two franchise unit operations from their first full year of operation through 1972. The data points for a time series are, in effect, bivariate observations, where the dependent variable, Y, is the value of the series (in our case sales in $ thousands) and the independent variable, X, is time. A straight line fitted to the time series will describe the long-term average change in sales, Y, associated with the passage of time.

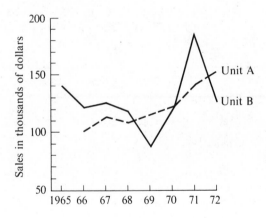

Figure 17.1   Sales for Two Units of a Franchise Operation, 1965–1972

We have available from Chapter 3 on association the equations for a "best fitting" straight line for describing a relationship between Y and X,†

$$\sum Y = \alpha \cdot N + \beta \cdot \sum X,$$

$$\sum XY = \alpha \cdot \sum X + \beta \cdot \sum X^2,$$

where α and β are the constants of the least-squares line in terms of the Y values. If

† As in Chapter 3, we treat the data here as describing a population relationship. The population is the particular historical series under study.

we designate the X variable in such a way that $\sum X = 0$, these equations are simplified. They become, respectively,

$$\sum Y = \alpha \cdot N + \beta \cdot 0,$$

$$\sum XY = \alpha \cdot 0 + \beta \cdot \sum X^2.$$

The first equation will then yield a single formula for α and the second a single formula for β. If our time points are evenly spaced and we select the mid-value of the time scale as the origin where $X = 0$, then $\sum X$ will be zero and we can use the equations

$$\blacksquare \qquad \alpha = \frac{\sum Y}{N}, \qquad \blacksquare \qquad \qquad \text{(17.1a)}$$

$$\blacksquare \qquad \beta = \frac{\sum XY}{\sum X^2}, \qquad \blacksquare \qquad \qquad \text{(17.1b)}$$

In Table 17.1 the scales for X are adjusted so that $\sum X = 0$ and the summations called for in Equations (17.1a and b) are calculated. The values for Y are sales in thousands of dollars from which Figure 17.1 was constructed.

Table 17.1   Data for Calculating Straight-Line Trends for Two Units of a Franchise Operation

	Unit A					Unit B			
Year	X	Y	X^2	XY	Year	X	Y	X^2	XY
1965	—	—	—	—	1965	−7	141	49	−987
1966	−3	101	9	−303	1966	−5	122	25	−610
1967	−2	113	4	−226	1967	−3	126	9	−378
1968	−1	109	1	−109	1968	−1	118	1	−118
1969	0	116	0	0	1969	1	87	1	87
1970	1	121	1	121	1970	3	119	9	357
1971	2	141	4	282	1971	5	187	25	935
1972	3	153	9	459	1972	7	126	49	882
Total		854	28	224			1026	168	168

Unit A's first full year of operation was 1966, so we have seven years of annual data. The mean year is 1969, so we let $X = 0$ in 1969 (the mid-value of the time scale is centered in 1969—that is, $X = 0$ on July 1, 1969). Then we can write the remaining years in terms of yearly deviations from 1969. This is the most convenient pattern whenever there is an odd number of yearly values in the time series.

For Unit B the series starts in 1965 and there are eight time series values. The mid-value of the time scale is $1968\frac{1}{2}$. This designates the year from July 1, 1968 to June 30, 1969, which has a mid-time point ($X = 0$) of January 1, 1969. Here it is convenient to rescale time as deviations from the mid-time point in one-half year (six-month) units. This achieves integer values of X that sum to zero, are as small as possible, and thus are easiest for hand calculation. This pattern will always work for annual series with an even number of values.

In interpreting and using the trend equation ($T = \alpha + \beta X$) it is necessary to remember the origin and units of the transformed time scale. The trend line for Unit A is found as follows, where T is the trend of annual sales.

$$\alpha = \frac{\sum Y}{N} = \frac{854}{7} = 122,$$

$$\beta = \frac{\sum XY}{\sum X^2} = \frac{224}{28} = 8.0;$$

$$T = 122.0 + 8X.$$

$$X = 0 \text{ in 1969 (that is, July 1, 1969)}.$$

$$X \text{ is in one-year units.}$$

For Unit B we calculate the trend line as

$$\alpha = \frac{\sum Y}{N} = \frac{1026}{8} = 128.25,$$

$$\beta = \frac{\sum XY}{\sum X^2} = \frac{168}{168} = 1.0;$$

$$T = 128.25 + 1.0X.$$

$$X = 0 \text{ in } 1968\tfrac{1}{2} \text{ (that is, January 1, 1969)}.$$

$$X \text{ is in one-half year units.}$$

The α constant is the level of the trend of the series when $X = 0$, that is, at the mid-value of the time series, and the constant β is the increment in the trend of the series per unit of X. In the case of Unit A the increment in trend is 8 ($ thousand) per year. For Unit B the trend increment is 1 ($ thousand) per six-month period. Of course, this increment can be expressed as 2 ($ thousand) per year for more direct comparison with the trend increment for Unit A. In fact, we might want to reexpress the entire trend equation for Unit B so that the trend equations for the two units could be directly compared. This can be done by moving the origin of the trend equation for Unit B up by one half-year period—that is, from $1968\tfrac{1}{2}$ ($X = 0$ at January 1, 1969) to 1969 ($X = 0$ at July 1, 1969) *and* restating β in yearly units. First we find T (1969) for Unit B from the originally calculated equation.

$$T(1969) = 128.25 + 1.0(1) = 129.25.$$

Then, using $T(1969)$ as the new origin, we express the trend equation with an annual increment by doubling the coefficient of X from the original equation. The result is

$$T = 129.25 + 2.0X.$$

$$X = 0 \text{ in 1969 (that is, July 1, 1969)}.$$

$$X \text{ is in one-year units.}$$

Now we can see directly that the trend value of Unit B's sales in 1969 was 7.25 ($ thousand) above Unit A (129.25 − 122.00), but the trend increment for Unit B (2.0) is only one-fourth of the trend increment for Unit A (8.0).

Cyclical-Irregular Relatives

The approach employed in time-series decomposition will now be illustrated with the sales series for Unit B. The variations of the actual series (Y) from the trend values are attributed to nontrend forces. For annual data, these forces are cyclical and irregular. In the multiplicative model the effects of cyclical and irregular forces (CI) are measured relative to the trend base (T). The cyclical-irregular relatives for annual data are

$$\blacksquare \qquad CI = \frac{Y}{T}. \qquad \blacksquare \qquad\qquad (17.2)$$

In Table 17.2 we calculate first the trend values and then show the cyclical-irregular relatives. We use the restated trend equation with the origin shifted to July 1, 1969. In Figure 17.2 the trend is shown along with the original series, and the cyclical-irregular relatives are shown in the second plot.

Table 17.2   Data for Calculating
Cyclical-Irregular
Relatives for Unit B

X	$Year$	Y	T	$CI = Y/T$
−4	1965	141	121.25	1.163
−3	1966	122	123.25	0.990
−2	1967	126	125.25	1.006
−1	1968	118	127.25	0.927
0	1969	87	129.25	0.673
1	1970	119	131.25	0.906
2	1971	187	133.25	1.403
3	1972	126	135.25	0.932

$T = 129.25 + 2.0X.$

$X = 0$ in 1969 (that is, July 1, 1969).

X is in one-year units.

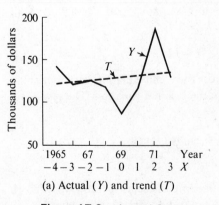

(a) Actual (Y) and trend (T)

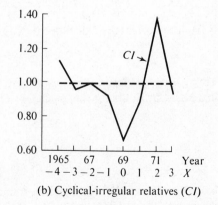

(b) Cyclical-irregular relatives (CI)

Figure 17.2   Actual Sales and Trend of Sales for Unit B and Cyclical-Irregular Relatives

In 1971 the actual sales are 40.3 percent above the trend of sales. This deviation is attributed to a combination of cyclical and irregular forces. Similarly the actual sales in 1972 are 6.8 percent below the trend, and the explanation lies with cyclical and irregular factors. It should be clear that statements like these can be made only on the presumption that the trend equation is an apt description of the long-term trend of the series.

Description of Cycles—Moving Averages

In order to describe the cyclical element in a time series, it is necessary to develop a set of relatives that does not contain the irregular element in the time series. Irregular changes are mostly of short duration, and a process of *smoothing* the original time series will tend to eliminate them. A short-term *moving average* is used to achieve this smoothing effect.

Table 17.3 Decomposition of a Time Series—Revenues of a Local Utility ($ Millions)

Year	Actual Y	Trend T	Three-year moving average MA	Cyclical relative MA/T	Irregular relative Y/MA
1956	20.8	19.9			
1957	22.5	20.1	21.4	1.06	1.05
1958	20.9	20.3	21.5	1.06	0.97
1959	21.1	20.5	20.7	1.01	1.02
1960	20.1	20.7	18.9	0.91	1.06
1961	15.6	20.9	17.6	0.84	0.89
1962	17.0	21.1	18.1	0.86	0.94
1963	21.7	21.3	20.2	0.95	1.07
1964	21.8	21.5	22.9	1.06	0.95
1965	25.1	21.7	23.6	1.09	1.06
1966	23.8	21.9	24.0	1.10	0.99
1967	23.2	22.1	23.0	1.04	1.01
1968	22.1	22.3	22.3	1.00	0.99
1969	21.6	22.5	21.9	0.97	0.99
1970	22.1	22.7	21.5	0.95	1.03
1971	20.8	22.9	22.7	0.99	0.92
1972	25.3	23.1			

Calculation of Trend

$X = 0$ in 1964.

X is in yearly units.

$$\alpha = \frac{\sum Y}{N} = \frac{365.5}{17} = 21.5,$$

$$\beta = \frac{\sum XY}{\sum X^2} = \frac{81.6}{408} = 0.2;$$

$$T = 21.5 + 0.2\ X.$$

Calculation of Moving Average

$$MA(1957) = \frac{20.8 + 22.5 + 20.9}{3} = 21.4$$

$$\vdots$$

$$MA(1971) = \frac{22.1 + 20.8 + 25.3}{3} = 22.7$$

In Table 17.3 we show an annual time series of revenues for a local utility company. A least squares trend has been fitted to the series and the calculated trend values are shown. In the third column we show for each time point the average of the time series values for the year preceding the time point, the year of the time point, and the year following. This is called a three-year moving average, and its effect is to smooth out the irregularities in the year-to-year changes in the original data.† The moving average is graphed in Figure 17.3 along with the trend values. It contains the upward drift of the trend as well as a wave-like movement, or cycle. It has dampened out much of the irregular variation of the original series. *The moving average is an attempt to describe trend-cycle in combination.*

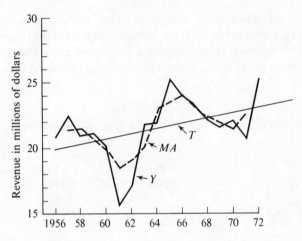

Figure 17.3   Actual (Y), Trend (T), and Three-Year Moving Average (MA) for Utility Series

The fourth column of Table 17.3 shows relatives of the moving average values to the trend values. These are the cyclical relatives in the multiplicative model for annual data.

$$\blacksquare \qquad C = \frac{MA}{T}. \qquad \blacksquare \qquad\qquad (17.3)$$

The logic of the multiplicative model is carried to a final step in the last column of Table 17.3. Here we show the irregular relatives. These are established *for annual data* by dividing the original values (Y) by the moving average for each time point.

$$\blacksquare \qquad I = \frac{Y}{MA}. \qquad \blacksquare \qquad\qquad (17.4)$$

This completes the decomposition; Figure 17.4 portrays graphically the several elements that have been separated out. Here we see a smooth long-term trend in the form of a straight line, and a wave-like cyclical variation. The irregular relatives form no discernible pattern, as indeed that term implies.

† The exact length of the moving average is a compromise. Too short a term will not sufficiently smooth out irregular variation, and too long a term will begin to smooth (or cancel out) a considerable part of the cyclical variation.

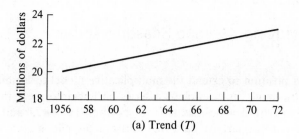

(a) Trend (*T*)

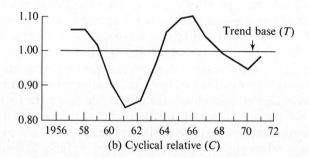

(b) Cyclical relative (*C*)

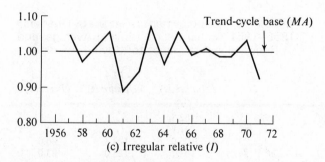

(c) Irregular relative (*I*)

Figure 17.4   Components of Variation for Utility Series

Time series decomposition using the multiplicative model can be summarized as follows for annual data. The model,

$$Y = T \times C \times I,$$

is implemented by calculating the elements

$$Y = T \times \frac{MA}{T} \times \frac{Y}{MA}.$$

The trend element, T, is found by fitting a simple (often straight-line) equation to the series. A short-term moving average is used to describe a combined trend-cycle level. This is often called the trend-cycle base. In terms of elements of variation represented we have in effect related the following

$$Y = T \times \frac{TC}{T} \times \frac{TCI}{TC}.$$

The Ratio to Moving Average Seasonal Index

We are now in a position to extend the multiplicative model to a time series containing seasonal variation. Our example is a series on unemployment in Arizona by quarters from 1956 to 1961. The original series is shown in Table 17.4 and Figure 17.5.

The element present in quarterly data that was not present in annual data is, of course, seasonal variation. Seasonal variation was defined as the typical pattern of variation within a calendar period of a year or less. The key to understanding the method for isolating the seasonal element is the idea that seasonal variations take place around a base level established by the two longer-term elements of variation—trend and cycle. Thus a first step is to determine this trend-cycle base. This is done by a moving average. The subsequent steps are designed to bring out the seasonal element and to separate it from the shorter-term irregular variations in the series.

The moving average used to establish the trend-cycle base covers a period of one year. This nets out the annual seasonal variation as well as irregular elements. In order to bypass a procedural detail (discussed later) we ask you to accept provisionally the moving-average values given in the second column of Table 17.4 and graphed in Figure 17.5. They contain a general upward drift, or trend, as well as a wave-like

Table 17.4 Unemployment in Arizona by Quarters,*
1956–1961 Actual Series, Moving Average, and
Ratios of Actual Series to Moving Average

		Y (Hundreds)	MA (Hundreds)	Y/MA (Percent)
1956	1	162.3		
	2	121.7		
	3	106.0	123.3	86.0
	4	103.7	125.0	83.0
1957	1	161.3	131.0	123.1
	2	136.0	144.0	94.4
	3	140.3	165.9	84.6
	4	173.0	193.5	89.4
1958	1	267.3	219.0	122.1
	2	250.7	232.7	107.7
	3	230.0	230.5	99.8
	4	192.7	217.1	88.8
1959	1	230.0	202.1	113.8
	2	180.7	193.4	93.4
	3	180.0	190.4	94.5
	4	173.0	190.6	90.8
1960	1	225.7	194.2	116.2
	2	186.3	205.6	90.6
	3	203.3	226.3	89.8
	4	241.0	250.4	96.2
1961	1	323.3	268.9	120.2
	2	282.0	275.6	102.3
	3	255.3		
	4	242.7		

* Source: Arizona Employment Security Commission.

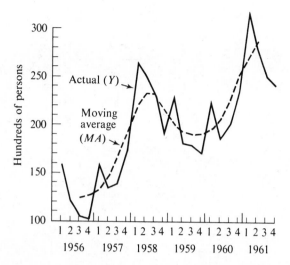

Figure 17.5 Unemployment in Arizona by Quarters, 1956–1961;
Actual Series and Moving Average

cyclical variation. The moving average line does not describe a within year seasonal pattern nor does it have short abrupt changes suggestive of irregular variation. In terms of kinds of variation present we say, as before,

$$MA = T \times C.$$

The central step in establishing the seasonal pattern is the calculation of ratios of the original series values (Y) to the moving average values (MA). The result of this step is

$$\frac{Y}{MA} = \frac{T \times C \times S \times I}{T \times C} = S \times I.$$

These values (Y/MA ratios) are shown in Table 17.4. We now have a set of values containing a combination of seasonal and irregular variation.

The final step in isolating the seasonal pattern is to average the Y/MA ratios for each term of the seasonal period. The purpose of this step is to net out the irregular elements present in individual Y/MA ratios. The calculation is shown in Table 17.5.

The seasonal index is the adjusted average of the Y/MA ratios for each quarter. The adjustment is made so that the average seasonal index will equal 100.0. The final seasonal indexes tell us that unemployment in the first quarter has averaged 19.9 percent above the trend-cycle base, second quarter unemployment has averaged 1.6 percent below the trend-cycle base, and so on. *The seasonal index is the average standing for each quarter's actual data relative to the trend-cycle base.*

Let us return now to the procedural problem that we postponed earlier—the detailed calculation of the moving average values. The problem stems from the fact that there is an even number of quarters in a year. Recall that when we calculated a moving average of annual data we used a three-term average. Any moving average value was properly centered at the second time point of the three included in its calculation. In the unemployment data, seasonal variation occurs over a period of four quarters. A four-quarter moving average is properly centered at the middle of the relevant period. For

Table 17.5 *Y/MA* Ratios by Quarters

Year	1	2	3	4	Total
		Quarter			
1956			86.0	83.0	
1957	123.1	94.4	84.6	89.4	
1958	122.1	107.7	99.8	88.8	
1959	113.8	93.4	94.5	90.8	
1960	116.2	90.6	89.8	96.2	
1961	120.2	102.3			
Total	595.4	488.4	454.7	448.2	
Average	119.1	97.7	90.9	89.6	397.3
Seasonal Index*	119.9	98.4	91.5	90.2	400.0

* (Average)(400.0/397.3)

the first four quarters, this would be June 30, 1956, (the end of the second quarter) and for the second four-quarter period, it would be September 30, 1956 (the end of the third quarter).

We wish to have moving average values whose time points correspond to those of the actual data. The reason for this is that we are going to calculate Y/MA ratios for each of the time points. The solution is to run a two-term average of the simple moving average. The average for the four quarters of 1956 is combined with the average for second quarter 1956 through first quarter 1957, and the result is properly centered half way between June 30, 1956, and September 30, 1956. This is August 15, the mid-point of the third quarter of 1956. The moving average shown in Table 17.4 is called a "two-of-a-four term" moving average. The easiest way to calculate such a moving average is as follows:

		Series Y	Four-quarter moving total	Two-term moving total	Moving average MA
1956	1	162.3			
	2	121.7	493.7		
	3	106.0	492.7	986.4	123.3 (=986.4/8)
	4	103.7	507.0	999.7	125.0 (=999.7/8)
1957	1	161.3	etc.	etc.	etc. etc.
	2	136.0			
		etc.			

Decomposition Including Seasonal Variation

Table 17.6 and Figure 17.6 show the complete decomposition of the quarterly unemployment series. First, a least squares trend was fitted to provide quarterly trend values.† This trend where T is the trend of unemployment at time X, is

† The calculations for this trend line are shown on page 370.

$$T = 201.9015 + 6.443X.$$

$X = 0$ in first quarter 1959.

X is in quarterly units.

Now we can isolate the cyclical component by relating the moving average values to the trend values. As before, the cyclical relatives are

$$C = \frac{MA}{T}.$$

The seasonal element is represented by the indexes we have determined for each quarter. It is a pattern of variation recurring each year as shown in (c) of Figure 17.6. As emphasized before, this pattern occurs around a base established by trend and cycle.

Finally we show a set of irregular relatives. These are viewed as relative variations from a base established by trend, cycle, and seasonal. They can be calculated by

$$\blacksquare \qquad I = \frac{Y}{MA \times S}. \qquad \blacksquare \qquad (17.5)$$

Table 17.6 Complete Decomposition of Arizona Unemployment Series

Period		Y (Hundreds)	T^* (Hundreds)	MA (Hundreds)	C (MA/T)	S	I [$Y/(MA \times S)$]
1956	1	162.3					
	2	121.7					
	3	106.0	137.5	123.3	0.90	0.915	0.94
	4	103.7	143.9	125.0	0.87	0.902	0.92
1957	1	161.3	150.4	131.0	0.87	1.199	1.03
	2	136.0	156.8	144.0	0.92	0.984	0.96
	3	140.3	163.2	165.9	1.02	0.915	0.92
	4	173.0	169.7	193.5	1.14	0.902	0.99
1958	1	267.3	176.1	219.0	1.24	1.199	1.02
	2	250.7	182.6	232.7	1.27	0.984	1.09
	3	230.0	189.0	230.5	1.22	0.915	1.09
	4	192.7	195.5	217.1	1.11	0.902	0.98
1959	1	230.0	201.9	202.1	1.00	1.199	0.95
	2	180.7	208.3	193.4	0.93	0.984	0.95
	3	180.0	214.8	190.4	0.87	0.915	1.03
	4	173.0	221.2	190.6	0.86	0.902	1.01
1960	1	225.7	227.7	194.2	0.85	1.199	0.97
	2	186.3	234.1	205.6	0.88	0.984	0.92
	3	203.3	240.6	226.3	0.94	0.915	0.98
	4	241.0	247.0	250.4	1.01	0.902	1.07
1961	1	323.3	253.4	268.9	1.06	1.199	1.00
	2	282.0	259.9	275.6	1.06	0.984	1.04
	3	255.3					
	4	242.7					

* The quarterly trend is $T = 201.9015 + 6.443X$, with $X = 0$ in 1st quarter, 1959.

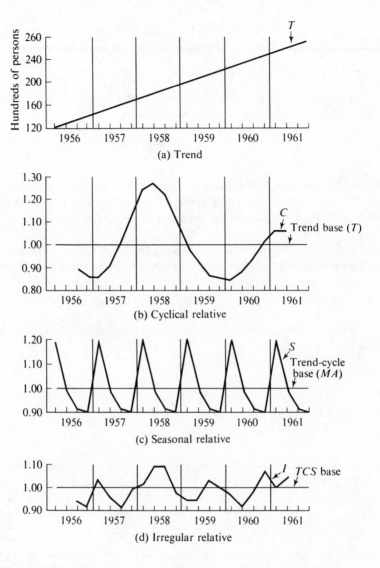

Figure 17.6   Components of Quarterly Unemployment Series

Our decomposition is now complete. The original series can be expressed entirely in terms of components that we have separated out. The model

$$Y = T \times C \times S \times I$$

is represented in the individual elements

$$Y = T \times \left(\frac{MA}{T}\right) \times S \times \left(\frac{Y}{MA \times S}\right).$$

Thus, we have calculated

Actual value = (Trend) × (Cyclical relative) × (Seasonal relative)

× (Irregular relative).

Refinements in Measuring Seasonal Variation

In determining seasonal indexes we have assumed that the seasonal pattern does not change over time. The assumption of a constant seasonal factor can be checked by looking at the sequence of Y/MA values for each term of the seasonal period. While individual Y/MA values contain irregular variation, this variation will not cause systematic drifts in the values for individual months (or quarters). Such drifts over a period of time would reflect a changing seasonal pattern. Methods for describing a changing seasonal pattern include a step whereby "trends" are fitted to the drifts of the Y/MA values in order to produce a different set of seasonal indexes for each year.

Computer programs have been devised to introduce refinements into the basic ratio-to-moving-average method. One program in widespread use is known as the BLS (Bureau of Labor Statistics) Seasonal Factor Method. The principal refinement is a modification of the moving average line so that it more fully reflects the complete cyclical variation (a one-year moving average actually tends to understate the total cyclical swing because it is not sensitive at the cycle turning points). This is done within a model that allows for changes in the seasonal pattern over time. The procedure is based on the idea that components or variations that are imperfectly reflected at early stages of the time series decomposition will show up in the final stage when the irregular relatives are calculated. These elements, particularly cyclical variation, can then be removed from the irregular relatives and put back into the moving average line where they belong.† The entire process of decomposition is then repeated using the revised trend-cycle base. The resulting new set of irregular relatives is examined for non-irregular components, and the process of reintroducing these into the moving average can be done again, or "iterated." Experience has shown that three iterations usually are sufficient to recapture all cyclical elements into the trend-cycle base. The moving average from the final iteration is used for the final estimates of the moving seasonal pattern. There are also procedures to prevent one-time extreme irregular variation (as might result from strikes, floods, domestic riots, etc.) from dominating portions of the moving average.

Summary

Time series decomposition proceeds from a model that regards a time series as the result of sets of forces that produce different patterns of variation. These different patterns of variation are known as *trend*, *cyclical* variation, and *seasonal* variation. The residual variation remaining when these patterns are identified and removed from the data is termed *irregular* variation. Simple equations fitted to the data are often adequate descriptions of trend. The key to describing the remaining elements of variation is the calculation of a short-term moving average which forms a trend-cycle base. In a multiplicative model, the ratios of moving average to trend will describe relative cyclical variation. Seasonal variation is described by finding the average ratio of a   period   moving   average   to   trend   for   each   term   in   the   seasonal

† Figure 17.6(d) of the irregular relatives for the Arizona unemployment series contains the kind of cyclical pattern referred to.

period. If desired, the entire decomposition may be completed by calculating irregular relatives from the ratios of original series values to a base level which represents the combined effect of all the other elements of variation present in the series.

See Self-Correcting Exercises 17.A

Exercises

1. From the following data on sales ($ millions) of a machine tool company, fit a least squares trend line to the five-year averages.

1950	9.8	1955	11.4	1960	11.4	1965	11.6
1951	13.6	1956	17.0	1961	11.7	1966	13.0
1952	14.4	1957	12.7	1962	12.3	1967	11.1
1953	8.7	1958	9.9	1963	10.4	1968	16.0
1954	8.2	1959	10.9	1964	7.0	1969	13.3
Average	10.9		12.4		10.6		13.0

2. Reexpress your trend equation from Exercise 1 in annual terms with 1960 as the year of origin where $X = 0$. Find trend values for 1950 and 1965. Plot the trend on a graph along with the original series.

3. Fit a five-year moving average to the data in Exercise 1. Plot the moving average line on a graph along with the original data and the trend line from Exercises 1 and 2. Comment on the components of time series variation present in each of the three lines.

4. Determine the cyclical-irregular relative and the cyclical relative for 1965 from your calculations in the first three exercises.

5. From the following data on quarterly sales of a merchandising firm ($ millions), determine seasonal indexes by the ratio to moving average method.

	Year					
Quarter	1967	1968	1969	1970	1971	1972
1		40	49	44	47	60
2		50	60	58	63	67
3	41	50	60	56	72	
4	54	70	73	70	80	

6. Determine a least squares trend equation from your moving average values for the third quarter for the data in Exercise 5. Express the trend as an equation with origin, $X = 0$, in the first quarter of 1970 and the trend increment in quarterly terms.

7. Plot the original data and moving averages from Exercise 5 and the trend line from Exercise 6 on the same graph. Comment on the elements of time series variation reflected by each of the three lines.

8. From the following data on discount rates (in quarter percents) of the New York Federal Reserve Bank, determine the least squares trend line.

	1953	1954	1955	1956	1957	1958	1959
Jan. 1		8	6	10	12	12	10
July 1	8	6	7	11	12	7	14

9. Plot the original data along with the trend line from Exercise 8. Comment on the deviations from trend.

10. From the following data on orders received by a newly established company, determine the least squares trend line. Do you think the least squares trend is a good description of growth in orders over the period?

Period	1	2	3	4	5	6	7	8	9	10	11	12
Orders	2	2	0	3	0	4	2	4	7	11	15	17

Applications of Time Series Analysis

Up to this point we have been concerned with analyzing the past levels of a time series, which led to the method of time series decomposition. We shift emphasis now, to use time series analysis to interpret current levels and forecast future levels of a time series. When using the past to interpret the present and foreshadow the future, it is wise to remember that present and future time "slices" are not random samples of past time periods. For example, there is no statistical justification for using past variations of actual data from a calculated trend to make probabilistic statements about future variations from trend. Indeed in our discussion of time series decomposition we did not even suggest that these variations were the result of sampling errors. While we search for patterns that have exhibited some degree of regularity in the past, there is no guarantee that changing circumstances will not destroy their usefulness for the present and future. The best insurance here is for the analyst to have a close familiarity with the business and economic environment relating to the time series at issue.

Adjustments for Seasonal Variation and Trend

One purpose for measuring seasonal variation is to enable persons to remove this element from short-term comparisons of levels of the time series. Suppose that you were in the position in early 1962 of evaluating the course of unemployment in Arizona during the preceding year. In 1961 a number of programs designed to reduce unemployment had in fact been inaugurated as part of the New Frontier of the Kennedy administration. In this light should you take satisfaction in the record shown in Table 17.7?

Table 17.7   Unemployment in Arizona during 1961

		Hundreds of persons	
		Unemployed	*Change from first quarter*
1961	first quarter	323.3	—
	second quarter	282.0	−41.3
	third quarter	255.3	−68.0
	fourth quarter	242.7	−80.6

A politician who boasted that unemployment had fallen by eight thousand persons since the first quarter of 1961 when the Kennedy administration began could not be accused of using false data. But we know from the seasonal pattern that unemployment is seasonally high in the first quarter and low in the fourth quarter. Thus there is a normal seasonal drop in unemployment between the first quarter and the fourth quarter.

Removing this normal seasonal variation from the actual unemployment figures can be helpful. The adjustment is accomplished by dividing the actual data by the seasonal relative.

$$\blacksquare \qquad Y_{\text{adjS}} = \frac{Y}{S}. \qquad \blacksquare \qquad (17.6)$$

The data adjusted for seasonal variation state the level of the series brought about by trend and cyclical and irregular variation. In Table 17.8 we repeat Table 17.7 using seasonally adjusted unemployment. The change in seasonally adjusted unemployment

Table 17.8   Seasonally Adjusted Unemployment in Arizona during 1961

		Hundreds of persons	
		Unemployed (*adjusted for seasonal variation*)	*Change from first quarter*
1961	first quarter	323.3/1.199 = 269.6	—
	second quarter	282.0/0.984 = 286.6	+17.0
	third quarter	255.3/0.915 = 279.0	+ 9.4
	fourth quarter	242.7/0.902 = 269.1	− 0.5

from first to fourth quarter is a reduction of only 50 persons. The period-to-period change, adjusted for seasonal variation, can be written

$$\blacksquare \qquad \Delta_{\text{adjS}} = \frac{Y_n}{S_n} - \frac{Y_0}{S_0}, \qquad \blacksquare \qquad (17.7)$$

where Δ is the capital Greek letter " delta "—often used where differences are involved —and

$$Y_n = \text{actual value, later period,}$$
$$Y_0 = \text{actual value, earlier period,}$$
$$S_n = \text{seasonal relative for later period,}$$
$$S_0 = \text{seasonal relative for earlier period.}$$

One might wish to correct a period-to-period change for trend. In fairness to our politician we should grant that there is a normally expected increase in unemployment in Arizona arising from growth in state population and perhaps other long-term factors. This increment is 6.44 hundred persons per quarter (see Table 17.6), or $3(6.44) = 19.32$ hundred over a three-quarter interval. The period-to-period change with the trend effect removed is

$$\Delta_{adjT} = Y_n - Y_0 - n\beta, \tag{17.8}$$

where n is the number of periods between the earlier and later values, and β is the trend increment per period. For first to fourth quarter above,

$$\Delta_{adjT} = 242.7 - 323.3 - 3(6.44)$$
$$= -80.6 - 19.3 = -99.9.$$

Unemployment adjusted for trend alone declined by about 10,000 persons. The fairest comparison in the present situation is to adjust for both seasonal variation and trend. The justification is that effects of recent political and economic programs, viewed against an established trend, will show up as either cyclical or irregular variation. To remove both the normal seasonal change and trend increment from a period-to-period comparison we can calculate

$$\blacksquare \qquad \Delta_{adjST} = \frac{Y_n}{S_n} - \frac{Y_0}{S_0} - n\beta. \qquad \blacksquare \tag{17.9}$$

For our example

$$\Delta_{adjST} = 269.1 - 269.6 - 19.3$$
$$= -0.5 - 19.3 = -19.8.$$

The change attributable to cyclical and irregular forces is a decrease in unemployment of 1980 persons. The New Frontier programs are one among many causal factors that might be advanced to explain this change. In correcting for past seasonal variation and trend we have at least removed some effects that are probably irrelevant to the politician's claim.

Forecasting a Time Series

We return now to the series on unemployment in Arizona and place ourselves in the position of having to forecast unemployment by quarters for 1962 on the basis of our analysis for 1956–1961. The complete forecasting model for the multiplicative case is

$$\hat{Y} = \hat{T} \times \hat{C} \times \hat{S}. \tag{17.10}$$

Here we place "hats" over the symbols for the elements of variation to emphasize that forecasts are estimates. Our first task is to forecast the trend element, for which we use the trend equation. Our trend measure for the series is a least squares straight line fitted to the average annual unemployment figures for the six years. This is done for convenience—the 24 quarterly figures given in Table 17.6 could just as well have been used. The results are given in Table 17.9.

Table 17.9 Data for Calculating Straight-Line Trend for Arizona Unemployment Using Annual Figures, 1956-1961

Year	Y	X	X^2	XY
1956	123.42	−5	25	−617.10
1957	152.67	−3	9	−458.01
1958	235.17	−1	1	−235.17
1959	190.92	1	1	190.92
1960	214.08	3	9	642.24
1961	275.83	5	25	1379.15
Totals	1192.09		70	902.03

$$\beta = \frac{\Sigma XY}{\Sigma X^2} = \frac{902.03}{70} = 12.886,$$

$$a = \frac{\Sigma Y}{N} = \frac{1192.09}{6} = 198.68;$$

$T = 198.68 + 12.886X.$

$X = 0$ in $1958\frac{1}{2}$ (that is, January 1, 1959).

X is in 6-month units.

Note that we have an annual series with an even number of values as was illustrated in Table 17.1 for Unit B. In the scale of X used in calculating the trend equation in Table 17.9, $X = 0$ in $1958\frac{1}{2}$. This is the year from July 1, 1958 to June 30, 1959, which has a mid-time point of January 1, 1959. Hence, $X = 0$ on January 1, 1959 and X is measured in six-month units, so $X = 1$ on July 1, 1959. We would like a trend expression with an origin, say of first quarter 1959. This is a centered time point of February 15, 1959, which is one-half quarter removed from January 1, 1959. Since the original trend increment comes out in six-month units, we must use $0.25 = X$ to find the trend for first quarter 1959.

$$T(1959-1) = 198.68 + 12.886(0.25)$$
$$= 198.68 + 3.2215 = 201.9015.$$

To express a quarterly trend we want the quarterly trend increment in unemployment. This is the six-month increment divided by 2, or 6.443. A quarterly trend equation is, then,

$$T = 201.9015 + 6.443X.$$

$X = 0$ in first quarter 1959.

X is in quarterly units.

Using this form of the trend equation, the forecast of trend for each quarter of 1962 is

$$T(1962-1) = 201.9015 + 6.443(12) = 279.2,$$
$$T(1962-2) = 201.9015 + 6.443(13) = 285.7,$$
$$T(1962-3) = 201.9015 + 6.443(14) = 292.1,$$
$$T(1962-4) = 201.9015 + 6.443(15) = 298.5.$$

In terms of the multiplicative model the next element to forecast is the cyclical relative. Even when description of the cyclical element is refined, it proves difficult to forecast. Past cycles are often nonuniform in magnitude and duration. Further, in the usual moving average methods, the cyclical ratios will terminate one-half a year prior to the end of the observed series. If we choose not to include a cyclical forecast, the result obtained is *as if the* cyclical relative remains a constant 1.0.

Applying the seasonal element is straightforward. If we are satisfied that the seasonal pattern is a stable one, the seasonal relatives are simply applied to the trend-cycle base forecast. Forecasting Arizona unemployment by quarters for 1962 (using a cyclical relative of 1.0) yields

$$\hat{T} \times \hat{C} \times \hat{S} = \hat{Y},$$

$$
\begin{array}{rl}
1962 \quad 1 & 279.2 \times 1.0 \times 1.199 = 334.8, \\
2 & 285.7 \times 1.0 \times 0.984 = 281.1, \\
3 & 292.1 \times 1.0 \times 0.915 = 267.3, \\
4 & 298.5 \times 1.0 \times 0.902 = 269.2.
\end{array}
$$

If cyclical ratios could be obtained through a fundamental analysis of economic conditions and prospects, they would replace the 1.0 values that we show above.

Index Numbers

The purpose of an index number is to allow the behavior of a group of variables to be reduced to a single variable. In this chapter the variables to be combined are *time* series, but other kinds of comparisons may also be made by means of index numbers.[†] While index numbers are averages, certain problems not generally associated with simple averages give time series index numbers a claim to special treatment.

Index numbers have been employed in business and public policy decisions since World War I. Periods of price inflation here and in Europe gave impetus to studies of price-level changes. Business contracts frequently contain clauses in which various recognized price indexes form the basis for cost and price adjustments. Farm price support programs have been keyed to the concept of "parity" since the early 1930's. The concept of parity involves two indexes, one an index of prices received by farmers for goods they sell, and one of prices paid by farmers for goods they buy. Wages of many workers are tied through "escalator" clauses to changes in the Consumer Price Index of the Bureau of Labor Statistics.

Relatives of Price, Quantity, and Value

Consider the data in Table 17.10 on average prices, quantities purchased, and resulting total expenditure per week on three breakfast items in 1968 and 1969 by a hypothetical young couple.

† For example IQ, or intelligence quotient, is a complex index number. An individual's scores in a number of intellective areas are reduced to a single number.

Comparisons may be made between the two years as to the price, quantity purchased, and total expenditure on each item by means of relatives. These relatives in the table tell us, for instance, that the price of eggs in 1969 was 90 percent of its price in 1968, that the value of expenditures on bacon in 1969 was 133.3 percent of the corresponding expenditure in 1968. But how do we say something meaningful about overall price change, overall quantity change, or overall change in expenditures for the family over the period. This is the purpose served by an index number. An index number is an overall percentage relative that encompasses in some manner the change in a group of related individual series.

Table 17.10   Average Prices Paid, Consumption and Expenditures per Week by a Young Couple for Three Breakfast Items, 1968 and 1969

Item	Unit	Price 1968	Price 1969	Consumption 1968	Consumption 1969	Expenditure 1968	Expenditure 1969
Bacon	(lb)	$0.75	$0.80	1.6	2.0	$1.20	$1.60
Eggs	(doz)	0.80	0.72	2.0	1.5	1.60	1.08
Coffee	(lb)	1.20	0.96	1.2	1.5	1.44	1.44

	Relatives		
	(p_1/p_0)	(q_1/q_0)	(v_1/v_0)
Bacon	1.067	1.25	1.333
Eggs	0.900	0.75	0.675
Coffee	0.800	1.25	1.000

$p_1 = $ 1969 price $q_1 = $ 1969 consumption $v_1 = $ 1969 expenditure
$p_0 = $ 1968 price $q_0 = $ 1968 consumption $v_0 = $ 1968 expenditure

Aggregate Price Indexes

One may seek a single combined or an overall relative from a summation of weighted original prices. The simplest such aggregate price index can be obtained by totaling the unit prices of a list of commodities in two years and striking the relative of the resulting totals. The computation of this index for the prices of the three breakfast items is shown in Table 17.11.

The value extension columns, which are summed, in this case are identical with the price columns. But this is only because what might be called *unit* weights have been employed; that is, each price per unit influences the index only to the extent of one unit. Consider the relative influence each commodity would have on the aggregate index if we changed the unit for the quotation of the price of eggs to the crate! The meaning of any aggregate index is restricted to the set of quantity weights used. The weights implicit in summing a list of unit prices are one unit of each of the listed commodities.

Table 17.11 Calculation of a Simple Aggregate Price Index for 1969 on 1968 as a Base for Three Breakfast Items

Item	Unit	1968			1969		
		Price p_0	Unit quantity q	Value extension $p_0 q$	Price p_1	Unit quantity q	Value extension $p_1 q$
Bacon	(lb)	0.75	1	0.75	0.80	1	0.80
Eggs	(doz)	0.80	1	0.80	0.72	1	0.72
Coffee	(lb)	1.20	1	1.20	0.96	1	0.96
				2.75			2.48

$$\text{Index} = \frac{\sum (p_1 q)}{\sum (p_0 q)} \times 100 = \frac{2.48}{2.75} \times 100 = 90.2.$$

The aggregate index discussed above is clear in meaning but generally not too useful. If all aggregate indexes are really weighted indexes, one should make certain that the weights are related to the purpose of the index. The quantities consumed per week are such a set of weights, because they indicate the importance of each item to our young couple. Let us apply the quantities of the base year, 1968, as shown in Table 17.12.

Table 17.12 Calculation of Weighted Aggregate Prices for Three Breakfast Items Employing 1968 Quantity Weights

Item	Unit	p_0	q_0	$p_0 q_0$	p_1	q_0	$p_1 q_0$
Bacon	(lb)	0.75	1.6	$1.20	$0.80	1.6	$1.28
Eggs	(doz)	0.80	2.0	1.60	0.72	2.0	1.44
Coffee	(lb)	1.20	1.2	1.44	0.96	1.2	1.152
				$4.24			$3.872

We find that the quantities of breakfast items consumed weekly in the base year (1968) cost the family $4.24 = $\sum (p_0 q_0)$. To purchase *these same quantities* at 1969 prices would cost the family $3.872 = $\sum (p_1 q_0)$. We have, in effect, priced a particular market basket of goods in both years. The relative price of the fixed basket of goods can form a useful index of prices.

$$PI = \frac{\$3.872}{\$4.240} \times 100 = 91.3.$$

The price of the fixed market basket of breakfast items in 1969 was 91.3 percent of the price in 1968, that is, 8.7 percent less.

The year used as a basis of comparison in an index number is called the base year. In our comparison above, 1968 was the base year. If the fixed quantity weights in a price index are those of the base year, the index is called a base-weighted price index.

Definition	A fixed base-weighted price index measures the relative price of a fixed base-year bill (or market basket) of goods. At any point in time the index compares the total price of the bill of goods to the total price of the same bill of goods in the base year.

$$\blacksquare \quad PI_n = \frac{\sum (p_n q_0)}{\sum (p_0 q_0)} \times 100. \quad \blacksquare \qquad (17.11)$$

It should be evident from the previous discussion that fixed weights (quantities) are basic to the aggregate method of construction of price indexes. Should these weights represent the quantities of the base year or those of the current year? In consumer price indexes this means should the index be designed to reflect a past consumption pattern or the most recent one? There are two possible weighted aggregate indexes, one embodying base year weights and the other current year weights. These are commonly called the Laspeyres index and the Paasche index, after economists who proposed them. The Laspeyres index is the one we calculated for the breakfast prices. The Paasche index is defined as

$$PI \text{ (Paasche)} = \frac{\sum (p_n q_n)}{\sum (p_0 q_n)} \times 100.$$

This index prices the current year's bill of goods in both the current and base year. It compares the current cost of what is consumed today with what it would have cost to consume these same things in the base year.

Most of the price indexes prepared by governmental agencies use a historic set of quantity weights, though the period they represent is not necessarily that of the base year. For example, during the 1960's the Bureau of Labor Statistics Consumer Price Index had a set of weights for 1960–1961, but the base year was the period 1957–1959. The CPI weights are based on an extensive survey of consumer expenditure patterns, and it is not practical to repeat this expensive survey each year in order to obtain weights for a Paasche index. In order to prevent the Consumer Price Index from reflecting in time an obsolete consumption pattern, a new expenditure study is carried out roughly every ten years. The indexes for the late 1970's will be based on a set of consumption weights for 1972–1973, and the index will have 1967 as the base year.

Weighted Mean of Relatives Price Index

An alternative formulation of the base weighted price index involves the weighting of price relatives by base-year values. The formula is

$$\blacksquare \quad PI_n = \frac{\sum \left[\frac{p_n}{p_0} (p_0 q_0) \right]}{\sum (p_0 q_0)} \times 100. \quad \blacksquare \qquad (17.12)$$

It is evident that the p_0 elements in the numerator cancel out. The result is $\sum (p_n q_0) / \sum (p_0 q_0)$, the base-weighted relative of aggregates just discussed. Why have another formula for the same index?

The answer is that value weights are often easier to obtain than quantity weights. A housewife is more likely to know how much she spends for milk than how many quarts her family consumes. A further reason is that in a comprehensive price index it is impossible to price every item consumers might buy. In developing the Consumer Price Index, the Bureau of Labor Statistics establishes price relatives for commodity classes (such as dry breakfast foods, men's shoes, etc.) by sampling price changes for selected items in selected establishments. Then these (class) relatives are weighted by the proportion of family budgets spent on the different product classes. The base weights are determined from the periodic studies of consumer expenditures mentioned earlier.

Aggregate Quantity Index

It would seem at first that the main difficulty in constructing quantity indexes is the different units in which quantities are measured. How does one, for example, add tons of steel, gallons of gasoline, and kilowatts of electric power in constructing an index of industrial production? Consider, however, a situation in which all the quantity units are the same. Such a situation might prevail for production of selected grain crops, which are commonly measured in bushels. Why not simply find the total bushels of output in each year for all the crops included and construct from this an index of agricultural production for these crops? The answer is that one bushel of corn and one bushel of wheat may not be of equal importance. The only way of measuring the importance of identical quantity units in a money exchange economy is to refer to the judgment of that economy in terms of unit prices. The selected unit prices will become weights in the quantity index in the same way that quantities become the weights in aggregate price indexes. The resulting weighted aggregate index will measure the relative change in *value* of production associated with changes in quantities produced, prices being held constant at levels prevailing at a particular time. The employment of a simple aggregate of quantities produced would amount to the application of equal price weights for all commodities.

Having recognized the need for price weights in an aggregate quantity index, let us return to our three breakfast items to develop the quantity index using base year (1968) prices as weights as shown in Table 17.13.

We find that the couple spent \$4.24 a week to obtain the quantities of breakfast items actually consumed in 1968, that is, $\sum (p_0 q_0) = \$4.24$. In 1969 the quantities consumed weekly changed somewhat from 1968. The quantities consumed in 1969, were

Table 17.13 Calculation of Weighted Aggregates for a Quantity Index of Three Breakfast Items Using Base Year (1968) Price Weights

Item	Unit	p_0	q_0	$p_0 q_0$	p_0	q_1	$p_0 q_1$
Bacon	(lb)	\$0.75	1.6	\$1.20	\$0.75	2.0	\$1.50
Eggs	(doz)	0.80	2.0	1.60	0.80	1.5	1.20
Coffee	(lb)	1.20	1.2	1.44	1.20	1.5	1.80
				\$4.24			\$4.50

they bought at 1968 prices, would have cost \$4.50. The effect of the changes in quantities consumed would have raised the couple's weekly cost of breakfasts even if prices had stayed the same.

$$QI = \frac{\$4.50}{\$4.24} \times 100 = 106.1.$$

The effect of changes in quantities consumed would have been to raise expenditures on breakfasts by 6.1 percent, holding prices constant at 1968 levels.

Definition | A fixed base-weighted quantity index measures the effect on value of changes in quantities between the base year and the given year. In measuring this effect, prices are held constant at base-year levels.

$$QI_n = \frac{\sum (p_0 q_n)}{\sum (p_0 q_0)} \times 100. \tag{17.13}$$

Price, Quantity, and Value

We have seen that a base-weighted price index measures the relative change in value attributable to price changes. Comparable quantity indexes measure the relative change in value attributable to quantity changes. Since value is the product of price and quantity, we might think that the aggregate change in value (measured by an index) should be the product of a price index and a quantity index. Let us examine this.

Our couple actually spent \$4.12 weekly on breakfasts in 1969 and \$4.24 in 1968 (see Table 17.10). The value index relates these two aggregates.

$$VI_n = \frac{\sum (p_n q_n)}{\sum (p_0 q_0)} \times 100. \tag{17.14}$$

For our data,

$$VI = \frac{\$4.12}{\$4.24} \times 100 = 97.2.$$

Our base-weighted price index tells us that the effect of price changes on a constant (base-year) bill of goods would have been to lower expenditures by 8.7 percent.

$$PI = \frac{\sum (p_1 q_0)}{\sum (p_0 q_0)} \times 100 = \frac{\$3.872}{\$4.24} \times 100 = 91.3.$$

The base-weighted quantity index tells us that the effect on expenditures of the quantity changes alone would have been to raise expenditures by 6.1 percent.

$$QI = \frac{\sum (p_0 q_1)}{\sum (p_0 q_0)} \times 100 = \frac{\$4.50}{\$4.24} \times 100 = 106.1.$$

The product of the price index and the quantity index is

$$(0.913)(1.061) = 0.969, \quad \text{or } 96.9 \text{ percent,}$$

which is quite close to the value index of 97.2 percent.

The base-weighted price and quantity indexes do a reasonable job of "factoring out" the effects of price and quantity changes on total value. This is important because price indexes are frequently used to adjust value series so that they become indicators of quantities. Adjustment of dollar income to an indication of real income is a case in point. Shown below is the median income of families in the United States for several different periods.

Year	Median family income
1950	$3319
1955	4421
1960	5620
1964	6569

If a family had these incomes in the different years, we might ask whether its real income level in terms of goods and services had also increased and by how much. The income figures are a value series, and a related price index is the B.L.S. Consumer Price Index, which measures the change in price of a fixed bill of goods purchased by urban wage earners and moderate income families. Below we show the income figures, V, along with the Consumer Price Index for the same years, P.† The base of the Consumer Price Index is the average for the period 1957–1959.

	V	P	Q	$1/P$
1950	$3319	83.8	$3961	1.193
1955	4421	93.3	4738	1.072
1960	5620	103.1	5451	0.970
1964	6569	108.1	6077	0.925

In the third column, under Q, we show the result of dividing the income figure by the Consumer Price Index for the same year. The result is called *real income in 1957–1959 dollars*. That is, it indicates the quantity of goods, (valued at 1957–1959 prices) that each year's income could purchase. Thus, while the median money income increased by $1199 dollars from 1955 to 1960, the typical basket of goods and services that could be purchased with the median income increased by only $713 dollars. The difference is attributable to price increases.

In the final column we show the reciprocal of the price index, commonly called the purchasing power of the dollar. The purchasing power of the dollar is 1.00 in the 1957–1959 base period. The purchasing power of the dollar was 19 percent greater in 1950 than in 1957–1959. That is, one dollar in 1950 purchased 19 percent more goods and services than one dollar in 1957–1959. Similarly, the 1964 dollar bought 7.5 percent less than the 1957–1959 dollar. The purchasing power of the dollar shows the change in buying power of a fixed number of dollars, that is, a constant money income in each year.

The major problem in the adjustment of any particular value series is the selection of a really suitable price index. Ideally the price index selected should belong to the same index number system as the existing value series and the desired quantity index.

† U.S. Bureau of the Census, *Pocket Data Book, USA 1967*, pp. 191, 196.

For instance, if one wished to adjust dollar sales in the chemical industry to reflect changes in quantities sold, the adjustment should be made with an index of prices in the chemical industry. The chemical segment of the BLS Wholesale Price Index might be the most applicable price index available. However, if one were interested in a similar adjustment of sales in a particular segment of the chemical industry, or for a particular firm in the industry, a price index of correspondingly limited scope would be called for. To use the index of prices in the entire industry would be to make an assumption that may not be justified—namely, that movements in a particular restricted group of prices were similar to those of the industry as a whole.

Shifting the Base of an Index

The base employed in an index is not the one we wish to use in all comparisons. For example the table below shows an index of farm output and an index of farm prices for selected years. The base of the output index is 1957–1959, but the base of the price index is 1910–1914. The historic base is used because of its connection with farm price support programs.

Year	Index of farm output (1957–59 = 100)	Index of farm prices (1910–14 = 100)
1950	86	258
1955	96	232
1960	106	238
1965	116	248

U.S. Bureau of the Census, *Pocket Data Book, USA 1967*, p. 213.

Suppose we wish to compare the changes in output with the price changes from 1950 on. It would be best to put both indexes on the same base. To change the base of each index to 1950, the indexes for each series are divided by the existing index for 1950, the desired new base year. For example, 86 is the 1950 index of the first series and the remaining three numbers in the series are divided by 86 to shift the base to 1950.

Statement	To change the base of an index series, the existing indexes are divided by the existing index for the desired new base period and stated in percentage form.

The series with 1950 as a base is

Year	Index of farm output	Index of farm prices
1950	100.0	100.0
1955	111.6	89.9
1960	123.3	92.2
1965	134.9	96.1

From these figures we can get a quicker appreciation of how changes in prices since 1950 compare with changes in output. We see, for example, that from 1950 to 1955 the drop in farm prices was of the same relative magnitude as the rise in farm output. Over the entire period from 1950 to 1965, however, output rose by 34.9 percent while prices fell only 3.9 percent.

Summary

The results of a time series decomposition are used to interpret current period-to-period changes in the time series as well as to forecast future values of the time series. In both instances trend and seasonal variation are most frequently taken into account because they are generally the more persistent and predictable elements. The purpose of an index number is to combine the behavior of a number of related series in a single series. The series of concern may be price series, quantity series, or value series. In a given context these are related by the identity that price times quantity equals value. Base-weighted price indexes measure the effect of changing prices on value, and corresponding quantity indexes measure the effect of changing quantities on value. The same fundamental identity is used to convert a value series to a quantity series, by dividing the value series by a related price index series. The most common application of this adjustment for price changes is in creating so-called " real income " series. The base of an index series can be shifted to a desired period by dividing the series by the value for the desired new base period.

See Self-Correcting Exercises 17.B

Exercises

1. For the year 1971, the quarterly sales figures from Exercise 5 of the previous exercises (page 366) were

    1st Q   47;   2nd Q   63;   3rd Q   72;   4th Q   80.

    Adjust each quarter's sales figures for seasonal variation and comment on the resulting period-to-period changes in the adjusted figures.

2. Adjust the change in sales from first to fourth quarter of 1971 in Exercise 1 for both seasonal variation and trend, and comment on the result.

3. Use your trend equation from Exercise 6, page 366, and your seasonal indexes from Exercise 5, page 366, to forecast sales for 1973. What major element of time series variation has been omitted from the forecast? How could this element be included?

4. From the following data (adapted from published sources) on shellfish catch and prices in the United States, construct a relative of aggregates price index for 1965 on 1950 as a base, using 1950 weights.

Variety	Catch (mil. lb)		Av. price per lb	
	1950	1965	1950	1965
Clams	40	70	$0.25	$0.23
Crabs	160	320	0.06	0.09
Lobsters	20	30	0.32	0.64
Oysters	80	50	0.24	0.50
Shrimp	200	240	0.18	0.32

5. From the data in Exercise 4, construct a weighted relative of aggregates quantity index for 1965, using 1950 as a base and employing base-year weights.

6. Do the price and quantity changes measured in Exercises 4 and 5 appear to account for the change in value of catch between 1950 and 1965?

7. From the following data (adapted from published sources) on prices and values of production of poultry products in the United States, construct a weighted average of relatives price index for 1970, using 1960 as a base. How would you interpret the resulting index?

Unit	Product	Price		Production value in 1960 ($ millions)
		1960	1970	
lb	Chickens	$0.120	$0.090	140
doz	Eggs	0.360	0.378	1850
lb	Turkeys	0.250	0.225	380

8. From the following data on purchases and prices of new and used cars in the United States, construct a base-weighted price index for 1970 with 1967 as a base.

	Purchases per 100 households		Average price		Annual expenditures
	New	Used	New	Used	(New & used)
1967	11.9	22.0	$2,731	$840	$30.4 billion
1970	11.1	19.6	3,018	959	33.3

9. Adjust the 1970 expenditure figure in Exercise 8 for the price change between 1967 and 1970. What does this adjusted figure tell you?

10. From the following indexes of farm output and prices received by farmers, construct an index of value of farm output using 1950 as a base.

	1950	1955	1960	1965
Farm output (1957–59 = 100)	86	96	106	116
Prices received (1910–14 = 100)	258	232	238	248

Time	Y	X	X^2	XY
1968	10	-2	4	-20
1969	18	-1	1	-18
1970	22	0	0	0
1971	20	1	1	20
1972	30	2	4	60
	100	0	10	42

$$\alpha = \frac{\Sigma X}{n} = \frac{100}{5} = 20 \qquad \beta = \frac{\Sigma XY}{\Sigma X^2} = \frac{42}{10} = 4.2$$

$$T_{70} = 20 + 4.2X$$

					110
Time	Y	X	X^2	XY	76
1969	18	-3	9	-54	34
1970	22	-1	1	-22	
1971	20	+1	1	+20	
1972	30	+3	9	+90	
	90	0	20	34	

$$\alpha = \frac{\Sigma X}{n} = \frac{90}{4} = 22.5 \qquad \beta = \frac{\Sigma XY}{\Sigma X^2} = \frac{34}{20} = 1.7$$

$$T_{70\frac{1}{2}} = 22.5 + 1.7X$$

$$T_{70} = 20.8 + 3.4X$$

Glossary of Equations

$$\alpha = \frac{\sum Y}{N}$$

$$\beta = \frac{\sum XY}{\sum X^2}$$

The equations for the constants of a least square trend line of the form $T = \alpha + \beta X$ when the X, or time, variable is arranged so that $\sum X$ over the N time values is equal to zero.

$$CI = \frac{Y}{T}$$

Cyclical-irregular relatives are obtained for annual data by relating the original series value to the trend value for each period.

$$C = \frac{MA}{T}$$

Cyclical relatives are obtained by relating moving-average values to trend values for corresponding periods. For annual series a short-term moving average is used, and for series containing seasonal variation the moving average must extend over the period of seasonal variation.

$$I = \frac{Y}{MA}$$

Irregular relatives for annual data are obtained by relating original series values to moving-average values for corresponding periods.

$$I = \frac{Y}{MA \times S}$$

Irregular relatives for series containing seasonal variation are obtained by relating original series values to the products of moving average and seasonal relative for corresponding periods.

$$Y_{adjS} = \frac{Y}{S}$$

Values of a series adjusted for seasonal variation are obtained by dividing each series value by the seasonal relative for the corresponding period.

$$\Delta_{adjS} = \frac{Y_n}{S_n} - \frac{Y_0}{S_0}$$

Adjustment of a period-to-period change for seasonal variation is made by calculating the difference in seasonally adjusted values for the periods.

$$\Delta_{adjT} = Y_n - Y_0 - n\beta$$

Adjustment of a period-to-period change for past trend is made by subtracting the trend increment over the period from the difference in values for the periods.

$$\Delta_{adjST} = \frac{Y_n}{S_n} - \frac{Y_0}{S_0} - n\beta$$

Adjustment of a period-to-period change for seasonal variation and trend is made by subtracting the trend adjustment from the difference in seasonally adjusted values for the periods.

$$\hat{Y} = \hat{T} \times \hat{C} \times \hat{S}$$

A forecast of a time series is the product of a trend forecast, cyclical relative forecast, and seasonal relative forecast.

$$PI_n = \frac{\sum (p_n q_0)}{\sum (p_0 q_0)} \times 100$$

A base-weighted price index is the percentage relative of the aggregate value of base year quantities at given year prices to the aggregate value of base-year quantities at base year prices.

$$PI_n = \frac{\sum \left[\frac{p_n}{p_0} (p_0 q_0) \right]}{\sum (p_0 q_0)} \times 100$$

An alternative formulation of the base-weighted price index in which base-year value weights are applied to price relatives for the given to the base year.

$$QI_n = \frac{\sum (p_0 q_n)}{\sum (p_0 q_0)} \times 100$$

A base-weighted quantity index is the percentage relative of aggregate value of given year quantities at base-year prices to the aggregate value of base-year quantities at base-year prices.

$$VI_n = \frac{\sum (p_n q_n)}{\sum (p_0 q_0)} \times 100$$

The value index is the percentage relative of aggregate value in the given year to aggregate value in the base year.

18

Bayesian Statistics

This chapter is an extension of the basic ideas of Bayesian decision making that appeared in Chapter 7. Recall that the unique thrust of Bayesian methods is the evaluation of alternative strategies from the standpoint of expected payoff. In the situations of this chapter, the relevant state of nature is a continuous variable. As in our discussion of classical inference, we will be concerned especially with means and proportions.

The first half of the chapter covers the elements of Bayesian methods needed to deal with continuous-state problems. Then we use these methods to evaluate strategies in situations in which the payoffs depend on continuous-state variables. Bayesian point estimation is presented as a special kind of strategy selection problem.

Extensions of Decision Elements

The Use of Regrets

We return now to the following decision problem, first encountered in Chapter 7.

Our firm (A) has developed a product that gives it a unique capability to exploit a new market. The key to the product is a possibly patentable device. However, Firm B holds a patent on a somewhat similar device, and it is not clear whether Firm A's patent application would be granted. If we enter the market and our device proves not patentable, we will be liable to a suit for infringement of B's patent. An alternative strategy is to purchase rights from Firm B for use of its device in our product. The payoff table was

| | Strategy | |
State	S_1 Apply for patent	S_2 Purchase rights
θ_1: Device Patentable	\$60,000	\$50,000
θ_2: Device not patentable	\$30,000	\$50,000

We considered various aspects of this problem, including the evaluation of a decision rule involving the advice of a lawyer as to patentability. In all cases we realized that there were no objective probabilities for the states. We used probabilities that represented the current subjective belief of the firm's officers. Although we introduced a method that can be used to find or verify such a belief, this method is often difficult to apply. Finally, we suggested a kind of sensitivity analysis that can help when no subjective probabilities can be formulated.

The absence of probabilities is often called *complete uncertainty*. An approach frequently used under complete uncertainty is based on the concept of *regrets*.

Definition	The **regret** associated with state *i* and strategy *j* is the difference between the maximum payoff given state *i* under any strategy and the payoff given state *i* and strategy *j*.

$$R_{ij} = (V_{\max}|\,\theta_i) - V_{ij}. \tag{18.1}$$

The key to regrets is to fix on states and look across strategies. For example, if the device is patentable and we adopt S_1, we will not regret our choice of strategy because the payoff under S_1 is the maximum given that state. The regret is zero. On the other hand if we adopt S_2 and the device later were to prove patentable, we will regret our choice of strategy because we could have done \$10,000 better the way things turned out. Similarly, if we purchase the rights and the device proves not patentable, we will have no regrets, but if the device proves not patentable after we apply for the patent, we will have a regret of \$50,000 − \$30,000 = \$20,000. The regret table is

| | Regrets from strategy | |
θ_i	S_1	S_2
θ_1	0	\$10,000
θ_2	\$20,000	0

If no probabilities of states are available, a strategy selection rule often recommended is the *minimax regret* criterion.

Rule	The minimax regret criterion is to select the strategy whose maximum regret is a minimum. That is

$$\text{minimize } (R_{\max}|\,S_j).$$

In the present case $R_{max}|\ S_1$ is \$20,000 and $R_{max}|\ S_2$ is \$10,000. The minimax regret criterion dictates selection of S_2, to purchase rights. Generally speaking, the minimax regret criterion is a rule that emphasizes avoidance of bad mistakes, as opposed to seeking great rewards. In this sense it is a conservative rule.

When probabilities of states are available, a criterion of minimizing expected regret leads to the same results as maximizing expected payoff. The use of regrets, then, becomes just an alternative way of looking at a problem. Consider our original payoff table with $P(\theta_1) = 0.75$. The units will be understood to be thousands of dollars.

		Payoffs	
θ_i	$P(\theta_i)$	S_1	S_2
θ_1	0.75	60	50
θ_2	0.25	30	50

$$EV(S_1) = 60(.75) + 30(.25) = 52.5,$$

$$EV(S_2) = 50(.75) + 50(.25) = 50.0,$$

$$EV(CP) = 60(.75) + 50(.25) = 57.5,$$

$$EVPI = EV(CP) - EV(S_{opt}) = 57.5 - 52.5 = 5.0.$$

Working with the regrets table we obtain the expected regrets for each strategy.

		Regrets	
θ_i	$P(\theta_i)$	S_1	S_2
θ_1	0.75	0	10
θ_2	0.25	20	0

$$ER(S_1) = 0(.75) + 20(.25) = 5.0,$$

$$ER(S_2) = 10(.75) + 0(.25) = 7.5.$$

Strategy 1 is optimal because it minimizes expected regret. Note that it is 2.5 thousand better than S_2, as was the case with expected payoffs. Note also that if we had perfect information we could switch to S_2 if the information indicated that the device was not patentable. By so doing, we would avoid the regret of 20 that would be incurred by staying with S_1. But our assessment of the probability that θ_2 is the true state is 0.25. Our expected gain (avoidance of regret) from perfect information is thus $20(0.25) = 5.0$. This is the expected regret of the optimal act. We are thus led to the important identity.

$$\blacksquare \qquad EVPI = ER(S_{opt}). \qquad \blacksquare \qquad\qquad (18.2)$$

The expected regret of the optimal strategy gives us directly the expected value of perfect information. Perhaps this is one reason why some people prefer to work with regrets rather than payoffs. The results in terms of strategy selection are the same, but the use of regrets continually reminds us of the improvement possible through more complete knowledge about states.

Continuous-State Variables

The distinguishing feature of this chapter compared with Chapter 7 is that the state of nature of concern to us is a continuous variable as opposed to a categorical one. In the example cited from Chapter 7 the possible states were device patentable and device not patentable. Consider now the same payoff table applied to a different problem. The relevant state of nature is the proportion of defective products in a production run. The state θ_1 is the condition that this parameter is 0.10 or less, and θ_2 is the state that this parameter exceeds 0.10. The critical value 0.10 represents a specification limit set by a buyer. The strategy S_1 is to send the lot of product out immediately to the buyer. If the lot meets his specification limit, we get an agreed price that gives us a net profit of $60,000. However, if the lot does not meet his specification an agreed penalty is assessed that reduces our net profit to $30,000. Strategy S_2 is to submit the lot now to 100 percent inspection. This guarantee that we will meet the buyer's specification will cost us $10,000, thus reducing our net profit to $50,000. This problem is the same as our preceding one except that the state variable is a continuous one.

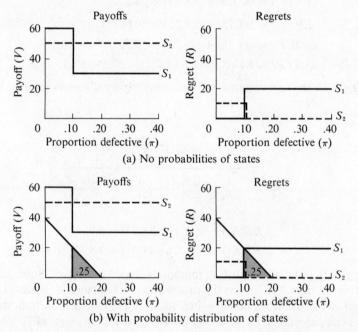

Figure 18.1   Payoffs and Regrets as Functions of a Continuous-State Variable

Figure 18.1(a) shows graphs of the payoffs and the corresponding regrets for each strategy as a function of π, the proportion of defective products in a production run. These are the graphs of (a) and (b) below:

| | (a) Payoffs under: | | | (b) Regrets under: | |
State	S_1	S_2	State	S_1	S_2
$\pi \leq 0.10$	60	50	$\pi \leq 0.10$	0	10
$\pi > 0.10$	30	50	$\pi > 0.10$	20	0

It is clear that to evaluate the strategies we must find $P(\pi \leq 0.10)$ and $P(\pi > 0.10)$, the relevant probabilities of states. In (b) of Figure 18.1 we insert a probability density function so located that $P(\pi \leq 0.10) = 0.75$. This is done for illustrative purposes only. A major section of the chapter will take up some methods for dealing with the probability distribution of a continuous-state variable. The particular parameters we will deal with are proportions and means. The methods are applications of Bayes' theorem whereby we combine sample evidence about the parameter with a prior probability distribution to obtain the posterior (to the sample) probability distribution of the parameter.

Linear Payoff Functions

In many situations the payoffs (and regrets) will not be step functions of the continuous-state variable. For example, consider a batch process that turns out lots of 1000 parts each for use in a further assembly. For each batch the processing department is credited $2,000 minus $2 for each part rejected by the assembly operation as defective. The processing department is considering inspecting all parts in any batch prior to shipment to the assembly department. The inspection operation will cost $400 per batch. The defective parts found can be repaired at $1 each. Thus, the processing department could be assured of shipping 100 percent good parts. If we let the "no inspection" strategy be S_1 and the "inspect all" strategy be S_2, the payoffs from each strategy can be expressed as a linear function of the proportion of defective parts in a batch.

$$V_1 = \$2,000 - \$2,000\pi,$$
$$V_2 = \$1,600 - \$1,000\pi.$$

Using these payoff functions, a payoff table for selected values of π has been calculated and is presented as part of Table 18.1. There we see that S_1 (no inspection) is the better strategy for listed π values from 0.00 to 0.35, while S_2 (inspect all) is the better strategy for the listed π values from 0.45 to 1.00. How should we decide between the strategies?

Table 18.1 Payoffs and Regrets for Selected Values of π

	Payoffs		Regrets	
π	V_1	V_2	R_1	R_2
0.00	$2000	$1600	$ 0	$400
0.05	1900	1550	0	350
0.15	1700	1450	0	250
0.25	1500	1350	0	150
0.35	1300	1250	0	50
0.45	1100	1150	50	0
0.55	900	1050	150	0
0.65	700	950	250	0
0.75	500	850	350	0
0.85	300	750	450	0
0.95	100	650	550	0
1.00	0	600	600	0

If no probabilities are available, we might consider selection by the minimax regret criterion. Remember that complete uncertainty means we really have no idea at all about what proportion of the lot is defective. The minimax regret criterion says to select the strategy whose maximum regret is a minimum. In the present problem this rule would indicate selecting S_2 (inspect all), because the maximum regret under Strategy 2 is $400, while the maximum regret under Strategy 1 is $600.

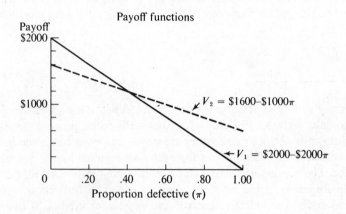

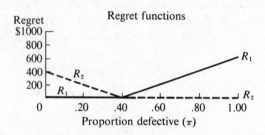

Figure 18.2   Linear Payoff and Regret Functions for Alternative Strategies

In Figure 18.2 we see the payoff functions and the regret functions as continuous functions rather than just the selected values included in Table 18.1. Note that at $\pi = 0.40$, we would be indifferent about the strategies.

As before, if a probability distribution of the relevant parameter (π in this case) can be determined, the expected payoffs of the alternative strategies can be assessed and a choice made accordingly. In Table 18.2 we show a uniform (equiprobable) distribution of π within each of ten intervals covering the range $0.0 \leq \pi \leq 1.0$. Then we use the payoff figures for π values centered in these intervals. In doing this we are creating a discrete approximation to the continuous functions. The expected values of the two strategies are

$$EV(S_1) = \sum [V_1 \cdot P(\pi)] = \$1000,$$
$$EV(S_2) = \sum [V_2 \cdot P(\pi)] = \$1100,$$

and Strategy 2 is optimal under the expected value criterion.

In the case of linear payoff functions the expected value of a strategy can be determined by substituting the expected value (mean) of the decision parameter in the

Table 18.2 Expected Payoffs of Alternative Strategies with Uniform Distribution of π

Interval	π	V_1	V_2	$P(\pi)$	$V_1 \cdot P(\pi)$	$V_2 \cdot P(\pi)$
.00– .10	.05	$1900	$1550	1/10	$ 190	$ 155
.10– .20	.15	1700	1450	1/10	170	145
.20– .30	.25	1500	1350	1/10	150	135
.30– .40	.35	1300	1250	1/10	130	125
.40– .50	.45	1100	1150	1/10	110	115
.50– .60	.55	900	1050	1/10	90	105
.60– .70	.65	700	950	1/10	70	95
.70– .80	.75	500	850	1/10	50	85
.80– .90	.85	300	750	1/10	30	75
.90–1.00	.95	100	650	1/10	10	65
					$1000	$1100

payoff equation. This follows from a theorem on expectations of linear functions. Our linear payoff equations are of the form $V = a + bX$, and it can be shown that

$$E(V) = a + b \cdot E(X),$$

or

$$\mu_V = a + b \cdot \mu_X.$$

In our example, $V_1 = \$2000 - \$2000\pi$. The mean of the probability distribution of π is $\sum [\pi \cdot P(\pi)] = 0.50$, and

$$EV(S_1) = \$2000 - \$2000\mu_\pi = \$2000 - \$2000(.50) = \$1000.$$

Similarly,

$$EV(S_2) = \$1600 - \$1000\mu_\pi = \$1600 - \$1000(.50) = \$1100.$$

Bayesian Posterior Distributions

We have just seen the role of the probability distribution of the decision parameter in evaluating alternative strategies. When we have sample evidence about a parameter, the prior distribution will have to be revised in accordance with the new information. We need, then, to examine some common situations. What we seek are formulas for the parameters of the posterior (revised) distribution under given conditions concerning the prior distribution and the sample evidence.

Posterior Distribution of a Proportion—Uniform Prior

Suppose an equiprobable prior belief about proportion of defective product in a lot is followed by the observation of two defectives out of six items. In Table 18.3 we have allocated prior probabilities of 1/10 to each of ten equal intervals over the range

Table 18.3 Revision of Uniform Prior Distribution of π Given Two
Successes in Six Trials

Interval	Midvalue π	Prior ① $P(\pi)$	Likelihood $P(E\mid\pi)$	Prior × Likelihood ② $P(\pi)\cdot P(E\mid\pi)$	Posterior distribution $P(\pi\mid E)$
.00– .10	.05	.10	.0305	.00305	.021
.10– .20	.15	.10	.1762	.01762	.123
.20– .30	.25	.10	.2966	.02966	.208
.30– .40	.35	.10	.3280	.03280	.230
.40– .50	.45	.10	.2780	.02780	.194
.50– .60	.55	.10	.1861	.01861	.130
.60– .70	.65	.10	.0951	.00951	.067
.70– .80	.75	.10	.0330	.00330	.023
.80– .90	.85	.10	.0055	.00055	.004
.90–1.00	.95	.10	.0001	.00001	.000
		1.00		.14291	1.000

(handwritten: $.021 = \dfrac{.00305}{.14291}$)

$\pi = 0.0$ to $\pi = 1.0$. For greater detail, especially if a computer program is devised, more intervals could be employed. Ten intervals will give quite good results, however.

In applying Bayes' theorem we first find (from the binomial table) the probability of two successes in six trials given each mid-value of π. The term given to these values of P(evidence | state) is *likelihood.* Then we find the joint probability of each state (mid-value) *and* the evidence of $r = 2$ out of $n = 6$. This is the product of the prior probability and the likelihood for each mid-value. The posterior distribution is found by rescaling the products of prior probabilities and likelihoods. This is, in effect, the step in Bayes' theorem where the numerator products are divided by the marginal probability of the evidence actually observed—the sum of the joint probability column. Recall the Bayes' theorem formula:

$$P(\theta_i\mid E) = \frac{P(\theta_i \cap E)}{P(E)}.$$

(handwritten margin note: as opposed to $\mu_\gamma = .50$)

The mean and variance of the posterior distribution in Table 18.3 can be calculated to be 0.375 and 0.026075, respectively. If continuous methods (from the calculus) are employed, the following general result will prevail.

Statement

The posterior probability distribution of a proportion predicated on a uniform (rectangular) prior probability distribution and the observation of r successes in n independent draws from the population will have the parameters

Mean: $\pi_B = \dfrac{r + 1}{n + 2}.$ (18.3a)

Variance: $\sigma_B^2 = \dfrac{(r + 1)(n - r + 1)}{(n + 2)^2(n + 3)}.$ (18.3b)

Here we use the symbols π_B and σ_B^2 (the subscript standing for Bayesian) to avoid confusion between the parameters of the prior and posterior distributions. In our example, with $r = 2$ and $n = 6$, the formulas yield

$$\pi_B = \frac{r+1}{n+2} = \frac{2+1}{6+2} = 0.375,$$

$$\sigma_B^2 = \frac{(r+1)(n-r+1)}{(n+2)^2(n+3)} = \frac{3(5)}{(8)^2(9)} = 0.026040.$$

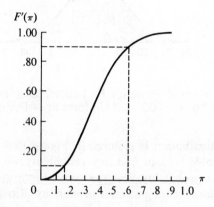

Figure 18.3   Posterior Distribution of π Given $r = 2$, $n = 6$, and Uniform Prior Distribution

Figure 18.3 shows the graph of the cumulated posterior probability distribution of π, determined from the final column of Table 18.3. We have used a smoothed cumulative graph. Any desired percentiles of the posterior distribution can be read from this graph. For example, it would appear that $\pi_{.10}$ is close to 0.17 and $\pi_{.90}$ is close to 0.60. This graphic method is applicable with any prior distribution that might be determined to represent the belief of a decision maker before the objective evidence is collected.

Note that in connection with Equations (18.3) we did not specify the form of the posterior probability distribution. We are now ready to do so.

Statement

With increasing sample size, the posterior probability distribution of a proportion predicated on a uniform (rectangular) prior probability distribution and the observation of r successes out of n independent draws from the population approaches a normal distribution with parameters

$$\text{Mean:} \quad \pi_B = p, \tag{18.4a}$$

$$\text{Variance:} \quad \sigma_B^2 = \frac{p(1-p)}{n}, \tag{18.4b}$$

where $p = r/n$.

For example, suppose 36 defectives out of 100 were observed. The posterior distribution of π given equal priors is approximately normal with

$$\pi_B = 36/100 = 0.36,$$

$$\sigma_B = \sqrt{\frac{.36(.64)}{100}} = 0.048.$$

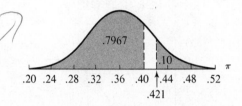

Figure 18.4   Normal Posterior Probability Distribution of π Given $r = 36$, $n = 100$, and Uniform Prior Distribution

This probability distribution is pictured in Figure 18.4. Now, with the standard normal probability table we can find any probability we are concerned with. For example, what is the probability that the true proportion is less than 0.40? The standard normal deviate for the posterior probability distribution of π is

$$z = \frac{\pi - \pi_B}{\sigma_B} ;$$

thus

$$P(\pi < .40) = P\left(z < \frac{.40 - .36}{.048}\right) = P(z < .83),$$

$$P(\pi < .40) = 0.7967.$$

We can also find any desired percentile of π. For example, to find $\pi_{.90}$,

$$\pi_P = \pi_B + z_P \sigma_B,$$

$$\pi_{.90} = .36 + 1.28(.048),$$

$$\pi_{.90} = 0.421.$$

As a guide to the size of sample needed to employ the above normal approximation with safety, Table 18.4 is provided.

Table 18.4   Minimum Sample Sizes for Normal Posterior Distribution of π Given a Uniform Prior Distribution†

Smaller of π_B or $1 - \pi_B$	0.01	0.02	0.05	0.10	0.20	0.30	0.40	0.50
Minimum sample size (n)	10,000	5,000	2,000	900	300	150	60	30

† This is the same minimum sample-size table given in Chapter 12 for using a normal distribution approximation for setting confidence limits for a true proportion. Note that establishment of a $1 - \alpha$ confidence interval by the method of Chapter 12 will yield a lower limit that is $\pi_{\alpha/2}$ and an upper limit that is $\pi_{1-\alpha/2}$ in the Bayesian procedure of this chapter.

Posterior Distribution of a Mean—Normal Prior

For an illustration of the Bayesian approach applied to a population mean we return to an example from Chapter 9 about the setting on a machine for filling 100-pound bags of cement. Each morning the foreman is accustomed to resetting the fill indicator on the machine. Long experience shows the standard deviation of weights of filled bags from the machine to be 0.9 pounds.

Suppose the foreman also knows that past daily resettings have produced a distribution of mean daily fills that has a mean of 100.1 pounds and a standard deviation of 0.3 pounds. A large number of bags are produced each day and here we are talking about a probability distribution of *mean* daily fills. To obtain the foregoing values, the foreman must have recorded the mean fill for each of a large number of days when he reset the machine and calculated the mean and standard deviation of these values. Suppose that he has also examined the frequency distribution of these mean fills and found a close correspondence with the normal distribution shape.

The above paragraph describes a prior probability distribution of mean daily fills. Since the mean fill on any day is a parameter, this is a probability distribution of the parameter, μ. Its mean is designated by μ_μ and its standard deviation by σ_μ. We have then, a normal prior distribution of μ with

$$\mu_\mu = 100.1, \qquad \sigma_\mu = 0.3.$$

Now suppose the foreman, after setting the machine this morning, takes a sample of $n = 36$ fills and finds a sample mean, $\overline{X}$, of 99.6 pounds. We are now in a position to combine the prior probability distribution of μ with the evidence provided by the sample mean in order to construct the posterior probability distribution of the mean fill for today.

Statement

Given a normal prior probability distribution of μ and a sample mean whose probability distribution given μ is normal, the posterior probability distribution of μ is normal with parameters

$$\blacksquare \qquad \mu_B = w\mu_\mu + (1 - w)\overline{X}, \qquad \blacksquare \qquad (18.5a)$$

$$\blacksquare \qquad \sigma_B^2 = w\sigma_\mu^2, \qquad \blacksquare \qquad (18.5b)$$

$$\text{where } w = \frac{\sigma_{\overline{X}}^2}{\sigma_{\overline{X}}^2 + n\sigma_\mu^2}. \qquad (18.5c)$$

In the filling machine example, $\mu_\mu = 100.1$, $\sigma_\mu = 0.3$, $\overline{X} = 99.6$, $n = 36$, and $\sigma_X = 0.9$. We have then

$$w = \frac{(.9)^2}{(.9)^2 + 36(.3)^2} = 0.2,$$

$$\mu_B = .2(100.1) + (1 - .2)99.6 = 99.7,$$

$$\sigma_B^2 = .2(.3)^2 = 0.018.$$

The posterior distribution of mean daily fill is normal with a mean of 99.7 and a standard deviation of $\sqrt{0.018} = 0.134$. The foreman can now state the probability, given the sample evidence, that today's mean fill will be less than 100.0 pounds. The standard normal deviate for any value of the mean is calculated from

$$z = \frac{\mu - \mu_B}{\sigma_B}.$$

For $\mu = 100.0$ pounds,

$$z(100.0) = \frac{100.0 - 99.7}{0.134} = 2.24.$$

Then,

$$P(\mu < 100.0) = P(z < 2.24) = 0.9874.$$

If the foreman is concerned to have today's mean fill equal or exceed 100.0 pounds, he had better reset the machine again and check a further sample.

While Equations 18.5 are the most convenient set for calculating μ_B and σ_B^2, the way they work is appreciated best by realizing that w and $(1 - w)$ are proportionate to the reciprocals of the variance of the prior distribution (σ_μ^2) and the variance of the sampling distribution of the mean $(\sigma_{\bar{X}}^2)$. The reciprocal of the variance can be taken as an index of the amount of information. The prior distribution of the population mean contains information about μ and so also does the sample mean. The posterior distribution of the population mean combines both these sources of information about μ. The reciprocal of its variance is the sum of the reciprocals of the variance of the prior and the variance of the sample mean.

$$\frac{1}{\sigma_B^2} = \frac{1}{\sigma_\mu^2} + \frac{1}{\sigma_{\bar{X}}^2}.$$

In the filling machine example, $\sigma_\mu^2 = (0.3)^2$ and $\sigma_{\bar{X}}^2 = (0.9)^2/36$. Thus

$$\frac{1}{\sigma_\mu^2} = \frac{1}{.09}, \quad \text{or } 11.1, \qquad \frac{1}{\sigma_{\bar{X}}^2} = \frac{1}{.81/36} = \frac{1}{.0225}, \quad \text{or } 44.4.$$

The sample contains four times as much information about μ as the prior ($\sigma_{\bar{X}}^2$ is one-fourth the size of σ_μ^2). Therefore in the determination of μ_B, the sample mean received a weight four times that of the mean of the prior distribution. The variance of the posterior distribution of μ is smaller than either σ_μ^2 or $\sigma_{\bar{X}}^2$, because the posterior distribution combines the information about μ contained in each. Using the reciprocal relationships,

$$\frac{1}{\sigma_B^2} = \frac{1}{\sigma_\mu^2} + \frac{1}{\sigma_{\bar{X}}^2}$$

$$= 11.1 + 44.4 = 55.5;$$

$$\sigma_B^2 = \frac{1}{55.5} = 0.018, \text{ as before.}$$

When σ_μ^2 is large compared to $\sigma_{\bar{X}}^2$, the prior distribution is said to be *diffuse*. In this case the information value, or weight attached to the prior distribution will be small and the parameters of the posterior distribution will be virtually determined by the sample information.

Statement	As the ratio of the variance of the sampling distribution of the mean to the variance of the prior distribution of the population mean approaches zero, the parameters of the posterior distribution of the population mean approach

$$\mu_B = \overline{X}, \tag{18.6a}$$

$$\sigma_B^2 = \frac{\sigma_X^2}{n}. \tag{18.6b}$$

A diffuse prior is one which has very close to equal probability densities over the range for μ covered by the posterior distribution. An example is pictured in Figure 18.5. While the prior appears flat, it is actually a normal distribution with a large standard deviation compared to the range of μ covered by $\pm 3\sigma_B$.

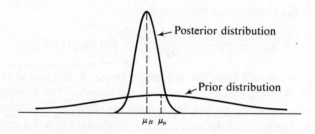

Figure 18.5   Schematic Representation of a Diffuse Normal Prior Distribution

Posterior Distribution of a Proportion—Normal Prior

Earlier we considered the posterior distribution of a proportion predicated on a uniform prior distribution. The case of the posterior distribution of a proportion when the prior distribution is normal resembles the situations we have just covered for means.

Statement	Given a normal prior distribution of a population proportion and sample evidence yielding r successes in n independent draws, the posterior distribution of the proportion will closely approximate a normal distribution with the following parameters under certain conditions.

$$\text{Mean:} \quad \pi_B = w\mu_\pi + (1 - w)p, \tag{18.7a}$$

$$\text{Variance:} \quad \sigma_B^2 = w\sigma_\pi^2, \tag{18.7b}$$

$$\text{where} \quad w = \frac{p(1 - p)}{p(1 - p) + n\sigma_\pi^2}. \tag{18.7c}$$

The condition for normality is that the ratio $\pi_B(1 - \pi_B)/\sigma_B^2$ must meet the requirement of the minimum sample size numbers given in Table 18.4.

Suppose a sales manager for Summa television sets holds a subjective probability distribution regarding the proportion of current buyers of his product who are purchasing to replace a set of the same brand. This distribution can be described as normal with a mean of 0.25 and a standard deviation of 0.03. For the prior distribution, then,

$$\mu_\pi = 0.25, \qquad \sigma_\pi^2 = (0.03)^2 = 0.0009.$$

A random sample of 100 recent purchasers reveals that 20 purchased to replace a set of the same brand. We have $p = r/n = 0.20$, and the posterior parameters

$$w = \frac{p(1 - p)}{p(1 - p) + n\sigma_\pi^2} = \frac{.20(.80)}{.20(.80) + 100(.0009)} = 0.64,$$

$$\pi_B = w\mu_\pi + (1 - w)p = .64(.25) + .36(.20) = 0.232,$$

$$\sigma_B^2 = w\sigma_\pi^2 = .64(.0009) = 0.000576.$$

Now we check the requirement mentioned earlier

$$\frac{\pi_B(1 - \pi_B)}{\sigma_B^2} = \frac{.232(.768)}{.000576} = 309 \text{ (approx.)}.$$

At 0.20 in the sample-size table the required number is 300, and at 0.30 the required number is 150. With $\pi_B = 0.232$ as above, the number 309 is above the minimal requirement. We can make probability statements about π or calculate percentiles of π using a normal posterior distribution with a mean of 0.232 and a standard deviation of $\sqrt{0.000576} = 0.024$.

If the prior normal probability distribution is diffuse, Equations (18.7) reduce to Equations (18.4) for the large-sample case with a uniform prior given earlier. These were

$$\pi_B = p, \qquad \sigma_B^2 = \frac{p(1 - p)}{n}.$$

Here again we must check that the condition $\pi_B(1 - \pi_B)/\sigma_B^2$ meets the sample-size requirement. In this case

$$\frac{\pi_B(1 - \pi_B)}{\sigma_B^2} = \frac{p(1 - p)}{p(1 - p)/n} = n.$$

This means that we can check the requirement by consulting Table 18.4 for the sample n, just as we did in connection with Equations (18.4).

Three major topics

Summary

The regret associated with a particular strategy given a state is the difference between the payoff from the optimal strategy given the state and the particular strategy given the same state. In the absence of probabilities of states the minimax regret rule is often used to select a strategy. When probabilities of states are available, minimizing expected regret is the same criterion for selecting a strategy as maximizing expected payoff.

The population mean and population proportion are examples of continuous-state variables. If payoffs from a strategy are a linear function of a continuous-state variable, the expected payoff from the strategy can be found by substituting the expected value of the state variable as the argument in the payoff function.

Evidence about the value of a continuous-state variable from a sample may be combined with a probability distribution expressing prior belief about the state by means of Bayes' theorem. In general when the prior belief is diffuse and the evidence is from a large sample, the posterior distribution of the continuous-state variable will be normal in shape. Formulas for the mean and variance of posterior distributions in particular situations were presented.

See Self-Correcting Exercises 18.A

Exercises

1. From the payoff table below, prepare a regrets table. Which strategy is optimal under the minimax regret criterion?

| | Strategies | | |
States	S_1	S_2	S_3
θ_1	80	20	0
θ_2	10	20	20
θ_3	5	10	15

2. Given $P(\theta_1) = 0.40$, $P(\theta_2) = 0.25$, and $P(\theta_3) = 0.35$ in Exercise 1, find the expected regret for each strategy. Show that the expected regret criterion aligns the strategies in the same way as expected payoff.

3. A manufacturer who turns out batches of a product for a particular customer makes a normal profit of $500 per batch. However, if there are more than 1 percent defectives in a batch, the probability is 0.6 that the customer will detect this and return the batch. The manufacturer then disposes of the batch at cost. At an additional cost of $100, the manufacturer can institute a screening process which guarantees that outgoing quality is no more than 1 percent defective. Which alternative minimizes the maximum regret?

4. If the probability is 0.7 that batches in Exercise 3 are not more than 1 percent defective, what is the optimal strategy? What is the expected value of perfect information (per batch)?

5. A manufacturer must decide whether to make and market a seasonal novelty which sells for $3.00 a unit. A special-purpose machine costing $10,000 will have to be completely written off against revenues from the novelty. The variable cost of manufacturing and marketing the novelty is $1.00 per unit. What would the manufacturer's regret be should he sell 3000 units? 5000 units? 6000 units?

6. Forty-two out of one-hundred randomly selected wage earners in a large firm were aware of the services available through an employee's credit union. Find the probability that a majority of employees in the firm were aware of the credit union services.

7. The season batting averages for rookies in a minor baseball league over the past five years was normally distributed with a mean of 0.250 and a standard deviation of 0.025. In his first 40 at bats, a new rookie gets 20 hits. Combine this information with the prior information to obtain the mean and standard deviation of the posterior distribution for the new rookie's season average.

8. A work crew filling orders in a warehouse was timed for 36 orders. Their mean time was 10.6 minutes, and the standard deviation of times was 4.5 minutes. Assume a diffuse normal prior and find the probability that the long-run mean for this crew exceeds 10.0 minutes.

9. Suppose it is known that the work crew in Exercise 8 belongs to a set of work crews whose mean order-filling times are normally distributed with a mean of 9.5 minutes and a standard deviation of 0.50. Find the mean and standard deviation of the posterior distribution of average time for the particular crew.

10. Given the information of Exercise 9, what is the probability that the long-run mean for the crew exceeds 10.0 minutes?

Evaluation of Strategies

Now that we have learned how to find Bayesian posterior distributions of a population mean and a population proportion in some common situations, we are ready to use these in Bayesian decision making. The principles are the same as in our earlier discussions. We must use the probability distribution of the states of nature (parameter) along with the payoffs from alternative strategies given states to determine the expected payoff of each strategy. We may wish to deal with regrets rather than payoffs, but the results will be equivalent. We will then select the strategy with the highest expected payoff or lowest expected regret.

Fixed Payoffs and Critical Parameter

In ordering a large supply of pine boards from a mill, a buyer has specified that no more than thirty percent of the boards contain knots exceeding a certain size. The buyer will pay a $3500 bonus for compliance with the specification. An inspection procedure to be conducted by the buyer has been agreed on. The mill has a supply of logs to use, but is uncertain about the quality of the resulting output. As a trial, 200 boards have been produced and 56 contained larger knots than desired.

The mill has two possible courses of action. One is to conclude that the process will meet the quality specification. If this strategy is adopted, no special costs are involved. The alternative strategy is to act as if the specification will not be met without taking further precautions. These further measures include a special sorting operation to remove low quality boards. The resulting output will meet the requirements, but the sorting operation costs an extra $500. The payoff table is as follows.

State of current process	Payoffs from strategy	
	Proceed as is S_1	Institute sorting S_2
$\pi \leq 0.30$	$3500	$3000
$\pi > 0.30$	0	$3000

All we need now to evaluate the strategies are the probabilities for the relevant states of the current process. We can assess this from the appropriate Bayesian posterior distribution. Assuming the prior distribution of π to be diffuse, the posterior distribution can be approximated as normal with the following parameters [see Equation (18.4)]:

$$\pi_B = p = 56/200 = 0.28,$$

$$\sigma_B^2 = \frac{p(1-p)}{n} = \frac{.28(.72)}{200} = 0.001008,$$

$$\sigma_B = \sqrt{0.001008} = 0.03175.$$

Then,

$$P(\pi \leq 0.30) = P\left(z \leq \frac{.30 - .28}{.03175}\right),$$

$$P(\pi \leq 0.30) = P(z \leq 0.63) = 0.74 \text{ (approx.).}$$

We can now evaluate the strategies:

$$EV(S_1) = \$3500(.74) + \$0(.26) = \$2590,$$

$$EV(S_2) = \$3000(.74) + \$3000(.26) = \$3000.$$

The preferred strategy is to institute sorting, S_2. The expected value of certain prediction is

$$EV(CP) = \$3500(.74) + \$3000(.26) = \$3370,$$

and the expected value of perfect information is

$$EVPI = EV(CP) - EV(S_{opt})$$
$$= \$3370 - \$3000 = \$370.$$

The evaluation can be carried out in terms of regrets. The regrets table, along with probabilities of states, follows.

State	Probability	Regrets	
		S_1	S_2
$\pi \leq 0.30$	0.74	0	$500
$\pi > 0.30$	0.26	$3000	0

$$ER(S_1) = \$0(.74) + \$3000(.26) = \$780.$$

$$ER(S_2) = \$500(.74) + \$0(.26) = \$370.$$

The optimal strategy is S_2 with an expected regret of \$370. This figure is also the expected value of perfect information.

The use of regrets is adaptable to situations where the consequences of decision errors are expressed in terms of relative "seriousness" or disutility rather than in money terms. Suppose that the sample of two defectives out of six products introduced on page 389 represents a pilot test of a new production method. At the current stage in development of the process, 20 percent or fewer defectives for the process is judged to be a good result and would lead to further development of the process along existing lines. On the other hand, if the process proportion defective exceeded 0.20, the R&D department would abandon attempts to improve the process along current lines. The error of abandoning a promising development is regarded as nine times as serious as the error of continuing a development that has little promise. The R&D department feels this way because the first type of error is not apt to be discovered later, but the second type will come to light when further experiments are conducted.

If we consult the cumulative posterior distribution drawn in Figure 18.3, we find that $P(\pi < 0.20) \approx 0.13$. We can then set up the regrets table along with probabilities of states and evaluate the strategies.

		Regrets	
		Continue	Abandon
State	Probability	S_1	S_2
$\pi \le 0.20$	0.13	0	9
$\pi > 0.20$	0.87	1	0

$$ER(S_1) = 0(.13) + 1(.87) = 0.87,$$
$$ER(S_2) = 9(.13) + 0(.87) = 1.17.$$

The optimal strategy is S_1, to continue development along existing lines. For S_2 to be optimal, the probability would have to be less than 0.10 that $\pi \le 0.20$. Because abandoning a promising avenue of development is relatively serious, we will not abandon a development unless the probability that it has promise is correspondingly small.

Linear Payoffs

A job shop operating under a profit center accounting system has a choice of two payment schedules for a job of machining 1000 parts for another division of the company. The first (S_1) is the regular schedule, which calls for a credit of \$5 per part plus \$2 per part for every minute that average machining time betters a standard of 30.0 minutes. The second (S_2), called an incentive schedule, calls for \$3 per part plus \$4 per part for each minute that average machining time betters the standard.

It is convenient first to express the payoffs in the form $V = a + b\mu$.

$$V_1 = 1000[\$5 + \$2(30.0 - \mu)],$$
$$V_2 = 1000[\$3 + \$4(30.0 - \mu)].$$

These can be rearranged into

$$V_1 = \$65,000 - \$2000\mu,$$
$$V_2 = \$123,000 - \$4000\mu.$$

These payoff equations are graphed in Figure 18.6. The point at which they cross is where $V_1 = V_2$. This is at $\mu = 29.0$, which can be found by equating the right-hand sides of the expressions for V_1 and V_2 and solving for μ.

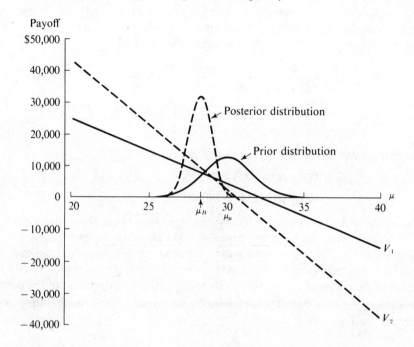

Figure 18.6 Prior and Posterior Distributions of μ in Relation to Linear Payoffs of Strategies

Let us suppose that this job is a new one for the machine shop. However, the shop manager has had experience with somewhat comparable jobs with similar time standards. He is able to translate this experience into a prior subjective probability distribution for mean time per part that has a mean of 30.2 minutes and a standard deviation of 1.5 minutes. The standard deviation of 1.5 minutes reflects considerable variation on past jobs from standard times, although on the average past jobs have come close to meeting standard times.

The manager can use his prior distribution to evaluate the strategies. Since his prior mean is 30.2, we already know that the regular schedule (S_1) will be better. However, let us find the expected values

$$EV(S_1) = \$65,000 - \$2000(30.2) = \$4,600,$$
$$EV(S_2) = \$123,000 - \$4000(30.2) = \$2,200.$$

Adopting the regular payment schedule is the best strategy if a decision must be made now. However, the manager asks and receives permission to machine 64 parts in order to check compliance with specifications. The mean time for this trial run was 27.8 minutes and the standard deviation of individual times was 6.0 minutes.

We now have

$$\text{Prior:} \quad \mu_\mu = 30.2, \qquad \sigma_\mu = 1.5;$$
$$\text{Sample:} \quad \overline{X} = 27.8, \qquad s_X = 6.0, \qquad n = 64.$$

While we do not know σ_X for the population, the sampling distribution of $(\overline{X} - \mu)/(s_X/\sqrt{n})$ is approximately normal because of the sample size. Therefore we can use Equation 18.5, with s_X in place of σ_X. To find the parameters of a normal posterior distribution of the population mean time, we find

$$w = \frac{\sigma_X^2}{\sigma_X^2 + n\sigma_\mu^2} \approx \frac{(6.0)^2}{(6.0)^2 + 64(1.5)^2} = 0.20,$$

$$\mu_B = w\mu_\mu + (1 - w)\overline{X} = .20(30.2) + .80(27.8) = 28.28,$$
$$\sigma_B^2 = w\sigma_\mu^2 = (.20)(1.5)^2 = 0.45.$$

The expected values of the two strategies are now

$$EV(S_1) = \$65,000 - \$2000(28.28) = \$8,440,$$
$$EV(S_2) = \$123,000 - \$4000(28.28) = \$9,880.$$

The results of the trial run have altered the manager's belief about mean machining time to the point that the incentive payment schedule is the better strategy.

While we did not use σ_B^2 in reevaluating the strategies, it does indicate how much "surer" the manager is about mean time than he was before the trial run. The standard deviation of the manager's prior was 1.5, while the standard deviation of the posterior distribution is $\sqrt{0.45} = 0.67$. This difference in the spread of the prior and the posterior can be seen in Figure 18.6.

Expected Value of Perfect Information

In introducing linear payoff functions (page 387) we dealt with a problem of whether to ship or hold for inspection and reprocessing batches of 1000 parts intended for use in an assembly. The payoff functions for the two strategies were

$$S_1, \text{ship:} \quad V_1 = \$2000 - \$2000\pi,$$
$$S_2, \text{hold:} \quad V_2 = \$1600 - \$1000\pi,$$

where

$$\pi = \text{proportion of defective parts.}$$

We then constructed a table showing the payoffs for values of π centered on intervals of 0.10 in length. From this we constructed a corresponding regrets table. Then we introduced a uniform prior distribution of proportion defective, that is, probability of 1/10 in each of the ten intervals. We then showed that the mean of the prior distribution of π, namely $\mu_\pi = 0.50$, could be inserted for π in the payoff functions to find the expected values of the strategies. Strategy 2, with an expected value of \$1100, turned out to be the better strategy under the uniform prior distribution. Table 18.5 shows again the uniform prior probability distribution of π along with the regrets under strategy 2, the optimal strategy.

In an earlier section on the use of regrets (page 385), we learned that the expected value of perfect information was the expected regret of the optimal strategy,

$$EVPI = ER(S_{opt}).$$

From Table 18.5 we are in a position to calculate $ER(S_{opt})$ from our discretized representation of the continuous uniform prior and the regrets function. We simply calculate the expected value of R_2, the regrets associated with the optimal strategy. The products of probabilities and regrets are entered in the final column of the table and summed to obtain the expected regret of the optimal strategy. This is the expected value of perfect information, which we find to be $80. This suggests that if the cost of a sample inspection is nominal, it may be wise to gather some sample evidence about the parameter.†

Table 18.5 Calculation of *EVPI* from Regrets of Optimal Strategy

Interval	π	$P(\pi)$	R_2	$P(\pi) \cdot R_2$
0.00–0.10	.05	.10	$350	$35
0.10–0.20	.15	.10	250	25
0.20–0.30	.25	.10	150	15
0.30–0.40	.35	.10	50	5
0.40–0.50	.45	.10	0	0
0.50–0.60	.55	.10	0	0
0.60–0.70	.65	.10	0	0
0.70–0.80	.75	.10	0	0
0.80–0.90	.85	.10	0	0
0.90–1.00	.95	.10	0	0

$$\$80$$

As an example of sample evidence that was combined with a uniform prior to produce a normal posterior distribution of π, we dealt on page 392 with the observation of 36 defectives out of 100 parts. This produced a posterior distribution with

$$\pi_B = p = 36/100 = 0.36,$$

$$\sigma_B = \sqrt{\frac{p(1-p)}{n}} = \sqrt{\frac{.36(.64)}{100}} = 0.048.$$

Suppose the sample result above occurred in sampling a particular batch of parts (with replacement). We would of course evaluate the strategies.

$$EV(S_1) = \$2000 - \$2000(.36) = \$1280,$$
$$EV(S_2) = \$1600 - \$1000(.36) = \$1240.$$

The optimal strategy is S_1, to ship out the lot. We might now want to find the expected value of perfect information, because it would give us a clue as to whether it

† An *EVSI* (expected value of sample information) calculation would better answer this question. We do not cover this calculation in this text for the continuous-state case, however.

would be worthwhile to gather even more evidence about the quality of the lot before making a final decision. Table 18.6 shows a method of discretizing a normal distribution in order to get an approximate figure for the expected value of perfect information.

Table 18.6 Discretizing a Normal Distribution of a Parameter in Order to Calculate *EVPI*

Standard interval	z	P	π	R_{opt}	$P \cdot R_{opt}$
−2.75 to −2.25	−2.5	.012	.240	$ 0	$ 0
−2.25 to −1.75	−2.0	.028	.264	0	0
−1.75 to −1.25	−1.5	.066	.288	0	0
−1.25 to −0.75	−1.0	.121	.312	0	0
−0.75 to −0.25	−0.5	.175	.336	0	0
−0.25 to 0.25	0.0	.197	.360	0	0
0.25 to 0.75	0.5	.175	.384	0	0
0.75 to 1.25	1.0	.121	.408	8	0.968
1.25 to 1.75	1.5	.066	.432	32	2.112
1.75 to 2.25	2.0	.028	.456	56	1.568
2.25 to 2.75	2.5	.012	.480	80	0.960
					$5.608

As with the uniform distribution, we decided that ten intervals would give us reasonable detail. Intervals half a standard deviation in length will provide this. The table shows such intervals with centered values expressed in terms of the standard normal deviate, z. In the column labeled P we have placed the standard normal probabilities for these intervals. This much of the table is uniform for all applications. To the right of the vertical line, we introduce the elements for our problem. First we find the values for the parameter (π) corresponding to the fixed z values. In our example we do this from

$$\pi = \pi_B + z\sigma_B = .36 + z(.048).$$

Then we enter the regrets of the optimal strategy. Our optimal strategy is S_1. Therefore, regrets occur when $V_2 - V_1$ is positive. Since $V_1 = \$2000 - \$2000\,\pi$ and $V_2 = \$1600 - \$1000\,\pi$, the regrets can be expressed as

$$R_{opt} = V_2 - V_1 = \$1000\pi - \$400$$

when $\pi > 0.40$. Now we need only multiply our standard set of probabilities by the corresponding centered regret values and sum the resulting products, to obtain the expected regret of $5.608. With perfect information valued at $5.61 it would seem wise to make the decision now to ship out the lot.

Bayesian Point Estimates as Decisions

Point estimation can be viewed as a decision to bet on one value of a random variable rather than some other value. At some later time a specific value of the random variable will occur, or become known. We will suffer a regret that depends on some function of the actual difference between our estimate and the true value. This outlook can be applied to any variable whose probability distribution is known. It can be applied to a variable representing single measurements (X) or to a mean viewed as a random variable. In the first case we make a decision to bet on a particular value of X. In the second we select some value of a mean (or proportion) as the best bet. The criterion of best bet is the bet that has the minimum expected regret.

An Inventory Problem

A health foods store receives delivery of fresh milk each morning. The milk, advertised as "fresh today," is sold to customers at $2.80 a gallon. The store pays $1.30 a gallon for the milk, and any milk not sold by the end of the day is delivered to a commercial user who pays $0.80 a gallon for it. For some time the store has been ordering 200 gallons a day, but it has never sold the entire quantity across the counter. Since it loses $0.50 on each unsold gallon, it is concerned about establishing an optimum order quantity. The manager has reviewed the daily sales figures over the last 100 business days. The frequency distribution of daily sales quantities appears normally distributed. Further, there appears to be no trend or within-the-week cyclical pattern in the daily sales figures when plotted as a time series. The mean daily sales is 120 gallons and the standard deviation of daily sales, 20 gallons.

The manager's problem is to determine the optimal daily order quantity. He accepts as his model for the probability distribution of daily demand a normal distribution with $\mu_X = 120$ and $\sigma_X = 20$. The economics of his problem are fairly clear. He has on hand at the beginning of the day a supply of gallons. For any given gallon sold he gains a gross margin of $2.80 - $1.30 = $1.50. On any gallon unsold he loses $1.30 - $0.80 = $0.50. Let us take a marginal approach that views the possible order quantities as the set of positive integers $\{k = 1, 2, 3, \ldots, 119, 120, 121, \ldots\}$. We then ask what is the *expected* net gain from adding the kth gallon to the number stocked.

If the actual amount demanded is equal to or greater than k, then the kth gallon will be sold and the store will gain $1.50. However, if actual demand is less than k, the kth gallon will remain unsold and the store will lose $0.50. We know the probabilities for the above events from the probability distribution of daily demand. Therefore, we can calculate for the kth gallon ordered.

Expected net gain = (gain if sold) $\cdot$ P(sell) + (gain if not sold) $\cdot$ P(not sold),

$$ENG = \$1.50 \cdot P(X \geq k) + (-\$0.50) \cdot P(X < k).$$

For example, for the 90th gallon stocked, that is, $k = 90$,†

$$z(90) = \frac{90 - 120}{20} = -1.5,$$

$$P(X < 90) = P(z < -1.5) = 0.0668,$$

$$P(X \geq 90) = 1 - .0668 = 0.9332,$$

$$ENG = \$1.50(.9332) + (-\$0.50)(.0668) = \$1.3664.$$

† In what follows we treat the demand variable as continuous rather than discrete.

The expected net gain from stocking the 90th gallon is about \$1.37. We should continue to add to the quantity stocked as long as the last (kth) unit stocked has a positive expected net gain. In the table below we summarize the calculation of expected net gain for several different values of k.

k (last unit stocked)	$z(k)$	$P(X < k)$	ENG
90	−1.5	.0668	\$ 1.3664
110	−0.5	.3085	\$ 0.8830
130	0.5	.6915	\$ 0.1170
150	1.5	.9332	\$−0.3664

We see here that the expected value (net gain) from stocking the 130th gallon is still positive, but the expected value of stocking the 150th gallon is negative. Somewhere between $k = 130$ and $k = 150$ the expected additional gain from stocking an incremental unit changes from positive to negative. This point can be found by setting ENG equal to zero and solving for $P(X < k)$. Using the symbol α_c for this critical level of $P(X < k)$, we find

$$0 = 1.50(1 - \alpha_c) + (-.50)\alpha_c,$$
$$0 = 1.50 - 1.50\alpha_c - .50\alpha_c,$$
$$\alpha_c = \frac{1.50}{2.00} = 0.75.$$

Now we can find the optimal order quantity, k_{opt}, represented by the α probability.

$$k_{opt} = \mu_X + z_{\alpha_c}\sigma_X,$$
$$k_{opt} = 120 + .67(20) = 133.4.$$

The optimal order quantity is 133.4 gallons. Of course, since the order quantity has to be in discrete gallons, we would place it at 133 or 134 gallons.

Bayesian Estimation

A reinterpretation and generalization of the health store's milk inventory problem allows us to appreciate a general problem in estimation. The store manager knows the probability distribution of a state variable, in this case demand on any given day. This is the probability distribution of X with $\mu_X = 120$ and $\sigma_X = 20$. His problem is to make an optimal point estimate of X when there are certain regrets (or cost consequences) associated with underestimating and with overestimating the actual X. The point estimate will be used as the order quantity. For each unit (gallon) by which this value overestimates actual demand he suffers a regret of \$0.50 (an unsold gallon). For each unit that he underestimates demand he suffers a regret of \$1.50, because he fails to stock a gallon on which he could have made a margin of \$1.50. The total expected regret is a minimum when the incremental expected regret from overestimating equals the incremental expected regret from underestimating. In the health food store example the regrets are linear functions of the amount of error between the point estimate (stock order quantity) and the actual demand level. At this point we need some symbols for the general case.

$$b_o = \text{regret per unit of overestimate,}$$
$$b_u = \text{regret per unit of underestimate.}$$

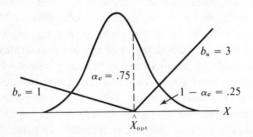

Figure 18.7   Linear Regret Functions and Location of Optimal Estimate

Figure 18.7 shows regret functions with b_o and b_u in a 1-to-3 ratio. Note that these regrets per unit are the slopes of the regret functions. The optimal estimate, $\hat{X}_{opt}$, is one for which the probability of overestimate is equal to

$$\alpha_c = \frac{b_u}{b_u + b_o}. \tag{18.8}$$

Thus, when the loss per unit of overestimate is equal to the loss per unit of underestimate ($b_o = b_u$), the median is the best Bayesian estimate. If the probability distribution of the random variable is symmetrical, the median will coincide with the mean. A common-sense interpretation of the general result is that if the penalty (per unit) of underestimation is, say, three times as serious as the penalty for overestimation, then we should be only one-third as willing to risk an underestimate as an overestimate. When the ratio of b_u to b_o is 3 to 1, $\hat{X}_{opt}$ is so located that the ratio of $1 - \alpha_c$ to α_c is 1 to 3. This was the situation in the health store inventory problem.

In the health store problem the distribution of X was normal and we found $\hat{X}_{opt}$ by

$$\hat{X}_{opt} = \mu_X + z_{\alpha_c} \sigma_X. \tag{18.9}$$

Where we wish to make an optimal Bayesian estimate of a mean based on a normal posterior probability distribution of the mean, we will have

$$\hat{\mu}_{opt} = \mu_B + z_{\alpha_c} \sigma_B. \tag{18.10}$$

Formulas for μ_B and σ_B were given in the section on posterior Bayesian distributions.

If the parameter of concern is a proportion and a normal posterior distribution applies, then

$$\hat{\pi}_{opt} = \pi_B + z_{\alpha_c} \sigma_B. \tag{18.11}$$

Formulas for the mean and standard deviation of normal posterior distributions of a proportion were given earlier.

Summary

The preceding sections have given some examples of applying Bayesian posterior distributions to the types of decision problems outlined in the first part of the chapter. These were "fixed payoffs defined by a critical value of the parameter" and "linear payoff" cases. We then illustrated the calculation of the expected value of perfect information. In this we discretized or "chunked up" the continuous distribution of the parameter and multiplied resulting interval probabilities by regret values for the midpoints of the intervals. We used the identity that the expected value of perfect information is equal to the expected regret of the optimal strategy.

Bayesian point estimation is equivalent to selecting a single value of a state parameter as the optimal strategy when there are cost consequences (regrets) associated with underestimation and overestimation. When these regrets are linear functions of the amount of underestimation and overestimation, an algebraic solution for the percentile location of the optimal estimate is available. In general the optimal (Bayesian) point estimate is the estimate that minimizes expected regret.

While we have dealt with model cases where payoffs are linear and the probability distribution of the parameter normal, the applicability of Bayesian statistics is not limited to these situations. The "discretized" calculations used at several points to illustrate basic concepts suggest how, with the aid of computational programs, the Bayesian methods can be applied to any payoff functions and any form of prior distribution. The prior probability distribution can be discretized by allocating probability to a number of equal intervals of the parameter. Payoff values are then found for the midpoints of the intervals. From this point, strategy evaluation proceeds as in "discrete-state" cases. If sample evidence is added, the posterior distribution can be calculated by applying Bayes' theorem for the discrete case. Reevaluation of strategies and calculation of the expected value of perfect information can also be done by "discretized" methods.

See Self-Correcting Exercises 18.B

Exercises

1. A large shipment of parts is awaiting sale. Buyer A has offered $20,000. Buyer B has offered $22,000 if 77.5 percent or more of the parts have no defects, and $15,000 if less than 77.5 percent are free from defects. The seller knows that a random sample of 400 parts has revealed 80 with defects. Which offer should the seller accept? What is the expected value of perfect information to the seller?

2. A task force on law enforcement in a city determined that the annual payoff from a program of burglary prevention measures would be $-\$400,000 + 2000\mu$, where μ is the average loss per burglary. Investigation of 100 recent burglaries found the average loss to be $220 and the standard deviation of losses was $150. What is the expected net gain from the program? What is the probability that the program would not show a net gain?

3. A company's staff figured that the return from an extensive coupon promotion would be $(-2 + 14\pi)$ millions of dollars, where π is the proportion of buyers taking advantage of the coupon. Their experience with other promotions led them to establish a normal prior distribution of π with a mean of 0.17 and a standard deviation of 0.04. Use the method of Table 18.6 to determine the expected value of perfect information.

4. The coupon promotion in Exercise 3 was tried out in a small test market. Of 400 buyers, 80 took advantage of the coupon. Revise the mean and variance of the prior distribution on the assumption that the test market results can be regarded as a random sample.

5. An office handling small claims figured the annual net return from a computer system that provided a check on the validity of claims would be $-\$120,000 + 100,000\mu$, where μ is the average dollars disallowed per claim. One hundred claims were examined in a random sample, using the same checks that the computer system would provide. The sample average disallowance per claim was $1.50 and the standard deviation of the sample values was $1.00. Should the office spend more time investigating the average disallowance per claim?

6. A baker located in an airport has kept track of requests for birthday cakes. He can bake them in the morning at a cost of $2.00, and sell them for $5.00. To protect his reputation for freshness, any cake left over at closing time would be donated to a nearby hospital. The following probability distribution for daily demand was prepared from request figures.

$X =$ demand	0	1	2	3	4	5
$P(X)$	.10	.25	.30	.20	.10	.05

Prepare a payoff table for the strategies "bake 1, 2, 3, 4, 5 cakes daily."

7. Find the optimal number of cakes to bake daily by calculating the expected values of the alternative strategies in Exercise 6.

8. Cumulate the probability distribution of demand in Exercise 6, and calculate the expected net gain from baking the first cake; the second cake; the third cake. Do these results agree with what you found in Exercise 7?

9. A lumber yard stocks and sells a certain variety of lumber on which its margin over costs (excluding storage costs) is 10 cents a board foot. Monthly storage costs are 1.2 cents a board foot. Monthly demand has a normal probability distribution with a mean of 20,000 board feet and a standard deviation of 3,000 board feet. If stocks are replenished monthly, what beginning-of-month stock level should be established?

10. A market research firm wishes to estimate average annual home furnishings expenditures in an area. A random sample of 225 households produces a mean expenditure figure of $186 and a sample standard deviation of $75. If the firm is twice as concerned to avoid overestimating as to avoid underestimating the true value, what is the optimal point estimate for them to make?

Glossary of Equations

$$R_{ij} = (V_{max}\mid \theta_i) - V_{ij}$$

The regret associated with state i and strategy j is the difference between the maximum payoff given state i under any strategy and the payoff given state i and strategy j.

$EVPI = ER(S_{opt})$

The expected value of perfect information is equal to the expected regret of the optimal strategy. The optimal strategy is the strategy that maximizes expected payoff (minimizes expected regret).

$$\pi_B = \frac{r+1}{n+2}$$

$$\sigma_B^2 = \frac{(r+1)(n-r+1)}{(n+2)^2(n+3)}$$

The mean and variance of the posterior distribution of a population proportion given a uniform prior probability and r successes in n independent draws from the population.

$$\pi_B = p$$

$$\sigma_B^2 = \frac{p(1-p)}{n}$$

The mean and variance of a normal approximation to the posterior distribution of a population proportion given r successes in n independent draws from the population ($p = r/n$). The prior distribution must be diffuse and the sample must meet certain size requirements.

$$\mu_B = w\mu_\mu + (1-w)\overline{X},$$

$$\sigma_B^2 = w\sigma_\mu^2$$

where $w = \dfrac{\sigma_X^2}{\sigma_X^2 + n\sigma_\mu^2}$

The mean and variance of the normal posterior distribution of a population mean given a normal prior distribution and a sample mean whose probability distribution, given the population mean (μ), is normal.

$$\mu_B = \overline{X}$$

$$\sigma_B^2 = \frac{\sigma_X^2}{n}$$

The mean and variance of a normal posterior distribution of the population mean given a *diffuse* normal prior distribution and a sample mean whose probability distribution, given the population mean (μ), is normal.

$$\pi_B = w\mu_\pi + (1-w)p$$

$$\sigma_B^2 = w\sigma_\pi^2$$

where $w = \dfrac{p(1-p)}{p(1-p) + n\sigma_\pi^2}$

The mean and variance of a normal posterior distribution of the population proportion given a normal prior distribution and r successes in n independent draws from the population. The posterior parameters must meet certain effective sample size requirements.

$$\alpha_c = \frac{b_u}{b_u + b_o}$$

The percentile location of the optimal estimate of a continuous-state variable is the ratio of regret per unit of underestimate to the sum of regret per unit of underestimate and regret per unit of overestimate. The regrets per unit error must be constants, not averages.

$$\hat{X}_{opt} = \mu_X + z_{\alpha_c}\sigma_X$$

$$\hat{\mu}_{opt} = \mu_B + z_{\alpha_c}\sigma_B$$

$$\hat{\pi}_{opt} = \pi_B + z_{\alpha_c}\sigma_B$$

Formulas for translating the percentile location, α_c, of an optimal estimate of a continuous state variable to a value of the variable when the (state) variable is normally distributed.

References

Chapter 1

1. Neiswanger, W. A., *Elementary Statistical Methods As Applied to Business and Economic Data*, revised ed. New York: The MacMillan Company, 1956, 281–294.
2. Richmond, S. B., *Statistical Analysis*, 2d ed. New York: The Ronald Press Company, 1964, 76–84.

Chapter 2

1. Chou, Y. L., *Statistical Analysis with Business and Economic Applications*. New York: Holt, Rinehart and Winston, Inc., 1969, 29–34.
2. Freund, J. E., *Modern Elementary Statistics*, 3d ed. Englewood Cliffs, N.J.: Prentice-Hall, Inc., 1967, 11–21.
3. Mendenhall, W., *Introduction to Probability and Statistics*, 3d ed. Belmont, Calif.: Wadsworth Publishing Company, Inc., 1971, 45–49.
4. Richmond, S. B., *Statistical Analysis*, 2d ed. New York: The Ronald Press Company, 1964, 71–73.
5. Clark, C. T. and L. L. Schkade, *Statistical Methods for Business Decisions*. Cincinnati, Ohio: South-Western Publishing Co., 1969, 44–47.

Chapter 3

1. Ezekiel, M., and K. A. Fox, *Methods of Correlation and Regression Analysis*, 3d ed. New York: John Wiley & Sons, Inc., 1959, 69–117.

Chapter 7

1. von Neumann, J., and O. Morgenstern, *Theory of Games and Economic Behavior*, 3d ed. Princeton, N.J.: Princeton University Press, 1953.

Chapter 9

1. Harnett, D. L., *Introduction to Statistical Methods*. Reading, Mass.: Addison-Wesley Publishing Company, 1970, 188–194.

Chapter 12

1. Peters, W. S., and G. W. Summers, *Statistical Analysis for Business Decisions*. Englewood Cliffs, N.J.: Prentice-Hall, Inc., 1968, Chapter 9.
2. Hays, W. L., *Statistics for Psychologists*. New York: Holt, Rinehart and Winston, Inc., 1963, Chapter 17.

Chapter 13

1. Cochran, W. G., and G. M. Cox, *Experimental Designs*, 2d ed. New York: John Wiley & Sons, Inc., 1957.
2. Mendenhall, W., *The Design and Analysis of Experiments*. Belmont, Calif.: Wadsworth Publishing Company, Inc., 1968.
3. Dixon, W. J., and F. J. Massey, Jr., *Introduction to Statistical Analysis*, 3d ed. New York: McGraw-Hill Book Company, 1969, a. 308–310, b. 313–316.
4. Ostle, B., *Statistics in Research*, 2d ed. Ames, Iowa: The Iowa State University Press, 1963, 310–312.
5. Peters, W. S., and G. W. Summers, *Statistical Analysis for Business Decisions*. Englewood Cliffs, N.J.: Prentice-Hall, Inc., 1968, 304–305.

Chapter 14

1. Ostle, B., *Statistics in Research*, 2d ed. Ames, Iowa: The Iowa State University Press, 1963, 170–176.
2. Peters, W. S., and G. W. Summers, *Statistical Analysis for Business Decisions*. Englewood Cliffs, N.J.: Prentice-Hall, Inc., 1968, 339–341.
3. Mood, A. M., and F. A. Graybill, *Introduction to the Theory of Statistics*, 2d ed. New York: McGraw-Hill Book Company, Inc., 1963, 198–203.
4. Ezekiel, M., and K. A. Fox, *Methods of Correlation and Regression Analysis*, 3d ed. New York: John Wiley & Sons, Inc., 1959, 69–117.

Chapter 15

1. Bradley, J. V., *Distribution-Free Statistical Tests*. Englewood Cliffs, N.J.: Prentice-Hall, Inc., 1968, b. 314.
2. Siegel, S., *Nonparametric Statistics for the Behavioral Sciences*. New York: McGraw-Hill Book Company, Inc., 1956.
3. Owen, D. B., *Handbook of Statistical Tables*. Reading, Mass.: Addison-Wesley Publishing Company, Inc., 1962, a. 420–422, b. 400–406.

Chapter 16

1. Deming, W. E., *Sample Design in Business Research*. New York: John Wiley & Sons, Inc., 1960, Chapter 5.

Chapter 17

1. Hamburg, M., *Statistical Analysis for Decision Making*. New York: Harcourt, Brace & World, Inc., 1970, Chapter 11.
2. Merrill, W. C., and K. A. Fox, *Introduction to Economic Statistics*. New York: John Wiley & Sons, Inc., 1970, Chapter 11.

Appendix

@ .08

two-tail | one-tail

total

.10   .05   1.28

.05   .025   1.645

.01   .025   .005   1.96

.05   .005   2.5?

.01   2.58

A: Areas under the Normal Distribution

For Negative Values of z

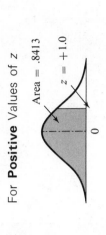

Area = .1587
z = −1.0

For Positive Values of z

Area = .8413
z = +1.0

For Positive Values of z

z To 1st Decimal	.00	.01	.02	.03	.04	.05	.06	.07	.08	.09
					Area					
0.0	.5000	.5040	.5080	.5120	.5160	.5199	.5239	.5279	.5319	.5359
0.1	.5398	.5438	.5478	.5517	.5557	.5596	.5636	.5675	.5714	.5753
0.2	.5793	.5832	.5871	.5910	.5948	.5987	.6026	.6064	.6103	.6141
0.3	.6179	.6217	.6255	.6293	.6331	.6368	.6406	.6443	.6480	.6517
0.4	.6554	.6591	.6628	.6664	.6700	.6736	.6772	.6808	.6844	.6879
0.5	.6915	.6950	.6985	.7019	.7054	.7088	.7123	.7157	.7190	.7224
0.6	.7257	.7291	.7324	.7357	.7389	.7422	.7454	.7486	.7517	.7549
0.7	.7580	.7611	.7642	.7673	.7703	.7734	.7764	.7794	.7823	.7852
0.8	.7881	.7910	.7939	.7967	.7995	.8023	.8051	.8078	.8106	.8133
0.9	.8159	.8186	.8212	.8238	.8264	.8289	.8315	.8340	.8365	.8389
1.0	.8413	.8438	.8461	.8485	.8508	.8531	.8554	.8577	.8599	.8621
1.1	.8643	.8665	.8686	.8708	.8729	.8749	.8770	.8790	.8810	.8830
1.2	.8849	.8869	.8888	.8907	.8925	.8944	.8962	.8980	.8997	.9015
1.3	.9032	.9049	.9066	.9082	.9099	.9115	.9131	.9147	.9162	.9177
1.4	.9192	.9207	.9222	.9236	.9251	.9265	.9278	.9292	.9306	.9319
1.5	.9332	.9345	.9357	.9370	.9382	.9394	.9406	.9418	.9430	.9441
1.6	.9452	.9463	.9474	.9485	.9495	.9505	.9515	.9525	.9535	.9545
1.7	.9554	.9564	.9573	.9582	.9591	.9599	.9608	.9616	.9625	.9633
1.8	.9641	.9649	.9656	.9664	.9671	.9678	.9686	.9693	.9700	.9706
1.9	.9713	.9719	.9726	.9732	.9738	.9744	.9750	.9756	.9762	.9767
2.0	.9772	.9778	.9783	.9788	.9793	.9798	.9803	.9808	.9812	.9817
2.1	.9821	.9826	.9830	.9834	.9838	.9842	.9846	.9850	.9854	.9857
2.2	.9861	.9865	.9868	.9871	.9874	.9878	.9881	.9884	.9887	.9890
2.3	.9893	.9896	.9898	.9901	.9904	.9906	.9909	.9911	.9913	.9916
2.4	.9918	.9920	.9922	.9924	.9926	.9928	.9930	.9932	.9934	.9936
2.5	.9938	.9940	.9941	.9943	.9944	.9946	.9948	.9949	.9951	.9952
2.6	.9953	.9955	.9956	.9957	.9958	.9960	.9961	.9962	.9963	.9964
2.7	.9965	.9966	.9967	.9968	.9969	.9970	.9971	.9972	.9973	.9974
2.8	.9974	.9975	.9976	.9977	.9977	.9978	.9979	.9979	.9980	.9981
2.9	.9981	.9982	.9982	.9983	.9984	.9984	.9985	.9985	.9986	.9986
3.0	.9986	.9987	.9987	.9988	.9988	.9988	.9989	.9989	.9990	.9990

Second Decimal

For Negative Values of z

z To 1st Decimal	.00	.01	.02	.03	.04	.05	.06	.07	.08	.09
					Area					
−3.0	.0014	.0013	.0013	.0012	.0012	.0011	.0011	.0011	.0010	.0010
−2.9	.0019	.0018	.0018	.0017	.0016	.0016	.0015	.0015	.0014	.0014
−2.8	.0026	.0025	.0024	.0023	.0023	.0022	.0021	.0021	.0020	.0019
−2.7	.0035	.0034	.0033	.0032	.0031	.0030	.0029	.0028	.0027	.0026
−2.6	.0047	.0045	.0044	.0043	.0041	.0040	.0039	.0038	.0037	.0036
−2.5	.0062	.0060	.0059	.0057	.0055	.0054	.0052	.0051	.0049	.0048
−2.4	.0082	.0080	.0078	.0075	.0073	.0071	.0069	.0068	.0066	.0064
−2.3	.0107	.0104	.0102	.0099	.0096	.0094	.0091	.0089	.0087	.0084
−2.2	.0139	.0136	.0132	.0129	.0126	.0122	.0119	.0116	.0113	.0110
−2.1	.0179	.0174	.0170	.0166	.0162	.0158	.0154	.0150	.0146	.0143
−2.0	.0228	.0222	.0217	.0212	.0207	.0202	.0197	.0192	.0188	.0183
−1.9	.0287	.0281	.0274	.0268	.0262	.0256	.0250	.0244	.0238	.0233
−1.8	.0359	.0352	.0344	.0336	.0329	.0322	.0314	.0307	.0300	.0294
−1.7	.0446	.0436	.0427	.0418	.0409	.0401	.0392	.0384	.0375	.0367
−1.6	.0548	.0537	.0526	.0516	.0505	.0495	.0485	.0475	.0465	.0455
−1.5	.0668	.0655	.0643	.0630	.0618	.0606	.0594	.0582	.0570	.0559
−1.4	.0808	.0793	.0778	.0764	.0749	.0735	.0722	.0708	.0694	.0681
−1.3	.0968	.0951	.0934	.0918	.0901	.0885	.0869	.0853	.0838	.0823
−1.2	.1151	.1131	.1112	.1093	.1075	.1056	.1038	.1020	.1003	.0985
−1.1	.1357	.1335	.1314	.1292	.1271	.1251	.1230	.1210	.1190	.1170
−1.0	.1587	.1562	.1539	.1515	.1492	.1469	.1446	.1423	.1401	.1379
−0.9	.1841	.1814	.1788	.1762	.1736	.1711	.1685	.1660	.1635	.1611
−0.8	.2119	.2090	.2061	.2033	.2005	.1977	.1949	.1922	.1894	.1867
−0.7	.2420	.2389	.2358	.2327	.2297	.2266	.2236	.2206	.2177	.2148
−0.6	.2743	.2709	.2676	.2643	.2611	.2578	.2546	.2514	.2483	.2451
−0.5	.3085	.3050	.3015	.2981	.2946	.2912	.2877	.2843	.2810	.2776
−0.4	.3446	.3409	.3372	.3336	.3300	.3264	.3228	.3192	.3156	.3121
−0.3	.3821	.3783	.3745	.3707	.3669	.3632	.3594	.3557	.3520	.3483
−0.2	.4207	.4168	.4129	.4090	.4052	.4013	.3974	.3936	.3897	.3859
−0.1	.4602	.4562	.4522	.4483	.4443	.4404	.4364	.4325	.4286	.4247
−0.0	.5000	.4960	.4920	.4880	.4840	.4801	.4761	.4721	.4681	.4641

Second Decimal

B: Areas under the *t* Distributions

t distribution for 2 d.f.

Area = .90

t = +1.886

0

d.f.	$t_{.005}$	$t_{.01}$	$t_{.025}$	$t_{.05}$	$t_{.10}$	$t_{.90}$	$t_{.95}$	$t_{.975}$	$t_{.99}$	$t_{.995}$
1	−63.657	−31.821	−12.706	−6.314	−3.078	3.078	6.314	12.706	31.821	63.657
2	− 9.925	− 6.965	− 4.303	−2.920	−1.886	1.886	2.920	4.303	6.965	9.925
3	− 5.841	− 4.541	− 3.182	−2.353	−1.638	1.638	2.353	3.182	4.541	5.841
4	− 4.604	− 3.747	− 2.776	−2.132	−1.533	1.533	2.132	2.776	3.747	4.604
5	− 4.032	− 3.365	− 2.571	−2.015	−1.476	1.476	2.015	2.571	3.365	4.032
6	− 3.707	− 3.143	− 2.447	−1.943	−1.440	1.440	1.943	2.447	3.143	3.707
7	− 3.499	− 2.998	− 2.365	−1.895	−1.415	1.415	1.895	2.365	2.998	3.499
8	− 3.355	− 2.896	− 2.306	−1.860	−1.397	1.397	1.860	2.306	2.896	3.355
9	− 3.250	− 2.821	− 2.262	−1.833	−1.383	1.383	1.833	2.262	2.821	3.250
10	− 3.169	− 2.764	− 2.228	−1.812	−1.372	1.372	1.812	2.228	2.764	3.169
11	− 3.106	− 2.718	− 2.201	−1.796	−1.363	1.363	1.796	2.201	2.718	3.106
12	− 3.055	− 2.681	− 2.179	−1.782	−1.356	1.356	1.782	2.179	2.681	3.055
13	− 3.012	− 2.650	− 2.160	−1.771	−1.350	1.350	1.771	2.160	2.650	3.012
14	− 2.977	− 2.624	− 2.145	−1.761	−1.345	1.345	1.761	2.145	2.624	2.977
15	− 2.947	− 2.602	− 2.131	−1.753	−1.341	1.341	1.753	2.131	2.602	2.947
16	− 2.921	− 2.583	− 2.120	−1.746	−1.337	1.337	1.746	2.120	2.583	2.921
17	− 2.898	− 2.567	− 2.110	−1.740	−1.333	1.333	1.740	2.110	2.567	2.898
18	− 2.878	− 2.552	− 2.101	−1.734	−1.330	1.330	1.734	2.101	2.552	2.878
19	− 2.861	− 2.539	− 2.093	−1.729	−1.328	1.328	1.729	2.093	2.539	2.861
20	− 2.845	− 2.528	− 2.086	−1.725	−1.325	1.325	1.725	2.086	2.528	2.845
21	− 2.831	− 2.518	− 2.080	−1.721	−1.323	1.323	1.721	2.080	2.518	2.831
22	− 2.819	− 2.508	− 2.074	−1.717	−1.321	1.321	1.717	2.074	2.508	2.819
23	− 2.807	− 2.500	− 2.069	−1.714	−1.319	1.319	1.714	2.069	2.500	2.807
24	− 2.797	− 2.492	− 2.064	−1.711	−1.318	1.318	1.711	2.064	2.492	2.797
25	− 2.787	− 2.485	− 2.060	−1.708	−1.316	1.316	1.708	2.060	2.485	2.787
26	− 2.779	− 2.479	− 2.056	−1.706	−1.315	1.315	1.706	2.056	2.479	2.779
27	− 2.771	− 2.473	− 2.052	−1.703	−1.314	1.314	1.703	2.052	2.473	2.771
28	− 2.763	− 2.467	− 2.048	−1.701	−1.313	1.313	1.701	2.048	2.467	2.763
29	− 2.756	− 2.462	− 2.045	−1.699	−1.311	1.311	1.699	2.045	2.462	2.756
30	− 2.750	− 2.457	− 2.042	−1.697	−1.310	1.310	1.697	2.042	2.457	2.750
40	− 2.704	− 2.423	− 2.021	−1.684	−1.303	1.303	1.684	2.021	2.423	2.704
60	− 2.660	− 2.390	− 2.000	−1.671	−1.296	1.296	1.671	2.000	2.390	2.660
120	− 2.617	− 2.358	− 1.980	−1.658	−1.289	1.289	1.658	1.980	2.358	2.617
∞	− 2.576	− 2.326	− 1.960	−1.645	−1.282	1.282	1.645	1.960	2.326	2.576

Table B is taken from Table IV of Fisher: *Statistical Methods for Research Workers*, published by Oliver & Boyd, Edinburgh, and used by permission of the author and publishers.

C: Areas under the Chi-Square Distributions

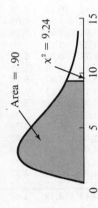

Chi-square distribution for 5 d.f.

Area = .90

$x^2 = 9.24$

$\mu = df$

$\sigma^2 = 2(df)$

d.f.	$\chi^2_{.005}$	$\chi^2_{.01}$	$\chi^2_{.025}$	$\chi^2_{.05}$	$\chi^2_{.10}$	$\chi^2_{.90}$	$\chi^2_{.95}$	$\chi^2_{.975}$	$\chi^2_{.99}$	$\chi^2_{.995}$
1	.000039	.00016	.00098	.0039	.0158	2.71	3.84	5.02	6.63	7.88
2	.0100	.0201	.0506	.1026	.2107	4.61	5.99	7.38	9.21	10.60
3	.0717	.115	.216	.352	.584	6.25	7.81	9.35	11.34	12.84
4	.207	.297	.484	.711	1.064	7.78	9.49	11.14	13.28	14.86
5	.412	.554	.831	1.15	1.61	9.24	11.07	12.83	15.09	16.75
6	.676	.872	1.24	1.64	2.20	10.64	12.59	14.45	16.81	18.55
7	.989	1.24	1.69	2.17	2.83	12.02	14.07	16.01	18.48	20.28
8	1.34	1.65	2.18	2.73	3.49	13.36	15.51	17.53	20.09	21.96
9	1.73	2.09	2.70	3.33	4.17	14.68	16.92	19.02	21.67	23.59
10	2.16	2.56	3.25	3.94	4.87	15.99	18.31	20.48	23.21	25.19
11	2.60	3.05	3.82	4.57	5.58	17.28	19.68	21.92	24.73	26.76
12	3.07	3.57	4.40	5.23	6.30	18.55	21.03	23.34	26.22	28.30
13	3.57	4.11	5.01	5.89	7.04	19.81	22.36	24.74	27.69	29.82
14	4.07	4.66	5.63	6.57	7.79	21.06	23.68	26.12	29.14	31.32
15	4.60	5.23	6.26	7.26	8.55	22.31	25.00	27.49	30.58	32.80
16	5.14	5.81	6.91	7.96	9.31	23.54	26.30	28.85	32.00	34.27
18	6.26	7.01	8.23	9.39	10.86	25.99	28.87	31.53	34.81	37.16
20	7.43	8.26	9.59	10.85	12.44	28.41	31.41	34.17	37.57	40.00
24	9.89	10.86	12.40	13.85	15.66	33.20	36.42	39.36	42.98	45.56
30	13.79	14.95	16.79	18.49	20.60	40.26	43.77	46.98	50.89	53.67
40	20.71	22.16	24.43	26.51	29.05	51.81	55.76	59.34	63.69	66.77
60	35.53	37.48	40.48	43.18	46.46	74.40	79.08	83.30	88.38	91.95
120	83.85	86.92	91.58	95.70	100.62	140.23	146.57	152.21	158.95	163.64

From *Introduction to Statistical Analysis*, 3rd edition, by Wilfrid J. Dixon and Frank J. Massey, Jr. Copyright © 1969 by McGraw-Hill, Inc. Used by permission of McGraw-Hill Book Company.

D: Values on the F Distributions

F distribution for 50 and 5 d.f.

$F = 4.44$ Area $= .05$

Values of F

Right tail of the distribution for $P = 0.05$ (light-face type), 0.01 (bold-face type)

m_2	$m_1 = $ Degrees of Freedom for Numerator											
	1	2	3	4	5	6	7	8	9	10	11	12
1	161	200	216	225	230	234	237	239	241	242	243	244
	4,052	**4,999**	**5,403**	**5,625**	**5,764**	**5,859**	**5,928**	**5,981**	**6,022**	**6,056**	**6,082**	**6,106**
2	18.51	19.00	19.16	19.25	19.30	19.33	19.36	19.37	19.38	19.39	19.40	19.41
	98.49	**99.01**	**99.17**	**99.25**	**99.30**	**99.33**	**99.34**	**99.36**	**99.38**	**99.40**	**99.41**	**99.42**
3	10.13	9.55	9.28	9.12	9.01	8.94	8.88	8.84	8.81	8.78	8.76	8.74
	34.12	**30.81**	**29.46**	**28.71**	**28.24**	**27.91**	**27.67**	**27.49**	**27.34**	**27.23**	**27.13**	**27.05**
4	7.71	6.94	6.59	6.39	6.26	6.16	6.09	6.04	6.00	5.96	5.93	5.91
	21.20	**18.00**	**16.69**	**15.98**	**15.52**	**15.21**	**14.98**	**14.80**	**14.66**	**14.54**	**14.45**	**14.37**
5	6.61	5.79	5.41	5.19	5.05	4.95	4.88	4.82	4.78	4.74	4.70	4.68
	16.26	**13.27**	**12.06**	**11.39**	**10.97**	**10.67**	**10.45**	**10.27**	**10.15**	**10.05**	**9.96**	**9.89**
6	5.99	5.14	4.76	4.53	4.39	4.28	4.21	4.15	4.10	4.06	4.03	4.00
	13.74	**10.92**	**9.78**	**9.15**	**8.75**	**8.47**	**8.26**	**8.10**	**7.98**	**7.87**	**7.79**	**7.72**
7	5.59	4.74	4.35	4.12	3.97	3.87	3.79	3.73	3.68	3.63	3.60	3.57
	12.25	**9.55**	**8.45**	**7.85**	**7.46**	**7.19**	**7.00**	**6.84**	**6.71**	**6.62**	**6.54**	**6.47**
8	5.32	4.46	4.07	3.84	3.69	3.58	3.50	3.44	3.39	3.34	3.31	3.28
	11.26	**8.65**	**7.59**	**7.01**	**6.63**	**6.37**	**6.19**	**6.03**	**5.91**	**5.82**	**5.74**	**5.67**
9	5.12	4.26	3.86	3.63	3.48	3.37	3.29	3.23	3.18	3.13	3.10	3.07
	10.56	**8.02**	**6.99**	**6.42**	**6.06**	**5.80**	**5.62**	**5.47**	**5.35**	**5.26**	**5.18**	**5.11**
10	4.96	4.10	3.71	3.48	3.33	3.22	3.14	3.07	3.02	2.97	2.94	2.91
	10.04	**7.56**	**6.55**	**5.99**	**5.64**	**5.39**	**5.21**	**5.06**	**4.95**	**4.85**	**4.78**	**4.71**
11	4.84	3.98	3.59	3.36	3.20	3.09	3.01	2.95	2.90	2.86	2.82	2.79
	9.65	**7.20**	**6.22**	**5.67**	**5.32** /	**5.07**	**4.88**	**4.74**	**4.63**	**4.54**	**4.46**	**4.40**
12	4.75	3.88	3.49	3.26	3.11	3.00	2.92	2.85	2.80	2.76	2.72	2.69
	9.33	**6.93**	**5.95**	**5.41**	**5.06**	**4.82**	**4.65**	**4.50**	**4.39**	**4.30**	**4.22**	**4.16**
13	4.67	3.80	3.41	3.18	3.02	2.92	2.84	2.77	2.72	2.67	2.63	2.60
	9.07	**6.70**	**5.74**	**5.20**	**4.86**	**4.62**	**4.44**	**4.30**	**4.19**	**4.10**	**4.02**	**3.96**
14	4.60	3.74	3.34	3.11	2.96	2.85	2.77	2.70	2.65	2.60	2.56	2.53
	8.86	**6.51**	**5.56**	**5.03**	**4.69**	**4.46**	**4.28**	**4.14**	**4.03**	**3.94**	**3.86**	**3.80**
15	4.54	3.68	3.29	3.06	2.90	2.79	2.70	2.64	2.59	2.55	2.51	2.48
	8.68	**6.36**	**5.42**	**4.89**	**4.56**	**4.32**	**4.14**	**4.00**	**3.89**	**3.80**	**3.73**	**3.67**
16	4.49	3.63	3.24	3.01	2.85	2.74	2.66	2.59	2.54	2.49	2.45	2.42
	8.53	**6.23**	**5.29**	**4.77**	**4.44**	**4.20**	**4.03**	**3.89**	**3.78**	**3.69**	**3.61**	**3.55**
17	4.45	3.59	3.20	2.96	2.81	2.70	2.62	2.55	2.50	2.45	2.41	2.38
	8.40	**6.11**	**5.18**	**4.67**	**4.34**	**4.10**	**3.93**	**3.79**	**3.68**	**3.59**	**3.52**	**3.45**
18	4.41	3.55	3.16	2.93	2.77	2.66	2.58	2.51	2.46	2.41	2.37	2.34
	8.28	**6.01**	**5.09**	**4.58**	**4.25**	**4.01**	**3.85**	**3.71**	**3.60**	**3.51**	**3.44**	**3.37**
19	4.38	3.52	3.13	2.90	2.74	2.63	2.55	2.48	2.43	2.38	2.34	2.31
	8.18	**5.93**	**5.01**	**4.50**	**4.17**	**3.94**	**3.77**	**3.63**	**3.52**	**3.43**	**3.36**	**3.30**
20	4.35	3.49	3.10	2.87	2.71	2.60	2.52	2.45	2.40	2.35	2.31	2.28
	8.10	**5.85**	**4.94**	**4.43**	**4.10**	**3.87**	**3.71**	**3.56**	**3.45**	**3.37**	**3.30**	**3.23**
21	4.32	3.47	3.07	2.84	2.68	2.57	2.49	2.42	2.37	2.32	2.28	2.25
	8.02	**5.78**	**4.87**	**4.37**	**4.04**	**3.81**	**3.65**	**3.51**	**3.40**	**3.31**	**3.24**	**3.17**
22	4.30	3.44	3.05	2.82	2.66	2.55	2.47	2.40	2.35	2.30	2.26	2.23
	7.94	**5.72**	**4.82**	**4.31**	**3.99**	**3.76**	**3.59**	**3.45**	**3.35**	**3.26**	**3.18**	**3.12**
23	4.28	3.42	3.03	2.80	2.64	2.53	2.45	2.38	2.32	2.28	2.24	2.20
	7.88	**5.66**	**4.76**	**4.26**	**3.94**	**3.71**	**3.54**	**3.41**	**3.30**	**3.21**	**3.14**	**3.07**
24	4.26	3.40	3.01	2.78	2.62	2.51	2.43	2.36	2.30	2.26	2.22	2.18
	7.82	**5.61**	**4.72**	**4.22**	**3.90**	**3.67**	**3.50**	**3.36**	**3.25**	**3.17**	**3.09**	**3.03**
25	4.24	3.38	2.99	2.76	2.60	2.49	2.41	2.34	2.28	2.24	2.20	2.16
	7.77	**5.57**	**4.68**	**4.18**	**3.86**	**3.63**	**3.46**	**3.32**	**3.21**	**3.13**	**3.05**	**2.99**
26	4.22	3.37	2.98	2.74	2.59	2.47	2.39	2.32	2.27	2.22	2.18	2.15
	7.72	**5.53**	**4.64**	**4.14**	**3.82**	**3.59**	**3.42**	**3.29**	**3.17**	**3.09**	**3.02**	**2.96**

$m_2 = $ Degrees of Freedom for Denominator

Reprinted by permission from *Statistical Methods*, 6th edition, by George W. Snedecor and William G. Cochran, © 1967 by The Iowa State University Press, Ames, Iowa.

D: Values on the F Distributions (continued)

$m_1 = $ Degrees of Freedom for Numerator												m_2
14	16	20	24	30	40	50	75	100	200	500	∞	
245	246	248	249	250	251	252	253	253	254	254	254	1
6,142	**6,169**	**6,208**	**6,234**	**6,258**	**6,286**	**6,302**	**6,323**	**6,334**	**6,352**	**6,361**	**6,366**	
19.42	19.43	19.44	19.45	19.46	19.47	19.47	19.48	19.49	19.49	19.50	19.50	2
99.43	**99.44**	**99.45**	**99.46**	**99.47**	**99.48**	**99.48**	**99.49**	**99.49**	**99.49**	**99.50**	**99.50**	
8.71	8.69	8.66	8.64	8.62	8.60	8.58	8.57	8.56	8.54	8.54	8.53	3
26.92	**26.83**	**26.69**	**26.60**	**26.50**	**26.41**	**26.35**	**26.27**	**26.23**	**26.18**	**26.14**	**26.12**	
5.87	5.84	5.80	5.77	5.74	5.71	5.70	5.68	5.66	5.65	5.64	5.63	4
14.24	**14.15**	**14.02**	**13.93**	**13.83**	**13.74**	**13.69**	**13.61**	**13.57**	**13.52**	**13.48**	**13.46**	
4.64	4.60	4.56	4.53	4.50	4.46	4.44	4.42	4.40	4.38	4.37	4.36	5
9.77	**9.68**	**9.55**	**9.47**	**9.38**	**9.29**	**9.24**	**9.17**	**9.13**	**9.07**	**9.04**	**9.02**	
3.96	3.92	3.87	3.84	3.81	3.77	3.75	3.72	3.71	3.69	3.68	3.67	6
7.60	**7.52**	**7.39**	**7.31**	**7.23**	**7.14**	**7.09**	**7.02**	**6.99**	**6.94**	**6.90**	**6.88**	
3.52	3.49	3.44	3.41	3.38	3.34	3.32	3.29	3.28	3.25	3.24	3.23	7
6.35	**6.27**	**6.15**	**6.07**	**5.98**	**5.90**	**5.85**	**5.78**	**5.75**	**5.70**	**5.67**	**5.65**	
3.23	3.20	3.15	3.12	3.08	3.05	3.03	3.00	2.98	2.96	2.94	2.93	8
5.56	**5.48**	**5.36**	**5.28**	**5.20**	**5.11**	**5.06**	**5.00**	**4.96**	**4.91**	**4.88**	**4.86**	
3.02	2.98	2.93	2.90	2.86	2.82	2.80	2.77	2.76	2.73	2.72	2.71	9
5.00	**4.92**	**4.80**	**4.73**	**4.64**	**4.56**	**4.51**	**4.45**	**4.41**	**4.36**	**4.33**	**4.31**	
2.86	2.82	2.77	2.74	2.70	2.67	2.64	2.61	2.59	2.56	2.55	2.54	10
4.60	**4.52**	**4.41**	**4.33**	**4.25**	**4.17**	**4.12**	**4.05**	**4.01**	**3.96**	**3.93**	**3.91**	
2.74	2.70	2.65	2.61	2.57	2.53	2.50	2.47	2.45	2.42	2.41	2.40	11
4.29	**4.21**	**4.10**	**4.02**	**3.94**	**3.86**	**3.80**	**3.74**	**3.70**	**3.66**	**3.62**	**3.60**	
2.64	2.60	2.54	2.50	2.46	2.42	2.40	2.36	2.35	2.32	2.31	2.30	12
4.05	**3.98**	**3.86**	**3.78**	**3.70**	**3.61**	**3.56**	**3.49**	**3.46**	**3.41**	**3.38**	**3.36**	
2.55	2.51	2.46	2.42	2.38	2.34	2.32	2.28	2.26	2.24	2.22	2.21	13
3.85	**3.78**	**3.67**	**3.59**	**3.51**	**3.42**	**3.37**	**3.30**	**3.27**	**3.21**	**3.18**	**3.16**	
2.48	2.44	2.39	2.35	2.31	2.27	2.24	2.21	2.19	2.16	2.14	2.13	14
3.70	**3.62**	**3.51**	**3.43**	**3.34**	**3.26**	**3.21**	**3.14**	**3.11**	**3.06**	**3.02**	**3.00**	
2.43	2.39	2.33	2.29	2.25	2.21	2.18	2.15	2.12	2.10	2.08	2.07	15
3.56	**3.48**	**3.36**	**3.29**	**3.20**	**3.12**	**3.07**	**3.00**	**2.97**	**2.92**	**2.89**	**2.87**	
2.37	2.33	2.28	2.24	2.20	2.16	2.13	2.09	2.07	2.04	2.02	2.01	16
3.45	**3.37**	**3.25**	**3.18**	**3.10**	**3.01**	**2.96**	**2.89**	**2.86**	**2.80**	**2.77**	**2.75**	
2.33	2.29	2.23	2.19	2.15	2.11	2.08	2.04	2.02	1.99	1.97	1.96	17
3.35	**3.27**	**3.16**	**3.08**	**3.00**	**2.92**	**2.86**	**2.79**	**2.76**	**2.70**	**2.67**	**2.65**	
2.29	2.25	2.19	2.15	2.11	2.07	2.04	2.00	1.98	1.95	1.93	1.92	18
3.27	**3.19**	**3.07**	**3.00**	**2.91**	**2.83**	**2.78**	**2.71**	**2.68**	**2.62**	**2.59**	**2.57**	
2.26	2.21	2.15	2.11	2.07	2.02	2.00	1.96	1.94	1.91	1.90	1.88	19
3.19	**3.12**	**3.00**	**2.92**	**2.84**	**2.76**	**2.70**	**2.63**	**2.60**	**2.54**	**2.51**	**2.49**	
2.23	2.18	2.12	2.08	2.04	1.99	1.96	1.92	1.90	1.87	1.85	1.84	20
3.13	**3.05**	**2.94**	**2.86**	**2.77**	**2.69**	**2.63**	**2.56**	**2.53**	**2.47**	**2.44**	**2.42**	
2.20	2.15	2.09	2.05	2.00	1.96	1.93	1.89	1.87	1.84	1.82	1.81	21
3.07	**2.99**	**2.88**	**2.80**	**2.72**	**2.63**	**2.58**	**2.51**	**2.47**	**2.42**	**2.38**	**2.36**	
2.18	2.13	2.07	2.03	1.98	1.93	1.91	1.87	1.84	1.81	1.80	1.78	22
3.02	**2.94**	**2.83**	**2.75**	**2.67**	**2.58**	**2.53**	**2.46**	**2.42**	**2.37**	**2.33**	**2.31**	
2.14	2.10	2.04	2.00	1.96	1.91	1.88	1.84	1.82	1.79	1.77	1.76	23
2.97	**2.89**	**2.78**	**2.70**	**2.62**	**2.53**	**2.48**	**2.41**	**2.37**	**2.32**	**2.28**	**2.26**	
2.13	2.09	2.02	1.98	1.94	1.89	1.86	1.82	1.80	1.76	1.74	1.73	24
2.93	**2.85**	**2.74**	**2.66**	**2.58**	**2.49**	**2.44**	**2.36**	**2.33**	**2.27**	**2.23**	**2.21**	
2.11	2.06	2.00	1.96	1.92	1.87	1.84	1.80	1.77	1.74	1.72	1.71	25
2.89	**2.81**	**2.70**	**2.62**	**2.54**	**2.45**	**2.40**	**2.32**	**2.29**	**2.23**	**2.19**	**2.17**	
2.10	2.05	1.99	1.95	1.90	1.85	1.82	1.78	1.76	1.72	1.70	1.69	26
2.86	**2.77**	**2.66**	**2.58**	**2.50**	**2.41**	**2.36**	**2.28**	**2.25**	**2.19**	**2.15**	**2.13**	

$m_2 = $ Degrees of Freedom for Denominator

D: Values on the F Distributions (continued)

m_2	$m_1 = $ Degrees of Freedom for Numerator											
	1	2	3	4	5	6	7	8	9	10	11	12
27	4.21	3.35	2.96	2.73	2.57	2.46	2.37	2.30	2.25	2.20	2.16	2.13
	7.68	**5.49**	**4.60**	**4.11**	**3.79**	**3.56**	**3.39**	**3.26**	**3.14**	**3.06**	**2.98**	**2.93**
28	4.20	3.34	2.95	2.71	2.56	2.44	2.36	2.29	2.24	2.19	2.15	2.12
	7.64	**5.45**	**4.57**	**4.07**	**3.76**	**3.53**	**3.36**	**3.23**	**3.11**	**3.03**	**2.95**	**2.90**
29	4.18	3.33	2.93	2.70	2.54	2.43	2.35	2.28	2.22	2.18	2.14	2.10
	7.60	**5.42**	**4.54**	**4.04**	**3.73**	**3.50**	**3.33**	**3.20**	**3.08**	**3.00**	**2.92**	**2.87**
30	4.17	3.32	2.92	2.69	2.53	2.42	2.34	2.27	2.21	2.16	2.12	2.09
	7.56	**5.39**	**4.51**	**4.02**	**3.70**	**3.47**	**3.30**	**3.17**	**3.06**	**2.98**	**2.90**	**2.84**
32	4.15	3.30	2.90	2.67	2.51	2.40	2.32	2.25	2.19	2.14	2.10	2.07
	7.50	**5.34**	**4.46**	**3.97**	**3.66**	**3.42**	**3.25**	**3.12**	**3.01**	**2.94**	**2.86**	**2.80**
34	4.13	3.28	2.88	2.65	2.49	2.38	2.30	2.23	2.17	2.12	2.08	2.05
	7.44	**5.29**	**4.42**	**3.93**	**3.61**	**3.38**	**3.21**	**3.08**	**2.97**	**2.89**	**2.82**	**2.76**
36	4.11	3.26	2.86	2.63	2.48	2.36	2.28	2.21	2.15	2.10	2.06	2.03
	7.39	**5.25**	**4.38**	**3.89**	**3.58**	**3.35**	**3.18**	**3.04**	**2.94**	**2.86**	**2.78**	**2.72**
38	4.10	3.25	2.85	2.62	2.46	2.35	2.26	2.19	2.14	2.09	2.05	2.02
	7.35	**5.21**	**4.34**	**3.86**	**3.54**	**3.32**	**3.15**	**3.02**	**2.91**	**2.82**	**2.75**	**2.69**
40	4.08	3.23	2.84	2.61	2.45	2.34	2.25	2.18	2.12	2.07	2.04	2.00
	7.31	**5.18**	**4.31**	**3.83**	**3.51**	**3.29**	**3.12**	**2.99**	**2.88**	**2.80**	**2.73**	**2.66**
42	4.07	3.22	2.83	2.59	2.44	2.32	2.24	2.17	2.11	2.06	2.02	1.99
	7.27	**5.15**	**4.29**	**3.80**	**3.49**	**3.26**	**3.10**	**2.96**	**2.86**	**2.77**	**2.70**	**2.64**
44	4.06	3.21	2.82	2.58	2.43	2.31	2.23	2.16	2.10	2.05	2.01	1.98
	7.24	**5.12**	**4.26**	**3.78**	**3.46**	**3.24**	**3.07**	**2.94**	**2.84**	**2.75**	**2.68**	**2.62**
46	4.05	3.20	2.81	2.57	2.42	2.30	2.22	2.14	2.09	2.04	2.00	1.97
	7.21	**5.10**	**4.24**	**3.76**	**3.44**	**3.22**	**3.05**	**2.92**	**2.82**	**2.73**	**2.66**	**2.60**
48	4.04	3.19	2.80	2.56	2.41	2.30	2.21	2.14	2.08	2.03	1.99	1.96
	7.19	**5.08**	**4.22**	**3.74**	**3.42**	**3.20**	**3.04**	**2.90**	**2.80**	**2.71**	**2.64**	**2.58**
50	4.03	3.18	2.79	2.56	2.40	2.29	2.20	2.13	2.07	2.02	1.98	1.95
	7.17	**5.06**	**4.20**	**3.72**	**3.41**	**3.18**	**3.02**	**2.88**	**2.78**	**2.70**	**2.62**	**2.56**
55	4.02	3.17	2.78	2.54	2.38	2.27	2.18	2.11	2.05	2.00	1.97	1.93
	7.12	**5.01**	**4.16**	**3.68**	**3.37**	**3.15**	**2.98**	**2.85**	**2.75**	**2.66**	**2.59**	**2.53**
60	4.00	3.15	2.76	2.52	2.37	2.25	2.17	2.10	2.04	1.99	1.95	1.92
	7.08	**4.98**	**4.13**	**3.65**	**3.34**	**3.12**	**2.95**	**2.82**	**2.72**	**2.63**	**2.56**	**2.50**
65	3.99	3.14	2.75	2.51	2.36	2.24	2.15	2.08	2.02	1.98	1.94	1.90
	7.04	**4.95**	**4.10**	**3.62**	**3.31**	**3.09**	**2.93**	**2.79**	**2.70**	**2.61**	**2.54**	**2.47**
70	3.98	3.13	2.74	2.50	2.35	2.23	2.14	2.07	2.01	1.97	1.93	1.89
	7.01	**4.92**	**4.08**	**3.60**	**3.29**	**3.07**	**2.91**	**2.77**	**2.67**	**2.59**	**2.51**	**2.45**
80	3.96	3.11	2.72	2.48	2.33	2.21	2.12	2.05	1.99	1.95	1.91	1.88
	6.96	**4.88**	**4.04**	**3.56**	**3.25**	**3.04**	**2.87**	**2.74**	**2.64**	**2.55**	**2.48**	**2.41**
100	3.94	3.09	2.70	2.46	2.30	2.19	2.10	2.03	1.97	1.92	1.88	1.85
	6.90	**4.82**	**3.98**	**3.51**	**3.20**	**2.99**	**2.82**	**2.69**	**2.59**	**2.51**	**2.43**	**2.36**
125	3.92	3.07	2.68	2.44	2.29	2.17	2.08	2.01	1.95	1.90	1.86	1.83
	6.84	**4.78**	**3.94**	**3.47**	**3.17**	**2.95**	**2.79**	**2.65**	**2.56**	**2.47**	**2.40**	**2.33**
150	3.91	3.06	2.67	2.43	2.27	2.16	2.07	2.00	1.94	1.89	1.85	1.82
	6.81	**4.75**	**3.91**	**3.44**	**3.14**	**2.92**	**2.76**	**2.62**	**2.53**	**2.44**	**2.37**	**2.30**
200	3.89	3.04	2.65	2.41	2.26	2.14	2.05	1.98	1.92	1.87	1.83	1.80
	6.76	**4.71**	**3.88**	**3.41**	**3.11**	**2.90**	**2.73**	**2.60**	**2.50**	**2.41**	**2.34**	**2.28**
400	3.86	3.02	2.62	2.39	2.23	2.12	2.03	1.96	1.90	1.85	1.81	1.78
	6.70	**4.66**	**3.83**	**3.36**	**3.06**	**2.85**	**2.69**	**2.55**	**2.46**	**2.37**	**2.29**	**2.23**
1000	3.85	3.00	2.61	2.38	2.22	2.10	2.02	1.95	1.89	1.84	1.80	1.76
	6.66	**4.62**	**3.80**	**3.34**	**3.04**	**2.82**	**2.66**	**2.53**	**2.43**	**2.34**	**2.26**	**2.20**
	3.84	2.99	2.60	2.37	2.21	2.09	2.01	1.94	1.88	1.83	1.79	1.75
	6.64	**4.60**	**3.78**	**3.32**	**3.02**	**2.80**	**2.64**	**2.51**	**2.41**	**2.32**	**2.24**	**2.18**

$m_2 = $ Degrees of Freedom for Denominator

D :   Values on the F Distributions (continued)

14	16	20	24	30	40	50	75	100	200	500	∞	m_2
						m_1 = Degrees of Freedom for Numerator						
2.08	2.03	1.97	1.93	1.88	1.84	1.80	1.76	1.74	1.71	1.68	1.67	27
2.83	**2.74**	**2.63**	**2.55**	**2.47**	**2.38**	**2.33**	**2.25**	**2.21**	**2.16**	**2.12**	**2.10**	
2.06	2.02	1.96	1.91	1.87	1.81	1.78	1.75	1.72	1.69	1.67	1.65	28
2.80	**2.71**	**2.60**	**2.52**	**2.44**	**2.35**	**2.30**	**2.22**	**2.18**	**2.13**	**2.09**	**2.06**	
2.05	2.00	1.94	1.90	1.85	1.80	1.77	1.73	1.71	1.68	1.65	1.64	29
2.77	**2.68**	**2.57**	**2.49**	**2.41**	**2.32**	**2.27**	**2.19**	**2.15**	**2.10**	**2.06**	**2.03**	
2.04	1.99	1.93	1.89	1.84	1.79	1.76	1.72	1,69	1.66	1.64	1.62	30
2.74	**2.66**	**2.55**	**2.47**	**2.38**	**2.29**	**2.24**	**2.16**	**2.13**	**2.07**	**2.03**	**2.01**	
2.02	1.97	1.91	1.86	1.82	1.76	1.74	1.69	1.67	1.64	1.61	1.59	32
2.70	**2.62**	**2.51**	**2.42**	**2.34**	**2.25**	**2.20**	**2.12**	**2.08**	**2.02**	**1.98**	**1.96**	
2.00	1.95	1.89	1.84	1.80	1.74	1.71	1.67	1.64	1.61	1.59	1.57	34
2.66	**2.58**	**2.47**	**2.38**	**2.30**	**2.21**	**2.15**	**2.08**	**2.04**	**1.98**	**1.94**	**1.91**	
1.98	1.93	1.87	1.82	1.78	1.72	1.69	1.65	1.62	1.59	1.56	1.55	36
2.62	**2.54**	**2.43**	**2.35**	**2.26**	**2.17**	**2.12**	**2.04**	**2.00**	**1.94**	**1.90**	**1.87**	
1.96	1.92	1.85	1.80	1.76	1.71	1.67	1.63	1.60	1.57	1.54	1.53	38
2.59	**2.51**	**2.40**	**2.32**	**2.22**	**2.14**	**2.08**	**2.00**	**1.97**	**1.90**	**1.86**	**1.84**	
1.95	1.90	1,84	1.79	1.74	1.69	1.66	1.61	1.59	1.55	1.53	1.51	40
2.56	**2.49**	**2.37**	**2.29**	**2.20**	**2.11**	**2.05**	**1.97**	**1.94**	**1.88**	**1.84**	**1.81**	
1.94	1.89	1.82	1.78	1.73	1.68	1.64	1.60	1.57	1.54	1.51	1.49	42
2.54	**2.46**	**2.35**	**2.26**	**2.17**	**2.08**	**2.02**	**1.94**	**1.91**	**1.85**	**1.80**	**1.78**	
1.92	1.88	1.81	1.76	1.72	1.66	1.63	1.58	1.56	1.52	1.50	1.48	44
2.52	**2.44**	**2.32**	**2.24**	**2.15**	**2.06**	**2.00**	**1.92**	**1.88**	**1.82**	**1.78**	**1.75**	
1.91	1.87	1.80	1.75	1.71	1.65	1.62	1.57	1.54	1.51	1.48	1.46	46
2.50	**2.42**	**2.30**	**2.22**	**2.13**	**2.04**	**1.98**	**1.90**	**1.86**	**1.80**	**1.76**	**1.72**	
1.90	1.86	1.79	1.74	1.70	1.64	1.61	1.56	1.53	1.50	1.47	1.45	48
2.48	**2.40**	**2.28**	**2.20**	**2.11**	**2.02**	**1.96**	**1.88**	**1.84**	**1.78**	**1.73**	**1.70**	
1.90	1.85	1.78	1.74	1.69	1.63	1.60	1.55	1.52	1.48	1.46	1.44	50
2.46	**2.39**	**2.26**	**2.18**	**2.10**	**2.00**	**1.94**	**1.86**	**1.82**	**1.76**	**1.71**	**1.68**	
1.88	1.83	1.76	1.72	1.67	1.61	1.58	1.52	1.50	1.46	1.43	1.41	55
2.43	**2.35**	**2.23**	**2.15**	**2.06**	**1.96**	**1.90**	**1.82**	**1.78**	**1.71**	**1.66**	**1.64**	
1.86	1.81	1.75	1.70	1.65	1.59	1.56	1.50	1.48	1.44	1.41	1.39	60
2.40	**2.32**	**2.20**	**2.12**	**2.03**	**1.93**	**1.87**	**1.79**	**1.74**	**1.68**	**1.63**	**1.60**	
1.85	1.80	1.73	1.68	1.63	1.57	1.54	1.49	1.46	1.42	1.39	1.37	65
2.37	**2.30**	**2.18**	**2.09**	**2.00**	**1.90**	**1.84**	**1.76**	**1.71**	**1.64**	**1.60**	**1.56**	
1.84	1.79	1.72	1.67	1.62	1.56	1.53	1.47	1.45	1.40	1.37	1.35	70
2.35	**2.28**	**2.15**	**2.07**	**1.98**	**1.88**	**1.82**	**1.74**	**1.69**	**1.62**	**1.56**	**1.53**	
1.82	1.77	1.70	1.65	1.60	1.54	1.51	1.45	1.42	1.38	1.35	1.32	80
2.32	**2.24**	**2.11**	**2.03**	**1.94**	**1.84**	**1.78**	**1.70**	**1.65**	**1.57**	**1.52**	**1.49**	
1.79	1.75	1.68	1.63	1.57	1.51	1.48	1.42	1.39	1.34	1.30	1.28	100
2.26	**2.19**	**2.06**	**1.98**	**1.89**	**1.79**	**1.73**	**1.64**	**1.59**	**1.51**	**1.46**	**1.43**	
1.77	1.72	1.65	1.60	1.55	1.49	1.45	1.39	1.36	1.31	1.27	1.25	125
2.23	**2.15**	**2.03**	**1.94**	**1.85**	**1.75**	**1.68**	**1.59**	**1.54**	**1.46**	**1.40**	**1.37**	
1.76	1.71	1.64	1.59	1.54	1.47	1.44	1.37	1.34	1.29	1.25	1.22	150
2.20	**2.12**	**2.00**	**1.91**	**1.83**	**1.72**	**1.66**	**1.56**	**1.51**	**1.43**	**1.37**	**1.33**	
1.74	1.69	1.62	1.57	1.52	1.45	1.42	1.35	1.32	1.26	1.22	1.19	200
2.17	**2.09**	**1.97**	**1.88**	**1.79**	**1.69**	**1.62**	**1.53**	**1.48**	**1.39**	**1.33**	**1.28**	
1.72	1.67	1.60	1.54	1.49	1.42	1.38	1.32	1.28	1.22	1.16	1.13	400
2.12	**2.04**	**1.92**	**1.84**	**1.74**	**1.64**	**1.57**	**1.47**	**1.42**	**1.32**	**1.24**	**1.19**	
1.70	1.65	1.58	1.53	1.47	1.41	1.36	1.30	1.26	1.19	1.13	1.08	1,000
2.09	**2.01**	**1.89**	**1.81**	**1.71**	**1.61**	**1.54**	**1.44**	**1.38**	**1.28**	**1.19**	**1.11**	
1.69	1.64	1.57	1.52	1.46	1.40	1.35	1.28	1.24	1.17	1.11	1.00	∞
2.07	**1.99**	**1.87**	**1.79**	**1.69**	**1.59**	**1.52**	**1.41**	**1.36**	**1.25**	**1.15**	**1.00**	

m_2 = Degrees of Freedom for Denominator

E: Binomial Distributions

π

n	r	.05	.10	.15	.20	.25	.30	.35	.40	.45	.50	.55	.60	.65	.70	.75	.80	.85	.90	.95
1	0	.9500	.9000	.8500	.8000	.7500	.7000	.6500	.6000	.5500	.5000	.4500	.4000	.3500	.3000 *	.2500	.2000	.1500	.1000	.0500
	1	.0500	.1000	.1500	.2000	.2500	.3000	.3500	.4000	.4500	.5000	.5500	.6000	.6500	.7000	.7500	.8000	.8500	.9000	.9500
2	0	.9025	.8100	.7225	.6400	.5625	.4900	.4225	.3600	.3025	.2500	.2025	.1600	.1225	.0900	.0625	.0400	.0225	.0100	.0025
	1	.0950	.1800	.2550	.3200	.3750	.4200	.4550	.4800	.4950	.5000	.4950	.4800	.4550	.4200	.3750	.3200	.2550	.1800	.0950
	2	.0025	.0100	.0225	.0400	.0625	.0900	.1225	.1600	.2025	.2500	.3025	.3600	.4225	.4900	.5625	.6400	.7225	.8100	.9025
3	0	.8574	.7290	.6141	.5120	.4219	.3430	.2746	.2160	.1664	.1250	.0911	.0640	.0429	.0270	.0156	.0080	.0034	.0010	.0001
	1	.1354	.2430	.3251	.3840	.4219	.4410	.4436	.4320	.4084	.3750	.3341	.2880	.2389	.1890	.1406	.0960	.0574	.0270	.0071
	2	.0071	.0270	.0574	.0960	.1406	.1890	.2389	.2880	.3341	.3750	.4084	.4320	.4436	.4410	.4219	.3840	.3251	.2430	.1354
	3	.0001	.0010	.0034	.0080	.0156	.0270	.0429	.0640	.0911	.1250	.1664	.2160	.2746	.3430	.4219	.5120	.6141	.7290	.8574
4	0	.8145	.6561	.5220	.4096	.3164	.2401	.1785	.1296	.0915	.0625	.0410	.0256	.0150	.0081	.0039	.0016	.0005	.0001	.0000
	1	.1715	.2916	.3685	.4096	.4219	.4116	.3845	.3456	.2995	.2500	.2005	.1536	.1115	.0756	.0469	.0256	.0115	.0036	.0005
	2	.0135	.0486	.0975	.1536	.2109	.2646	.3105	.3456	.3675	.3750	.3675	.3456	.3105	.2646	.2109	.1536	.0975	.0486	.0135
	3	.0005	.0036	.0115	.0256	.0469	.0756	.1115	.1536	.2005	.2500	.2995	.3456	.3845	.4116	.4219	.4096	.3685	.2916	.1715
	4	.0000	.0001	.0005	.0016	.0039	.0081	.0150	.0256	.0410	.0625	.0915	.1296	.1785	.2401	.3164	.4096	.5220	.6561	.8145
5	0	.7738	.5905	.4437	.3277	.2373	.1681	.1160	.0778	.0503	.0312	.0185	.0102	.0053	.0024	.0010	.0003	.0001	.0000	.0000
	1	.2036	.3280	.3915	.4096	.3955	.3602	.3124	.2592	.2059	.1562	.1128	.0768	.0488	.0284	.0146	.0064	.0022	.0004	.0000
	2	.0214	.0729	.1382	.2048	.2637	.3087	.3364	.3456	.3369	.3125	.2757	.2304	.1811	.1323	.0879	.0512	.0244	.0081	.0011
	3	.0011	.0081	.0244	.0512	.0879	.1323	.1811	.2304	.2757	.3125	.3369	.3456	.3364	.3087	.2637	.2048	.1382	.0729	.0214
	4	.0000	.0004	.0022	.0064	.0146	.0284	.0488	.0768	.1128	.1562	.2059	.2592	.3124	.3602	.3955	.4096	.3915	.3280	.2036
	5	.0000	.0000	.0001	.0003	.0010	.0024	.0053	.0102	.0185	.0312	.0503	.0778	.1160	.1681	.2373	.3277	.4437	.5905	.7738
6	0	.7351	.5314	.3771	.2621	.1780	.1176	.0754	.0467	.0277	.0156	.0083	.0041	.0018	.0007	.0002	.0001	.0000	.0000	.0000
	1	.2321	.3543	.3993	.3932	.3560	.3025	.2437	.1866	.1359	.0938	.0609	.0369	.0205	.0102	.0044	.0015	.0004	.0001	.0000
	2	.0305	.0984	.1762	.2458	.2966	.3241	.3280	.3110	.2780	.2344	.1861	.1382	.0951	.0595	.0330	.0154	.0055	.0012	.0001
	3	.0021	.0146	.0415	.0819	.1318	.1852	.2355	.2765	.3032	.3125	.3032	.2765	.2355	.1852	.1318	.0819	.0415	.0146	.0021
	4	.0001	.0012	.0055	.0154	.0330	.0595	.0951	.1382	.1861	.2344	.2780	.3110	.3280	.3241	.2966	.2458	.1762	.0984	.0305
	5	.0000	.0001	.0004	.0015	.0044	.0102	.0205	.0369	.0609	.0938	.1359	.1866	.2437	.3025	.3560	.3932	.3993	.3543	.2321
	6	.0000	.0000	.0000	.0001	.0002	.0007	.0018	.0041	.0083	.0156	.0277	.0467	.0754	.1176	.1780	.2621	.3771	.5314	.7351
7	0	.6983	.4783	.3206	.2097	.1335	.0824	.0490	.0280	.0152	.0078	.0037	.0016	.0006	.0002	.0001	.0000	.0000	.0000	.0000
	1	.2573	.3720	.3960	.3670	.3115	.2471	.1848	.1306	.0872	.0547	.0320	.0172	.0084	.0036	.0013	.0004	.0001	.0000	.0000
	2	.0406	.1240	.2097	.2753	.3115	.3177	.2985	.2613	.2140	.1641	.1172	.0774	.0466	.0250	.0115	.0043	.0012	.0002	.0000
	3	.0036	.0230	.0617	.1147	.1730	.2269	.2679	.2903	.2918	.2734	.2388	.1935	.1442	.0972	.0577	.0287	.0109	.0026	.0002
	4	.0002	.0026	.0109	.0287	.0577	.0972	.1442	.1935	.2388	.2734	.2918	.2903	.2679	.2269	.1730	.1147	.0617	.0230	.0036
	5	.0000	.0002	.0012	.0043	.0115	.0250	.0466	.0774	.1172	.1641	.2140	.2613	.2985	.3177	.3115	.2753	.2097	.1240	.0406
	6	.0000	.0000	.0001	.0004	.0013	.0036	.0084	.0172	.0320	.0547	.0872	.1306	.1848	.2471	.3115	.3670	.3960	.3720	.2573
	7	.0000	.0000	.0000	.0000	.0001	.0002	.0006	.0016	.0037	.0078	.0152	.0280	.0490	.0824	.1335	.2097	.3206	.4783	.6983

E: Binomial Distributions (continued)

π

n	r	.05	.10	.15	.20	.25	.30	.35	.40	.45	.50	.55	.60	.65	.70	.75	.80	.85	.90	.95
8	0	.6634	.4305	.2725	.1678	.1001	.0576	.0319	.0168	.0084	.0039	.0017	.0007	.0002	.0001	.0000	.0000	.0000	.0000	.0000
	1	.2793	.3826	.3847	.3355	.2670	.1977	.1373	.0896	.0548	.0312	.0164	.0079	.0033	.0012	.0004	.0001	.0000	.0000	.0000
	2	.0515	.1488	.2376	.2936	.3115	.2965	.2587	.2090	.1569	.1094	.0703	.0413	.0217	.0100	.0038	.0011	.0002	.0000	.0000
	3	.0054	.0331	.0839	.1468	.2076	.2541	.2786	.2787	.2568	.2188	.1719	.1239	.0808	.0467	.0231	.0092	.0026	.0004	.0000
	4	.0004	.0046	.0185	.0459	.0865	.1361	.1875	.2322	.2627	.2734	.2627	.2322	.1875	.1361	.0865	.0459	.0185	.0046	.0004
	5	.0000	.0004	.0026	.0092	.0231	.0467	.0808	.1239	.1719	.2188	.2568	.2787	.2786	.2541	.2076	.1468	.0839	.0331	.0054
	6	.0000	.0000	.0002	.0011	.0038	.0100	.0217	.0413	.0703	.1094	.1569	.2090	.2587	.2965	.3115	.2936	.2376	.1488	.0515
	7	.0000	.0000	.0000	.0001	.0004	.0012	.0033	.0079	.0164	.0312	.0548	.0896	.1373	.1977	.2670	.3355	.3847	.3826	.2793
	8	.0000	.0000	.0000	.0000	.0000	.0001	.0002	.0007	.0017	.0039	.0084	.0168	.0319	.0576	.1001	.1678	.2725	.4305	.6634
9	0	.6302	.3874	.2316	.1342	.0751	.0404	.0207	.0101	.0046	.0020	.0008	.0003	.0001	.0000	.0000	.0000	.0000	.0000	.0000
	1	.2985	.3874	.3679	.3020	.2253	.1556	.1004	.0605	.0339	.0176	.0083	.0035	.0013	.0004	.0001	.0000	.0000	.0000	.0000
	2	.0629	.1722	.2597	.3020	.3003	.2668	.2162	.1612	.1110	.0703	.0407	.0212	.0098	.0039	.0012	.0003	.0000	.0000	.0000
	3	.0077	.0446	.1069	.1762	.2336	.2668	.2716	.2508	.2119	.1641	.1160	.0743	.0424	.0210	.0087	.0028	.0006	.0001	.0000
	4	.0006	.0074	.0283	.0661	.1168	.1715	.2194	.2508	.2600	.2461	.2128	.1672	.1181	.0735	.0389	.0165	.0050	.0008	.0000
	5	.0000	.0008	.0050	.0165	.0389	.0735	.1181	.1672	.2128	.2461	.2600	.2508	.2194	.1715	.1168	.0661	.0283	.0074	.0006
	6	.0000	.0001	.0006	.0028	.0087	.0210	.0424	.0743	.1160	.1641	.2119	.2508	.2716	.2668	.2336	.1762	.1069	.0446	.0077
	7	.0000	.0000	.0000	.0003	.0012	.0039	.0098	.0212	.0407	.0703	.1110	.1612	.2162	.2668	.3003	.3020	.2597	.1722	.0629
	8	.0000	.0000	.0000	.0000	.0001	.0004	.0013	.0035	.0083	.0176	.0339	.0605	.1004	.1556	.2253	.3020	.3679	.3874	.2985
	9	.0000	.0000	.0000	.0000	.0000	.0000	.0001	.0003	.0008	.0020	.0046	.0101	.0207	.0404	.0751	.1342	.2316	.3874	.6302
10	0	.5987	.3487	.1969	.1074	.0563	.0282	.0135	.0060	.0025	.0010	.0003	.0001	.0000	.0000	.0000	.0000	.0000	.0000	.0000
	1	.3151	.3874	.3474	.2684	.1877	.1211	.0725	.0403	.0207	.0098	.0042	.0016	.0005	.0001	.0000	.0000	.0000	.0000	.0000
	2	.0746	.1937	.2759	.3020	.2816	.2335	.1757	.1209	.0763	.0439	.0229	.0106	.0043	.0014	.0004	.0001	.0000	.0000	.0000
	3	.0105	.0574	.1298	.2013	.2503	.2668	.2522	.2150	.1665	.1172	.0746	.0425	.0212	.0090	.0031	.0008	.0001	.0000	.0000
	4	.0010	.0112	.0401	.0881	.1460	.2001	.2377	.2508	.2384	.2051	.1596	.1115	.0689	.0368	.0162	.0055	.0012	.0001	.0000
	5	.0001	.0015	.0085	.0264	.0584	.1029	.1536	.2007	.2340	.2461	.2340	.2007	.1536	.1029	.0584	.0264	.0085	.0015	.0001
	6	.0000	.0001	.0012	.0055	.0162	.0368	.0689	.1115	.1596	.2051	.2384	.2508	.2377	.2001	.1460	.0881	.0401	.0112	.0010
	7	.0000	.0000	.0001	.0008	.0031	.0090	.0212	.0425	.0746	.1172	.1665	.2150	.2522	.2668	.2503	.2013	.1298	.0574	.0105
	8	.0000	.0000	.0000	.0001	.0004	.0014	.0043	.0106	.0229	.0439	.0763	.1209	.1757	.2335	.2816	.3020	.2759	.1937	.0746
	9	.0000	.0000	.0000	.0000	.0000	.0001	.0005	.0016	.0042	.0098	.0207	.0403	.0725	.1211	.1877	.2684	.3474	.3874	.3151
	10	.0000	.0000	.0000	.0000	.0000	.0000	.0000	.0001	.0003	.0010	.0025	.0060	.0135	.0282	.0563	.1074	.1969	.3487	.5987
11	0	.5688	.3138	.1673	.0859	.0422	.0198	.0088	.0036	.0014	.0005	.0002	.0000	.0000	.0000	.0000	.0000	.0000	.0000	.0000
	1	.3293	.3835	.3248	.2362	.1549	.0932	.0518	.0266	.0125	.0054	.0021	.0007	.0002	.0000	.0000	.0000	.0000	.0000	.0000
	2	.0867	.2131	.2866	.2953	.2581	.1998	.1395	.0887	.0513	.0269	.0126	.0052	.0018	.0005	.0001	.0000	.0000	.0000	.0000
	3	.0137	.0710	.1517	.2215	.2581	.2568	.2254	.1774	.1259	.0806	.0462	.0234	.0102	.0037	.0011	.0002	.0000	.0000	.0000
	4	.0014	.0158	.0536	.1107	.1721	.2201	.2428	.2365	.2060	.1611	.1128	.0701	.0379	.0173	.0064	.0017	.0003	.0000	.0000
	5	.0001	.0025	.0132	.0388	.0803	.1321	.1830	.2207	.2360	.2256	.1931	.1471	.0985	.0566	.0268	.0097	.0023	.0003	.0000
	6	.0000	.0003	.0023	.0097	.0268	.0566	.0985	.1471	.1931	.2256	.2360	.2207	.1830	.1321	.0803	.0388	.0132	.0025	.0001
	7	.0000	.0000	.0003	.0017	.0064	.0173	.0379	.0701	.1128	.1611	.2060	.2365	.2428	.2201	.1721	.1107	.0536	.0158	.0014
	8	.0000	.0000	.0000	.0002	.0011	.0037	.0102	.0234	.0462	.0806	.1259	.1774	.2254	.2568	.2581	.2215	.1517	.0710	.0137
	9	.0000	.0000	.0000	.0000	.0001	.0005	.0018	.0052	.0126	.0269	.0513	.0887	.1395	.1998	.2581	.2953	.2866	.2131	.0867
	10	.0000	.0000	.0000	.0000	.0000	.0000	.0002	.0007	.0021	.0054	.0125	.0266	.0518	.0932	.1549	.2362	.3248	.3835	.3293
	11	.0000	.0000	.0000	.0000	.0000	.0000	.0000	.0000	.0002	.0005	.0014	.0036	.0088	.0198	.0422	.0859	.1673	.3138	.5688

E: Binomial Distributions (continued)

π

n	r	.05	.10	.15	.20	.25	.30	.35	.40	.45	.50	.55	.60	.65	.70	.75	.80	.85	.90	.95
12	0	.5404	.2824	.1422	.0687	.0317	.0138	.0057	.0022	.0008	.0002	.0001	.0000	.0000	.0000	.0000	.0000	.0000	.0000	.0000
	1	.3413	.3766	.3012	.2062	.1267	.0712	.0368	.0174	.0075	.0029	.0010	.0003	.0001	.0000	.0000	.0000	.0000	.0000	.0000
	2	.0988	.2301	.2924	.2835	.2323	.1678	.1088	.0639	.0339	.0161	.0068	.0025	.0008	.0002	.0000	.0000	.0000	.0000	.0000
	3	.0173	.0852	.1720	.2362	.2581	.2397	.1954	.1419	.0923	.0537	.0277	.0125	.0048	.0015	.0004	.0001	.0000	.0000	.0000
	4	.0021	.0213	.0683	.1329	.1936	.2311	.2367	.2128	.1700	.1208	.0762	.0420	.0199	.0078	.0024	.0005	.0001	.0000	.0000
	5	.0002	.0038	.0193	.0532	.1032	.1585	.2039	.2270	.2225	.1934	.1489	.1009	.0591	.0291	.0115	.0033	.0006	.0005	.0000
	6	.0000	.0005	.0040	.0155	.0401	.0792	.1281	.1766	.2124	.2256	.2124	.1766	.1281	.0792	.0401	.0155	.0040	.0038	.0002
	7	.0000	.0000	.0006	.0033	.0115	.0291	.0591	.1009	.1489	.1934	.2225	.2270	.2039	.1585	.1032	.0532	.0193	.0213	.0021
	8	.0000	.0000	.0001	.0005	.0024	.0078	.0199	.0420	.0762	.1208	.1700	.2128	.2367	.2311	.1936	.1329	.0683	.0852	.0173
	9	.0000	.0000	.0000	.0001	.0004	.0015	.0048	.0125	.0277	.0537	.0923	.1419	.1954	.2397	.2581	.2362	.1720	.2301	.0988
	10	.0000	.0000	.0000	.0000	.0000	.0002	.0008	.0025	.0068	.0161	.0339	.0639	.1088	.1678	.2323	.2835	.2924	.3766	.3413
	11	.0000	.0000	.0000	.0000	.0000	.0000	.0001	.0003	.0010	.0029	.0075	.0174	.0368	.0712	.1267	.2062	.3012	.3559	.3593
	12	.0000	.0000	.0000	.0000	.0000	.0000	.0000	.0000	.0001	.0002	.0008	.0022	.0057	.0138	.0317	.0687	.1422	.2824	.5404
13	0	.5133	.2542	.1209	.0550	.0238	.0097	.0037	.0013	.0004	.0001	.0000	.0000	.0000	.0000	.0000	.0000	.0000	.0000	.0000
	1	.3512	.3672	.2774	.1787	.1029	.0540	.0259	.0113	.0045	.0016	.0005	.0001	.0000	.0000	.0000	.0000	.0000	.0000	.0000
	2	.1109	.2448	.2937	.2680	.2059	.1388	.0836	.0453	.0220	.0095	.0036	.0012	.0003	.0001	.0000	.0000	.0000	.0000	.0000
	3	.0214	.0997	.1900	.2457	.2517	.2181	.1651	.1107	.0660	.0349	.0162	.0065	.0022	.0006	.0001	.0000	.0000	.0000	.0000
	4	.0028	.0277	.0838	.1535	.2097	.2337	.2222	.1845	.1350	.0873	.0495	.0243	.0101	.0034	.0009	.0001	.0000	.0000	.0000
	5	.0003	.0055	.0266	.0691	.1258	.1803	.2154	.2214	.1989	.1571	.1089	.0656	.0336	.0142	.0047	.0011	.0001	.0000	.0000
	6	.0000	.0008	.0063	.0230	.0559	.1030	.1546	.1968	.2169	.2095	.1775	.1312	.0833	.0442	.0186	.0058	.0011	.0001	.0000
	7	.0000	.0001	.0011	.0058	.0186	.0442	.0833	.1312	.1775	.2095	.2169	.1968	.1546	.1030	.0559	.0230	.0063	.0008	.0000
	8	.0000	.0000	.0001	.0011	.0047	.0142	.0336	.0656	.1089	.1571	.1989	.2214	.2154	.1803	.1258	.0691	.0266	.0055	.0003
	9	.0000	.0000	.0000	.0001	.0009	.0034	.0101	.0243	.0495	.0873	.1350	.1845	.2222	.2337	.2097	.1535	.0838	.0277	.0028
	10	.0000	.0000	.0000	.0000	.0001	.0006	.0022	.0065	.0162	.0349	.0660	.1107	.1651	.2181	.2517	.2457	.1900	.0997	.0214
	11	.0000	.0000	.0000	.0000	.0000	.0001	.0003	.0012	.0036	.0095	.0220	.0453	.0836	.1388	.2059	.2680	.2937	.2448	.1109
	12	.0000	.0000	.0000	.0000	.0000	.0000	.0000	.0001	.0005	.0016	.0045	.0113	.0259	.0540	.1029	.1787	.2774	.3672	.3512
	13	.0000	.0000	.0000	.0000	.0000	.0000	.0000	.0000	.0000	.0001	.0004	.0013	.0037	.0097	.0238	.0550	.1209	.2542	.5133
14	0	.4877	.2288	.1028	.0440	.0178	.0068	.0024	.0008	.0002	.0001	.0000	.0000	.0000	.0000	.0000	.0000	.0000	.0000	.0000
	1	.3593	.3559	.2539	.1539	.0832	.0407	.0181	.0073	.0027	.0009	.0002	.0001	.0000	.0000	.0000	.0000	.0000	.0000	.0000
	2	.1229	.2570	.2912	.2501	.1802	.1134	.0634	.0317	.0141	.0056	.0019	.0005	.0001	.0000	.0000	.0000	.0000	.0000	.0000
	3	.0259	.1142	.2056	.2501	.2402	.1943	.1366	.0845	.0462	.0222	.0093	.0033	.0010	.0002	.0000	.0000	.0000	.0000	.0000
	4	.0037	.0349	.0998	.1720	.2202	.2290	.2022	.1549	.1040	.0611	.0312	.0136	.0049	.0014	.0003	.0000	.0000	.0000	.0000
	5	.0004	.0078	.0352	.0860	.1468	.1963	.2178	.2066	.1701	.1222	.0762	.0408	.0183	.0066	.0018	.0003	.0000	.0000	.0000
	6	.0000	.0013	.0093	.0322	.0734	.1262	.1759	.2066	.2088	.1833	.1398	.0918	.0510	.0232	.0082	.0020	.0003	.0000	.0000
	7	.0000	.0002	.0019	.0092	.0280	.0618	.1082	.1574	.1952	.2095	.1952	.1574	.1082	.0618	.0280	.0092	.0019	.0002	.0000
	8	.0000	.0000	.0003	.0020	.0082	.0232	.0510	.0918	.1398	.1833	.2088	.2066	.1759	.1262	.0734	.0322	.0093	.0013	.0000
	9	.0000	.0000	.0000	.0003	.0018	.0066	.0183	.0408	.0762	.1222	.1701	.2066	.2178	.1963	.1468	.0860	.0352	.0078	.0004
	10	.0000	.0000	.0000	.0000	.0003	.0014	.0049	.0136	.0312	.0611	.1040	.1549	.2022	.2290	.2202	.1720	.0998	.0349	.0037
	11	.0000	.0000	.0000	.0000	.0000	.0002	.0010	.0033	.0093	.0222	.0462	.0845	.1366	.1943	.2402	.2501	.2056	.1142	.0259
	12	.0000	.0000	.0000	.0000	.0000	.0000	.0001	.0005	.0019	.0056	.0141	.0317	.0634	.1134	.1802	.2501	.2912	.2570	.1229
	13	.0000	.0000	.0000	.0000	.0000	.0000	.0000	.0001	.0002	.0009	.0027	.0073	.0181	.0407	.0832	.1539	.2539	.3559	.3593
	14	.0000	.0000	.0000	.0000	.0000	.0000	.0000	.0000	.0000	.0001	.0002	.0008	.0024	.0068	.0178	.0440	.1028	.2288	.4877

E: Binomial Distributions (continued)

π

n	r	.05	.10	.15	.20	.25	.30	.35	.40	.45	.50	.55	.60	.65	.70	.75	.80	.85	.90	.95
15	0	.4633	.2059	.0874	.0352	.0134	.0047	.0016	.0005	.0001	.0000	.0000	.0000	.0000	.0000	.0000	.0000	.0000	.0000	.0000
	1	.3658	.3432	.2312	.1319	.0668	.0305	.0126	.0047	.0016	.0005	.0001	.0000	.0000	.0000	.0000	.0000	.0000	.0000	.0000
	2	.1348	.2669	.2856	.2309	.1559	.0916	.0476	.0219	.0090	.0032	.0010	.0003	.0001	.0000	.0000	.0000	.0000	.0000	.0000
	3	.0307	.1285	.2184	.2501	.2252	.1700	.1110	.0634	.0318	.0139	.0052	.0016	.0004	.0001	.0000	.0000	.0000	.0000	.0000
	4	.0049	.0428	.1156	.1876	.2252	.2186	.1792	.1268	.0780	.0417	.0191	.0074	.0024	.0006	.0001	.0000	.0000	.0000	.0000
	5	.0006	.0105	.0449	.1032	.1651	.2061	.2123	.1859	.1404	.0916	.0515	.0245	.0096	.0030	.0007	.0001	.0000	.0000	.0000
	6	.0000	.0019	.0132	.0430	.0917	.1472	.1906	.2066	.1914	.1527	.1048	.0612	.0298	.0116	.0034	.0007	.0001	.0000	.0000
	7	.0000	.0003	.0030	.0138	.0393	.0811	.1319	.1771	.2013	.1964	.1647	.1181	.0710	.0348	.0131	.0035	.0005	.0000	.0000
	8	.0000	.0000	.0005	.0035	.0131	.0348	.0710	.1181	.1647	.1964	.2013	.1771	.1319	.0811	.0393	.0138	.0030	.0003	.0000
	9	.0000	.0000	.0001	.0007	.0034	.0116	.0298	.0612	.1048	.1527	.1914	.2066	.1906	.1472	.0917	.0430	.0132	.0019	.0006
	10	.0000	.0000	.0000	.0001	.0007	.0030	.0096	.0245	.0515	.0916	.1404	.1859	.2123	.2061	.1651	.1032	.0449	.0105	.0049
	11	.0000	.0000	.0000	.0000	.0001	.0006	.0024	.0074	.0191	.0417	.0780	.1268	.1792	.2186	.2252	.1876	.1156	.0428	.0307
	12	.0000	.0000	.0000	.0000	.0000	.0001	.0004	.0016	.0052	.0139	.0318	.0634	.1110	.1700	.1559	.2501	.2184	.1285	.1348
	13	.0000	.0000	.0000	.0000	.0000	.0000	.0001	.0003	.0010	.0032	.0090	.0219	.0476	.0916	.0668	.2309	.2856	.2669	.3658
	14	.0000	.0000	.0000	.0000	.0000	.0000	.0000	.0000	.0001	.0005	.0016	.0047	.0126	.0305	.0134	.1319	.2312	.3432	.4633
	15	.0000	.0000	.0000	.0000	.0000	.0000	.0000	.0000	.0000	.0000	.0001	.0005	.0016	.0047	.0100	.0352	.0874	.2059	.4633
16	0	.4401	.1853	.0743	.0281	.0100	.0033	.0010	.0003	.0001	.0000	.0000	.0000	.0000	.0000	.0000	.0000	.0000	.0000	.0000
	1	.3706	.3294	.2097	.1126	.0535	.0228	.0087	.0030	.0009	.0002	.0001	.0000	.0000	.0000	.0000	.0000	.0000	.0000	.0000
	2	.1463	.2745	.2775	.2111	.1336	.0732	.0353	.0150	.0056	.0018	.0005	.0001	.0000	.0000	.0000	.0000	.0000	.0000	.0000
	3	.0359	.1423	.2285	.2463	.2079	.1465	.0888	.0468	.0215	.0085	.0029	.0008	.0002	.0000	.0000	.0000	.0000	.0000	.0000
	4	.0061	.0514	.1311	.2001	.2252	.2040	.1553	.1014	.0572	.0278	.0115	.0040	.0011	.0002	.0000	.0000	.0000	.0000	.0000
	5	.0008	.0137	.0555	.1201	.1802	.2099	.2008	.1623	.1123	.0667	.0337	.0142	.0049	.0013	.0002	.0000	.0000	.0000	.0000
	6	.0001	.0028	.0180	.0550	.1101	.1649	.1982	.1983	.1684	.1222	.0755	.0392	.0167	.0056	.0014	.0002	.0000	.0000	.0000
	7	.0000	.0004	.0045	.0197	.0524	.1010	.1524	.1889	.1969	.1746	.1318	.0840	.0442	.0185	.0058	.0012	.0001	.0000	.0000
	8	.0000	.0001	.0009	.0055	.0197	.0487	.0923	.1417	.1812	.1964	.1812	.1417	.0923	.0487	.0197	.0055	.0009	.0001	.0000
	9	.0000	.0000	.0001	.0012	.0058	.0185	.0442	.0840	.1318	.1746	.1969	.1889	.1524	.1010	.0524	.0197	.0045	.0004	.0000
	10	.0000	.0000	.0000	.0002	.0014	.0056	.0167	.0392	.0755	.1222	.1684	.1983	.1982	.1649	.1101	.0550	.0180	.0028	.0001
	11	.0000	.0000	.0000	.0000	.0002	.0013	.0049	.0142	.0337	.0667	.1123	.1623	.2008	.2099	.1802	.1201	.0555	.0137	.0008
	12	.0000	.0000	.0000	.0000	.0000	.0002	.0011	.0040	.0115	.0278	.0572	.1014	.1553	.2040	.2079	.2001	.1311	.0514	.0061
	13	.0000	.0000	.0000	.0000	.0000	.0000	.0002	.0008	.0029	.0085	.0215	.0468	.0888	.1465	.2079	.2463	.2285	.1423	.0359
	14	.0000	.0000	.0000	.0000	.0000	.0000	.0000	.0001	.0005	.0018	.0056	.0150	.0353	.0732	.1336	.2111	.2775	.2745	.1463
	15	.0000	.0000	.0000	.0000	.0000	.0000	.0000	.0000	.0001	.0002	.0009	.0030	.0087	.0228	.0535	.1126	.2097	.3294	.3706
	16	.0000	.0000	.0000	.0000	.0000	.0000	.0000	.0000	.0000	.0000	.0001	.0003	.0010	.0033	.0100	.0281	.0743	.1853	.4401

F: Random Rectangular Numbers

10 09 73 25 33	76 52 01 35 86	34 67 35 48 76	80 95 90 91 17	39 29 27 49 45
37 54 20 48 05	64 89 47 42 96	24 80 52 40 37	20 63 61 04 02	00 82 29 16 65
08 42 26 89 53	19 64 50 93 03	23 20 90 25 60	15 95 33 47 64	35 08 03 36 06
99 01 90 25 29	09 37 67 07 15	38 31 13 11 65	88 67 67 43 97	04 43 62 76 59
12 80 79 99 70	80 15 73 61 47	64 03 23 66 53	98 95 11 68 77	12 17 17 68 33
66 06 57 47 17	34 07 27 68 50	36 69 73 61 70	65 81 33 98 85	11 19 92 91 70
31 06 01 08 05	45 57 18 24 06	35 30 34 26 14	86 79 90 74 39	23 40 30 97 32
85 26 97 76 02	02 05 16 56 92	68 66 57 48 18	73 05 38 52 47	16 62 38 85 79
63 57 33 21 35	05 32 54 70 48	90 55 35 75 48	28 46 82 87 09	83 49 12 56 24
73 79 64 57 53	03 52 96 47 78	35 80 83 42 82	60 93 52 03 44	35 27 38 84 35
98 52 01 77 67	14 90 56 86 07	22 10 94 05 58	60 97 09 34 33	50 50 07 39 98
11 80 50 54 31	39 80 82 77 32	50 72 56 82 48	29 40 52 42 01	52 77 56 78 51
83 45 29 96 34	06 28 89 80 83	13 74 67 00 78	18 47 54 06 10	68 71 17 78 17
88 68 54 02 00	86 50 75 84 01	36 76 66 79 51	90 36 47 64 93	29 60 91 10 62
99 59 46 73 48	87 51 76 49 69	91 82 60 89 28	93 78 56 13 68	23 47 83 41 13
65 48 11 76 74	17 46 85 09 50	58 04 77 69 74	73 03 95 71 86	40 21 81 65 44
80 12 43 56 35	17 72 70 80 15	45 31 82 23 74	21 11 57 82 53	14 38 55 37 63
74 35 09 98 17	77 40 27 72 14	43 23 60 02 10	45 52 16 42 37	96 28 60 26 55
69 91 62 68 03	66 25 22 91 48	36 93 68 72 03	76 62 11 39 90	94 40 05 64 18
09 89 32 05 05	14 22 56 85 14	46 42 75 67 88	96 29 77 88 22	54 38 21 45 98
91 49 91 45 23	68 47 91 76 86	46 16 28 35 54	94 75 08 99 23	37 08 92 00 48
80 33 69 45 98	26 94 03 68 58	70 29 73 41 35	53 14 03 33 40	42 05 08 23 41
44 10 48 19 49	85 15 74 79 54	32 97 92 65 75	57 60 04 08 81	22 22 20 64 13
12 55 07 37 42	11 10 00 20 40	12 86 07 46 97	96 64 48 94 39	28 70 72 58 15
63 60 64 93 29	16 50 53 44 84	40 21 95 25 63	43 65 17 70 82	07 20 73 17 90
61 19 69 04 46	26 45 74 77 74	51 92 43 37 29	65 39 45 95 93	42 58 26 05 27
15 47 44 52 66	95 27 07 99 53	59 36 78 38 48	82 39 61 01 18	33 21 15 94 66
94 55 72 85 73	67 89 75 43 87	54 62 24 44 31	91 19 04 25 92	92 92 74 59 73
42 48 11 62 13	97 34 40 87 21	16 86 84 87 67	03 07 11 20 59	25 70 14 66 70
23 52 37 83 17	73 20 88 98 37	68 93 59 14 16	26 25 22 96 63	05 52 28 25 62
04 49 35 24 94	75 24 63 38 24	45 86 25 10 25	61 96 27 93 35	65 33 71 24 72
00 54 99 76 54	64 05 18 81 59	96 11 96 38 96	54 69 28 23 91	23 28 72 95 29
35 96 31 53 07	26 89 80 93 54	33 35 13 54 62	77 97 45 00 24	90 10 33 93 33
59 80 80 83 91	45 42 72 68 42	83 60 94 97 00	13 02 12 48 92	78 56 52 01 06
46 05 88 52 36	01 39 09 22 86	77 28 14 40 77	93 91 08 36 47	70 61 74 29 41
32 17 90 05 97	87 37 92 52 41	05 56 70 70 07	86 74 31 71 57	85 39 41 18 38
69 23 46 14 06	20 11 74 52 04	15 95 66 00 00	18 74 39 24 23	07 11 89 63 38
19 56 54 14 30	01 75 87 53 79	40 41 92 15 85	66 67 43 68 06	84 96 28 52 07
45 15 51 49 38	19 47 60 72 46	43 66 79 45 43	59 04 79 00 33	20 82 66 95 41
94 86 43 19 94	36 16 81 08 51	34 88 88 15 53	01 54 03 54 56	05 01 45 11 76
98 08 62 48 26	45 24 02 84 04	44 99 90 88 96	39 09 47 34 07	35 44 13 18 80
33 18 51 62 32	41 94 15 09 49	89 43 54 85 81	88 69 54 19 94	37 54 87 30 43
80 95 10 04 06	96 38 27 07 74	20 15 12 33 87	25 01 62 52 98	94 62 46 11 71
79 75 24 91 40	71 96 12 82 96	69 86 10 25 91	74 85 22 05 39	00 38 75 95 79
18 63 33 25 37	98 14 50 65 71	31 01 02 46 74	05 45 56 14 27	77 93 89 19 36
74 02 94 39 02	77 55 73 22 70	97 79 01 71 19	52 52 75 80 21	80 81 45 17 48
54 17 84 56 11	80 99 33 71 43	05 33 51 29 69	56 12 71 92 55	36 04 09 03 24
11 66 44 98 83	52 07 98 48 27	59 38 17 15 39	09 97 33 34 40	88 46 12 33 56
48 32 47 79 28	31 24 96 47 10	02 29 53 68 70	32 30 75 75 46	15 02 00 99 94
69 07 49 41 38	87 63 79 19 76	35 58 40 44 01	10 51 82 16 15	01 84 87 69 38

F: Random Rectangular Numbers (*continued*)

```
09 18 82 00 97    32 82 53 95 27    04 22 08 63 04    83 38 98 73 74    64 27 85 80 44
90 04 58 54 97    51 98 15 06 54    94 93 88 19 97    91 87 07 61 50    68 47 66 46 59
73 18 95 02 07    47 67 72 62 69    62 29 06 44 64    27 12 46 70 18    41 36 18 27 60
75 76 87 64 90    20 97 18 17 49    90 42 91 22 72    95 37 50 58 71    93 82 34 31 78
54 01 64 40 56    66 28 13 10 03    00 68 22 73 98    20 71 45 32 95    07 70 61 78 13

08 35 86 99 10    78 54 24 27 85    13 66 15 88 73    04 61 89 75 53    31 22 30 84 20
28 30 60 32 64    81 33 31 05 91    40 51 00 78 93    32 60 46 04 75    94 11 90 18 40
53 84 08 62 33    81 59 41 36 28    51 21 59 02 90    28 46 66 87 95    77 76 22 07 91
91 75 75 37 41    61 61 36 22 69    50 26 39 02 12    55 78 17 65 14    83 48 34 70 55
89 41 59 26 94    00 38 75 83 91    12 60 71 76 46    48 94 97 23 06    94 54 13 74 08

77 51 30 38 20    86 83 42 99 01    68 41 48 27 74    51 90 81 39 80    72 89 35 55 07
19 50 23 71 74    69 97 92 02 88    55 21 02 97 73    74 28 77 52 51    65 34 46 74 15
21 81 85 93 13    93 27 88 17 57    04 68 67 31 56    07 08 28 50 46    31 85 33 84 52
51 47 46 64 99    68 10 72 36 21    94 04 99 13 45    42 83 60 91 91    08 00 74 54 49
99 55 96 83 31    62 53 52 41 70    69 77 71 28 30    74 81 97 81 42    43 86 07 28 34

33 71 34 80 07    93 58 47 28 69    51 92 66 47 21    58 30 32 98 22    93 17 49 39 72
85 27 48 68 93    11 30 32 92 70    28 83 43 41 37    73 51 59 04 00    71 14 84 36 43
84 13 38 96 40    44 03 55 21 66    73 85 27 00 91    61 22 26 05 61    62 32 71 84 23
56 73 21 62 34    17 39 59 61 31    10 12 39 16 22    85 49 65 75 60    81 60 41 88 80
65 13 85 68 06    87 64 88 52 61    34 31 36 58 61    45 87 52 10 69    85 64 44 72 77

38 00 10 21 76    81 71 91 17 11    71 60 29 29 37    74 21 96 40 49    65 58 44 96 98
37 40 29 63 97    01 30 47 75 86    56 27 11 00 86    47 32 46 26 05    40 03 03 74 38
97 12 54 03 48    87 08 33 14 17    21 81 53 92 50    75 23 76 20 47    15 50 12 95 78
21 82 64 11 34    47 14 33 40 72    64 63 88 59 02    49 13 90 64 41    03 85 65 45 52
73 13 54 27 42    95 71 90 90 35    85 79 47 42 96    08 78 98 81 56    64 69 11 92 02

07 63 87 79 29    03 06 11 80 72    96 20 74 41 56    23 82 19 95 38    04 71 36 69 94
60 52 88 34 41    07 95 41 98 14    59 17 52 06 95    05 53 35 21 39    61 21 20 64 55
83 59 63 56 55    06 95 89 29 83    05 12 80 97 19    77 43 35 37 83    92 30 15 04 98
10 85 06 27 46    99 59 91 05 07    13 49 90 63 19    53 07 57 18 39    06 41 01 93 62
39 82 09 89 52    43 62 26 31 47    64 42 18 08 14    43 80 00 93 51    31 02 47 31 67

59 58 00 64 78    75 56 97 88 00    88 83 55 44 86    23 76 80 61 56    04 11 10 84 08
38 50 80 73 41    23 79 34 87 63    90 82 29 70 22    17 71 90 42 07    95 95 44 99 53
30 69 27 06 68    94 68 81 61 27    56 19 68 00 91    82 06 76 34 00    05 46 26 92 00
65 44 39 56 59    18 28 82 74 37    49 63 22 40 41    08 33 76 56 76    96 29 99 08 36
27 26 75 02 64    13 19 27 22 94    07 47 74 46 06    17 98 54 89 11    97 34 13 03 58

91 30 70 69 91    19 07 22 42 10    36 69 95 37 28    28 82 53 57 93    28 97 66 62 52
68 43 49 46 88    84 47 31 36 22    62 12 69 84 08    12 84 38 25 90    09 81 59 31 46
48 90 81 58 77    54 74 52 45 91    35 70 00 47 54    83 82 45 26 92    54 13 05 51 60
06 91 34 51 97    42 67 27 86 01    11 88 30 95 28    63 01 19 89 01    14 97 44 03 44
10 45 51 60 19    14 21 03 37 12    91 34 23 78 21    88 32 58 08 51    43 66 77 08 83

12 88 39 73 43    65 02 76 11 84    04 28 50 13 92    17 97 41 50 77    90 71 22 67 69
21 77 83 09 76    38 80 73 69 61    31 64 94 20 96    63 28 10 20 23    08 81 64 74 49
19 52 35 95 15    65 12 25 96 59    86 28 36 82 58    69 57 21 37 98    16 43 59 15 29
67 24 55 26 70    35 58 31 65 63    79 24 68 66 86    76 46 33 42 22    26 65 59 08 02
60 58 44 73 77    07 50 03 79 92    45 13 42 65 29    26 76 08 36 37    41 32 64 43 44

53 85 34 13 77    36 06 69 48 50    58 83 87 38 59    49 36 47 33 31    96 24 04 36 42
24 63 73 87 36    74 38 48 93 42    52 62 30 79 92    12 36 91 86 01    03 74 28 38 73
83 08 01 24 51    38 99 22 28 15    07 75 95 17 77    97 37 72 75 85    51 97 23 78 67
16 44 42 43 34    36 15 19 90 73    27 49 37 09 39    85 13 03 25 52    54 84 65 47 59
60 79 01 81 57    57 17 86 57 62    11 16 17 85 76    45 81 95 29 79    65 13 00 48 60
```

G :   Values of $w = \frac{1}{2} \ln \frac{1+r}{1-r}$

r	.00	.01	.02	.03	.04	.05	.06	.07	.08	.09
.0	.00000	.01000	.02000	.03001	.04002	.05004	.06007	.07012	.08017	.09024
.1	.10034	.11045	.12058	.13074	.14093	.15114	.16139	.17167	.18198	.19234
.2	.20273	.21317	.22366	.23419	.24477	.25541	.26611	.27686	.28768	.29857
.3	.30952	.32055	.33165	.34283	.35409	.36544	.37689	.38842	.40006	.41180
.4	.42365	.43561	.44769	.45990	.47223	.48470	.49731	.51007	.52298	.53606
.5	.54931	.56273	.57634	.59014	.60415	.61838	.63283	.64752	.66246	.67767
.6	.69315	.70892	.72500	.74142	.75817	.77530	.79281	.81074	.82911	.84795
.7	.86730	.88718	.90764	.92873	.95048	.97295	.99621	1.02033	1.04537	1.07143
.8	1.09861	1.12703	1.15682	1.18813	1.22117	1.25615	1.29334	1.33308	1.37577	1.42192
.9	1.47222	1.52752	1.58902	1.65839	1.73805	1.83178	1.94591	2.09229	2.29756	2.64665

For negative values of r put a minus sign in front of the tabled numbers.

From *Introduction to Statistical Analysis*, 3rd edition, by Wilfrid J. Dixon and Frank J. Massey, Jr. Copyright © 1969 by McGraw-Hill, Inc. Used by permission of McGraw-Hill Book Company.

H: Wilcoxon's Rank-Sum Test

Critical Lower Tail Values of W
[Largest value of W' for which $Pr(W \le W') \le \alpha$]

$n_1 = 1$

n_2	.005	.010	.025	.050	$2\mu_W$
			α		
3					5
4					6
5					7
6					8
7					9
8					10
9					11
10					12
11					13
12					14
13					15
14					16
15					17
16					18
17					19
18					20
19				1	21
20				1	22

$n_1 = 2$

n_2	.005	.010	.025	.050	$2\mu_W$
			α		
3					12
4					14
5				3	16
6				3	18
7				3	20
8			3	4	22
9			3	4	24
10			3	4	26
11			3	4	28
12			4	5	30
13		3	4	5	32
14		3	4	6	34
15		3	4	6	36
16		3	4	6	38
17		3	5	6	40
18		3	5	7	42
19	3	4	5	7	44
20	3	4	5	7	46

$n_1 = 3$

n_2	.005	.010	.025	.050	$2\mu_W$
			α		
3				6	21
4				6	24
5			6	7	27
6			7	8	30
7		6	7	8	33
8		6	8	9	36
9	6	7	8	10	39
10	6	7	9	10	42
11	6	7	9	11	45
12	7	8	10	11	48
13	7	8	10	12	51
14	7	8	11	13	54
15	8	9	11	13	57
16	8	9	12	14	60
17	8	10	12	15	63
18	8	10	13	15	66
19	9	10	13	16	69
20	9	11	14	17	72

$n_1 = 4$

n_2	.005	.010	.025	.050	$2\mu_W$
			α		
4			10	11	36
5		10	11	12	40
6	10	11	12	13	44
7	10	11	13	14	48
8	11	12	14	15	52
9	11	13	14	16	56
10	12	13	15	17	60
11	12	14	16	18	64
12	13	15	17	19	68
13	13	15	18	20	72
14	14	16	19	21	76
15	15	17	20	22	80
16	15	17	21	24	84
17	16	18	21	25	88
18	16	19	22	26	92
19	17	19	23	27	96
20	18	20	24	28	100

From Table 1, L. R. Verdooren, "Extended Tables of Critical Values for Wilcoxon's Test Statistic", *Biometrika*, 50 (1963), 177–186. Used with permission of the author and editor.

H.   Wilcoxon's Rank-Sum Test (*continued*)

$n_1 = 5$

n_2	.005	.010	.025	.050	$2\mu_W$
5	15	16	17	19	55
6	16	17	18	20	60
7	16	18	20	21	65
8	17	19	21	23	70
9	18	20	22	24	75
10	19	21	23	26	80
11	20	22	24	27	85
12	21	23	26	28	90
13	22	24	27	30	95
14	22	25	28	31	100
15	23	26	29	33	105
16	24	27	30	34	110
17	25	28	32	35	115
18	26	29	33	37	120
19	27	30	34	38	125
20	28	31	35	40	130

$n_1 = 6$

n_2	.005	.010	.025	.050	$2\mu_W$
6	23	24	26	28	78
7	24	25	27	29	84
8	25	27	29	31	90
9	26	28	31	33	96
10	27	29	32	35	102
11	28	30	34	37	108
12	30	32	35	38	114
13	31	33	37	40	120
14	32	34	38	42	126
15	33	36	40	44	132
16	34	37	42	46	138
17	36	39	43	47	144
18	37	40	45	49	150
19	38	41	46	51	156
20	39	43	48	53	162

$n_1 = 7$

n_2	.005	.010	.025	.050	$2\mu_W$
7	32	34	36	39	105
8	34	35	38	41	112
9	35	37	40	43	119
10	37	39	42	45	126
11	38	40	44	47	133
12	40	42	46	49	140
13	41	44	48	52	147
14	43	45	50	54	154
15	44	47	52	56	161
16	46	49	54	58	168
17	47	51	56	61	175
18	49	52	58	63	182
19	50	54	60	65	189
20	52	56	62	67	196
21	53	58	64	69	203
22	55	59	66	72	210
23	57	61	68	74	217
24	58	63	70	76	224
25	60	64	72	78	231

$n_1 = 8$

n_2	.005	.010	.025	.050	$2\mu_W$
8	43	45	49	51	136
9	45	47	51	54	144
10	47	49	53	56	152
11	49	51	55	59	160
12	51	53	58	62	168
13	53	56	60	64	176
14	54	58	62	67	184
15	56	60	65	69	192
16	58	62	67	72	200
17	60	64	70	75	208
18	62	66	72	77	216
19	64	68	74	80	224
20	66	70	77	83	232
21	68	72	79	85	240
22	70	74	81	88	248
23	71	76	84	90	256
24	73	78	86	93	264
25	75	81	89	96	272

H :   Wilcoxon's Rank-Sum Test (*continued*)

$n_1 = 9$

n_2	α .005	.010	.025	.050	$2\mu_W$
9	56	59	62	66	171
10	58	61	65	69	180
11	61	63	68	72	189
12	63	66	71	75	198
13	65	68	73	78	207
14	67	71	76	81	216
15	69	73	79	84	225
16	72	76	82	87	234
17	74	78	84	90	243
18	76	81	87	93	252
19	78	83	90	96	261
20	81	85	93	99	270
21	83	88	95	102	279
22	85	90	98	105	288
23	88	93	101	108	297
24	90	95	104	111	306
25	92	99	107	114	315

$n_1 = 10$

n_2	α .005	.010	.025	.050	$2\mu_W$
10	71	74	78	82	210
11	73	77	81	86	220
12	76	79	84	89	230
13	79	82	88	92	240
14	81	85	91	96	250
15	84	88	94	99	260
16	86	91	97	103	270
17	89	93	100	106	280
18	92	96	103	110	290
19	94	99	107	113	300
20	97	102	110	117	310
21	99	105	113	120	320
22	102	108	116	123	330
23	105	110	119	127	340
24	107	113	122	130	350
25	110	116	126	134	360

$n_1 = 11$

n_2	α .005	.010	.025	.050	$2\mu_W$
11	87	91	96	100	253
12	90	94	99	104	264
13	93	97	103	108	275
14	96	100	106	112	286
15	99	103	110	116	297
16	102	107	113	120	308
17	105	110	117	123	319
18	108	113	121	127	330
19	111	116	124	131	341
20	114	119	128	135	352
21	117	123	131	139	363
22	120	126	135	143	374
23	123	129	139	147	385
24	126	132	142	151	396
25	129	136	146	155	407

$n_1 = 12$

n_2	α .005	.010	.025	.050	$2\mu_W$
12	105	109	115	120	300
13	109	113	119	125	312
14	112	116	123	129	324
15	115	120	127	133	336
16	119	124	131	138	348
17	122	127	135	142	360
18	125	131	139	146	372
19	129	134	143	150	384
20	132	138	147	155	396
21	136	142	151	159	408
22	139	145	155	163	420
23	142	149	159	168	432
24	146	153	163	172	444
25	149	156	167	176	456

J: Squares and Square Roots

How to Find Square Roots

1. If the number contains more than 3 significant digits, round it to just 3 significant digits. Find the significant digits under column N.

2. Move the decimal point either left or right an *even* number of places until a number from 1 to 100 is found. If the result is less than 10, use the column under $\sqrt{N}$. If the result is greater than 10, use the column under $\sqrt{10N}$.

3. For the appropriate entry under either $\sqrt{N}$ or $\sqrt{10N}$, move the decimal point *half* as many places in the *opposite* direction as you moved it in step 2.

Example A Find $\sqrt{12345}$.

Step 1: Change 12345 to 12300.

Step 2: Change 12300 to 1.23 by moving the decimal point 4 places left. For the row with digits 1.23 under N in the table use the entry under $\sqrt{10N}$, which is 1.10905.

Step 3: Move the decimal 2 places *right* in 1.10905 to get 110.905, the square root of 12345 as accurately as is possible from the Table.

Example B Find $\sqrt{0.0093}$.

Step 2: Change 0.0093 to 93 by moving the decimal point 4 places right. For the row with digits 9.30 under N in the table, use the entry under $\sqrt{10N}$ which is 9.64365.

Step 3: Move the decimal 2 places *left* in 9.64365 to get 0.0964365, the square root of 0.0093 to 7 decimal places.

N	N^2	$\sqrt{N}$	$\sqrt{10N}$	N	N^2	$\sqrt{N}$	$\sqrt{10N}$
1.00	1.0000	1.00000	3.16228	1.50	2.2500	1.22474	3.87298
1.01	1.0201	1.00499	3.17805	1.51	2.2801	1.22882	3.88587
1.02	1.0404	1.00995	3.19374	1.52	2.3104	1.23288	3.89872
1.03	1.0609	1.01489	3.20936	1.53	2.3409	1.23693	3.91152
1.04	1.0816	1.01980	3.22490	1.54	2.3716	1.24097	3.92428
1.05	1.1025	1.02470	3.24037	1.55	2.4025	1.24499	3.93700
1.06	1.1236	1.02956	3.25576	1.56	2.4336	1.24900	3.94968
1.07	1.1449	1.03441	3.27109	1.57	2.4649	1.25300	3.96232
1.08	1.1664	1.03923	3.28634	1.58	2.4964	1.25698	3.97492
1.09	1.1881	1.04403	3.30151	1.59	2.5281	1.26095	3.98748
1.10	1.2100	1.04881	3.31662	1.60	2.5600	1.26491	4.00000
1.11	1.2321	1.05357	3.33167	1.61	2.5921	1.26886	4.01248
1.12	1.2544	1.05830	3.34664	1.62	2.6244	1.27279	4.02492
1.13	1.2769	1.06301	3.36155	1.63	2.6569	1.27671	4.03733
1.14	1.2996	1.06771	3.37639	1.64	2.6896	1.28062	4.04969
1.15	1.3225	1.07238	3.39116	1.65	2.7225	1.28452	4.06202
1.16	1.3456	1.07703	3.40588	1.66	2.7556	1.28841	4.07431
1.17	1.3689	1.08167	3.42053	1.67	2.7889	1.29228	4.08656
1.18	1.3924	1.08628	3.43511	1.68	2.8224	1.29615	4.09878
1.19	1.4161	1.09087	3.44964	1.69	2.8561	1.30000	4.11096
1.20	1.4400	1.09545	3.46410	1.70	2.8900	1.30384	4.12311
1.21	1.4641	1.10000	3.47851	1.71	2.9241	1.30767	4.13521
1.22	1.4884	1.10454	3.49285	1.72	2.9584	1.31149	4.14729
1.23	1.5129	1.10905	3.50714	1.73	2.9929	1.31529	4.15933
1.24	1.5376	1.11355	3.52136	1.74	3.0276	1.31909	4.17133
1.25	1.5625	1.11803	3.53553	1.75	3.0625	1.32288	4.18330
1.26	1.5876	1.12250	3.54965	1.76	3.0976	1.32665	4.19524
1.27	1.6129	1.12694	3.56371	1.77	3.1329	1.33041	4.20714
1.28	1.6384	1.13137	3.57771	1.78	3.1684	1.33417	4.21900
1.29	1.6641	1.13578	3.59166	1.79	3.2041	1.33791	4.23084
1.30	1.6900	1.14018	3.60555	1.80	3.2400	1.34164	4.24264
1.31	1.7161	1.14455	3.61939	1.81	3.2761	1.34536	4.25441
1.32	1.7424	1.14891	3.63318	1.82	3.3124	1.34907	4.26615
1.33	1.7689	1.15326	3.64692	1.83	3.3489	1.35277	4.27785
1.34	1.7956	1.15758	3.66060	1.84	3.3856	1.35647	4.28952
1.35	1.8225	1.16190	3.67423	1.85	3.4225	1.36015	4.30116
1.36	1.8496	1.16619	3.68782	1.86	3.4596	1.36382	4.31277
1.37	1.8769	1.17047	3.70135	1.87	3.4969	1.36748	4.32435
1.38	1.9044	1.17473	3.71484	1.88	3.5344	1.37113	4.33590
1.39	1.9321	1.17898	3.72827	1.89	3.5721	1.37477	4.34741
1.40	1.9600	1.18322	3.74166	1.90	3.6100	1.37840	4.35890
1.41	1.9881	1.18743	3.75500	1.91	3.6481	1.38203	4.37035
1.42	2.0164	1.19164	3.76829	1.92	3.6864	1.38564	4.38178
1.43	2.0449	1.19583	3.78153	1.93	3.7249	1.38924	4.39318
1.44	2.0736	1.20000	3.79473	1.94	3.7636	1.39284	4.40454
1.45	2.1025	1.20416	3.80789	1.95	3.8025	1.39642	4.41588
1.46	2.1316	1.20830	3.82099	1.96	3.8416	1.40000	4.42719
1.47	2.1609	1.21244	3.83406	1.97	3.8809	1.40357	4.43847
1.48	2.1904	1.21655	3.84708	1.98	3.9204	1.40712	4.44972
1.49	2.2201	1.22066	3.86005	1.99	3.9601	1.41067	4.46094

J: Squares and Square Roots (continued)

N	N²	√N	√10N
2.00	4.0000	1.41421	4.47214
2.01	4.0401	1.41774	4.48330
2.02	4.0804	1.42127	4.49444
2.03	4.1209	1.42478	4.50555
2.04	4.1616	1.42829	4.51664
2.05	4.2025	1.43178	4.52769
2.06	4.2436	1.43527	4.53872
2.07	4.2849	1.43875	4.54973
2.08	4.3264	1.44222	4.56070
2.09	4.3681	1.44568	4.57165
2.10	4.4100	1.44914	4.58258
2.11	4.4521	1.45258	4.59347
2.12	4.4944	1.45602	4.60435
2.13	4.5369	1.45945	4.61519
2.14	4.5796	1.46287	4.62601
2.15	4.6225	1.46629	4.63681
2.16	4.6656	1.46969	4.64758
2.17	4.7089	1.47309	4.65833
2.18	4.7524	1.47648	4.66905
2.19	4.7961	1.47986	4.67974
2.20	4.8400	1.48324	4.69042
2.21	4.8841	1.48661	4.70106
2.22	4.9284	1.48997	4.71169
2.23	4.9729	1.49332	4.72229
2.24	5.0176	1.49666	4.73286
2.25	5.0625	1.50000	4.74342
2.26	5.1076	1.50333	4.75395
2.27	5.1529	1.50665	4.76445
2.28	5.1984	1.50997	4.77493
2.29	5.2441	1.51327	4.78539
2.30	5.2900	1.51658	4.79583
2.31	5.3361	1.51987	4.80625
2.32	5.3824	1.52315	4.81664
2.33	5.4289	1.52643	4.82701
2.34	5.4756	1.52971	4.83735
2.35	5.5225	1.53297	4.84768
2.36	5.5696	1.53623	4.85798
2.37	5.6169	1.53948	4.86826
2.38	5.6644	1.54272	4.87852
2.39	5.7121	1.54596	4.88876
2.40	5.7600	1.54919	4.89898
2.41	5.8081	1.55242	4.90918
2.42	5.8564	1.55563	4.91935
2.43	5.9049	1.55885	4.92950
2.44	5.9536	1.56205	4.93964
2.45	6.0025	1.56525	4.94975
2.46	6.0516	1.56844	4.95984
2.47	6.1009	1.57162	4.96991
2.48	6.1504	1.57480	4.97996
2.49	6.2001	1.57797	4.98999

N	N²	√N	√10N
2.50	6.2500	1.58114	5.00000
2.51	6.3001	1.58430	5.00999
2.52	6.3504	1.58745	5.01996
2.53	6.4009	1.59060	5.02991
2.54	6.4516	1.59374	5.03984
2.55	6.5025	1.59687	5.04975
2.56	6.5536	1.60000	5.05964
2.57	6.6049	1.60312	5.06952
2.58	6.6564	1.60624	5.07937
2.59	6.7081	1.60935	5.08920
2.60	6.7600	1.61245	5.09902
2.61	6.8121	1.61555	5.10882
2.62	6.8644	1.61864	5.11859
2.63	6.9169	1.62173	5.12835
2.64	6.9696	1.62481	5.13809
2.65	7.0225	1.62788	5.14782
2.66	7.0756	1.63095	5.15752
2.67	7.1289	1.63401	5.16720
2.68	7.1824	1.63707	5.17687
2.69	7.2361	1.64012	5.18652
2.70	7.2900	1.64317	5.19615
2.71	7.3441	1.64621	5.20577
2.72	7.3984	1.64924	5.21536
2.73	7.4529	1.65227	5.22494
2.74	7.5076	1.65529	5.23450
2.75	7.5625	1.65831	5.24404
2.76	7.6176	1.66132	5.25357
2.77	7.6729	1.66433	5.26308
2.78	7.7284	1.66733	5.27257
2.79	7.7841	1.67033	5.28205
2.80	7.8400	1.67332	5.29150
2.81	7.8961	1.67631	5.30094
2.82	7.9524	1.67929	5.31037
2.83	8.0089	1.68226	5.31977
2.84	8.0656	1.68523	5.32917
2.85	8.1225	1.68819	5.33854
2.86	8.1796	1.69115	5.34790
2.87	8.2369	1.69411	5.35724
2.88	8.2944	1.69706	5.36656
2.89	8.3521	1.70000	5.37587
2.90	8.4100	1.70294	5.38516
2.91	8.4681	1.70587	5.39444
2.92	8.5264	1.70880	5.40370
2.93	8.5849	1.71172	5.41295
2.94	8.6436	1.71464	5.42218
2.95	8.7025	1.71756	5.43139
2.96	8.7616	1.72047	5.44059
2.97	8.8209	1.72337	5.44977
2.98	8.8804	1.72627	5.45894
2.99	8.9401	1.72916	5.46809

N	N²	√N	√10N
3.00	9.0000	1.73205	5.47723
3.01	9.0601	1.73494	5.48635
3.02	9.1204	1.73781	5.49545
3.03	9.1809	1.74069	5.50454
3.04	9.2416	1.74356	5.51362
3.05	9.3025	1.74642	5.52268
3.06	9.3636	1.74929	5.53173
3.07	9.4249	1.75214	5.54076
3.08	9.4864	1.75499	5.54977
3.09	9.5481	1.75784	5.55878
3.10	9.6100	1.76068	5.56776
3.11	9.6721	1.76352	5.57674
3.12	9.7344	1.76635	5.58570
3.13	9.7969	1.76918	5.59464
3.14	9.8596	1.77200	5.60357
3.15	9.9225	1.77482	5.61249
3.16	9.9856	1.77764	5.62139
3.17	10.0489	1.78045	5.63028
3.18	10.1124	1.78326	5.63915
3.19	10.1761	1.78606	5.64801
3.20	10.2400	1.78885	5.65685
3.21	10.3041	1.79165	5.66569
3.22	10.3684	1.79444	5.67450
3.23	10.4329	1.79722	5.68331
3.24	10.4976	1.80000	5.69210
3.25	10.5625	1.80278	5.70088
3.26	10.6276	1.80555	5.70964
3.27	10.6929	1.80831	5.71839
3.28	10.7584	1.81108	5.72713
3.29	10.8241	1.81384	5.73585
3.30	10.8900	1.81659	5.74456
3.31	10.9561	1.81934	5.75326
3.32	11.0224	1.82209	5.76194
3.33	11.0889	1.82483	5.77062
3.34	11.1556	1.82757	5.77927
3.35	11.2225	1.83030	5.78792
3.36	11.2896	1.83303	5.79655
3.37	11.3569	1.83576	5.80517
3.38	11.4244	1.83848	5.81378
3.39	11.4921	1.84120	5.82237
3.40	11.5600	1.84391	5.83095
3.41	11.6281	1.84662	5.83952
3.42	11.6964	1.84932	5.84808
3.43	11.7649	1.85203	5.85662
3.44	11.8336	1.85472	5.86515
3.45	11.9025	1.85742	5.87367
3.46	11.9716	1.86011	5.88218
3.47	12.0409	1.86279	5.89067
3.48	12.1104	1.86548	5.89915
3.49	12.1801	1.86815	5.90762

N	N²	√N	√10N
3.50	12.2500	1.87083	5.91608
3.51	12.3201	1.87350	5.92453
3.52	12.3904	1.87617	5.93296
3.53	12.4609	1.87883	5.94138
3.54	12.5316	1.88149	5.94979
3.55	12.6025	1.88414	5.95819
3.56	12.6736	1.88680	5.96657
3.57	12.7449	1.88944	5.97495
3.58	12.8164	1.89209	5.98331
3.59	12.8881	1.89473	5.99166
3.60	12.9600	1.89737	6.00000
3.61	13.0321	1.90000	6.00833
3.62	13.1044	1.90263	6.01664
3.63	13.1769	1.90526	6.02495
3.64	13.2496	1.90788	6.03324
3.65	13.3225	1.91050	6.04152
3.66	13.3956	1.91311	6.04979
3.67	13.4689	1.91572	6.05805
3.68	13.5424	1.91833	6.06630
3.69	13.6161	1.92094	6.07454
3.70	13.6900	1.92354	6.08276
3.71	13.7641	1.92614	6.09098
3.72	13.8384	1.92873	6.09918
3.73	13.9129	1.93132	6.10737
3.74	13.9876	1.93391	6.11555
3.75	14.0625	1.93649	6.12372
3.76	14.1376	1.93907	6.13188
3.77	14.2129	1.94165	6.14003
3.78	14.2884	1.94422	6.14817
3.79	14.3641	1.94679	6.15630
3.80	14.4400	1.94936	6.16441
3.81	14.5161	1.95192	6.17252
3.82	14.5924	1.95448	6.18061
3.83	14.6689	1.95704	6.18870
3.84	14.7456	1.95959	6.19677
3.85	14.8225	1.96214	6.20484
3.86	14.8996	1.96469	6.21289
3.87	14.9769	1.96723	6.22093
3.88	15.0544	1.96977	6.22896
3.89	15.1321	1.97231	6.23699
3.90	15.2100	1.97484	6.24500
3.91	15.2881	1.97737	6.25300
3.92	15.3664	1.97990	6.26099
3.93	15.4449	1.98242	6.26897
3.94	15.5236	1.98494	6.27694
3.95	15.6025	1.98746	6.28490
3.96	15.6816	1.98997	6.29285
3.97	15.7609	1.99249	6.30079
3.98	15.8408	1.99499	6.30872
3.99	15.9201	1.99750	6.31664

N	N²	√N	√10N
4.00	16.0000	2.00000	6.32456
4.01	16.0801	2.00250	6.33246
4.02	16.1604	2.00499	6.34035
4.03	16.2409	2.00749	6.34823
4.04	16.3216	2.00998	6.35610
4.05	16.4025	2.01246	6.36396
4.06	16.4836	2.01494	6.37181
4.07	16.5649	2.01742	6.37966
4.08	16.6464	2.01990	6.38749
4.09	16.7281	2.02237	6.39531
4.10	16.8100	2.02485	6.40312
4.11	16.8921	2.02731	6.41093
4.12	16.9744	2.02978	6.41872
4.13	17.0569	2.03224	6.42651
4.14	17.1396	2.03470	6.43428
4.15	17.2225	2.03715	6.44205
4.16	17.3056	2.03961	6.44981
4.17	17.3889	2.04206	6.45755
4.18	17.4724	2.04450	6.46529
4.19	17.5561	2.04695	6.47302
4.20	17.6400	2.04939	6.48074
4.21	17.7241	2.05183	6.48845
4.22	17.8084	2.05426	6.49615
4.23	17.8929	2.05670	6.50384
4.24	17.9776	2.05913	6.51153
4.25	18.0625	2.06155	6.51920
4.26	18.1476	2.06398	6.52687
4.27	18.2329	2.06640	6.53452
4.28	18.3184	2.06882	6.54217
4.29	18.4041	2.07123	6.54981
4.30	18.4900	2.07364	6.55744
4.31	18.5761	2.07605	6.56506
4.32	18.6624	2.07846	6.57267
4.33	18.7489	2.08087	6.58027
4.34	18.8356	2.08327	6.58787
4.35	18.9225	2.08567	6.59545
4.36	19.0096	2.08806	6.60303
4.37	19.0969	2.09045	6.61060
4.38	19.1844	2.09284	6.61816
4.39	19.2721	2.09523	6.62571
4.40	19.3600	2.09762	6.63325
4.41	19.4481	2.10000	6.64078
4.42	19.5364	2.10238	6.64831
4.43	19.6249	2.10476	6.65582
4.44	19.7136	2.10713	6.66333
4.45	19.8025	2.10950	6.67083
4.46	19.8916	2.11187	6.67832
4.47	19.9809	2.11424	6.68581
4.48	20.0704	2.11660	6.69328
4.49	20.1601	2.11896	6.70075
4.50	20.2500	2.12132	6.70820
4.51	20.3401	2.12368	6.71565
4.52	20.4304	2.12603	6.72309
4.53	20.5209	2.12838	6.73053
4.54	20.6116	2.13073	6.73795
4.55	20.7025	2.13307	6.74537
4.56	20.7936	2.13542	6.75278
4.57	20.8849	2.13776	6.76018
4.58	20.9764	2.14009	6.76757
4.59	21.0681	2.14243	6.77495
4.60	21.1600	2.14476	6.78233
4.61	21.2521	2.14709	6.78970
4.62	21.3444	2.14942	6.79706
4.63	21.4369	2.15174	6.80441
4.64	21.5296	2.15407	6.81175
4.65	21.6225	2.15639	6.81909
4.66	21.7156	2.15870	6.82642
4.67	21.8089	2.16102	6.83374
4.68	21.9024	2.16333	6.84105
4.69	21.9961	2.16564	6.84836
4.70	22.0900	2.16795	6.85565
4.71	22.1841	2.17025	6.86294
4.72	22.2784	2.17256	6.87023
4.73	22.3729	2.17486	6.87750
4.74	22.4676	2.17715	6.88477
4.75	22.5625	2.17945	6.89202
4.76	22.6576	2.18174	6.89928
4.77	22.7529	2.18403	6.90652
4.78	22.8484	2.18632	6.91375
4.79	22.9441	2.18861	6.92098
4.80	23.0400	2.19089	6.92820
4.81	23.1361	2.19317	6.93542
4.82	23.2324	2.19545	6.94262
4.83	23.3289	2.19773	6.94982
4.84	23.4256	2.20000	6.95701
4.85	23.5225	2.20227	6.96419
4.86	23.6196	2.20454	6.97137
4.87	23.7169	2.20681	6.97854
4.88	23.8144	2.20907	6.98570
4.89	23.9121	2.21133	6.99285
4.90	24.0100	2.21359	7.00000
4.91	24.1081	2.21585	7.00714
4.92	24.2064	2.21811	7.01427
4.93	24.3049	2.22036	7.02140
4.94	24.4036	2.22261	7.02851
4.95	24.5025	2.22486	7.03562
4.96	24.6016	2.22711	7.04273
4.97	24.7009	2.22935	7.04982
4.98	24.8004	2.23159	7.05691
4.99	24.9001	2.23383	7.06399
5.00	25.0000	2.23607	7.07107
5.01	25.1001	2.23830	7.07814
5.02	25.2004	2.24054	7.08520
5.03	25.3009	2.24277	7.09225
5.04	25.4016	2.24499	7.09930
5.05	25.5025	2.24722	7.10634
5.06	25.6036	2.24944	7.11337
5.07	25.7049	2.25167	7.12039
5.08	25.8064	2.25389	7.12741
5.09	25.9081	2.25610	7.13442
5.10	26.0100	2.25832	7.14143
5.11	26.1121	2.26053	7.14843
5.12	26.2144	2.26274	7.15542
5.13	26.3169	2.26495	7.16240
5.14	26.4196	2.26716	7.16938
5.15	26.5225	2.26936	7.17635
5.16	26.6256	2.27156	7.18331
5.17	26.7289	2.27376	7.19027
5.18	26.8324	2.27596	7.19722
5.19	26.9361	2.27816	7.20417
5.20	27.0400	2.28035	7.21110
5.21	27.1441	2.28254	7.21803
5.22	27.2484	2.28473	7.22496
5.23	27.3529	2.28692	7.23187
5.24	27.4576	2.28910	7.23878
5.25	27.5625	2.29129	7.24569
5.26	27.6676	2.29347	7.25259
5.27	27.7729	2.29565	7.25948
5.28	27.8784	2.29783	7.26636
5.29	27.9841	2.30000	7.27324
5.30	28.0900	2.30217	7.28011
5.31	28.1961	2.30434	7.28697
5.32	28.3024	2.30651	7.29383
5.33	28.4089	2.30868	7.30068
5.34	28.5156	2.31084	7.30753
5.35	28.6225	2.31301	7.31437
5.36	28.7296	2.31517	7.32120
5.37	28.8369	2.31733	7.32803
5.38	28.9444	2.31948	7.33485
5.39	29.0521	2.32164	7.34166
5.40	29.1600	2.32379	7.34847
5.41	29.2681	2.32594	7.35527
5.42	29.3764	2.32809	7.36205
5.43	29.4849	2.33024	7.36885
5.44	29.5936	2.33238	7.37564
5.45	29.7025	2.33452	7.38241
5.46	29.8116	2.33666	7.38918
5.47	29.9209	2.33880	7.39594
5.48	30.0304	2.34094	7.40270
5.49	30.1401	2.34307	7.40945
5.50	30.2500	2.34521	7.41620
5.51	30.3601	2.34734	7.42294
5.52	30.4704	2.34947	7.42967
5.53	30.5809	2.35160	7.43640
5.54	30.6916	2.35372	7.44312
5.55	30.8025	2.35584	7.44983
5.56	30.9136	2.35797	7.45654
5.57	31.0249	2.36008	7.46324
5.58	31.1364	2.36220	7.46994
5.59	31.2481	2.36432	7.47663
5.60	31.3600	2.36643	7.48331
5.61	31.4721	2.36854	7.48999
5.62	31.5844	2.37065	7.49667
5.63	31.6969	2.37276	7.50333
5.64	31.8096	2.37487	7.50999
5.65	31.9225	2.37697	7.51665
5.66	32.0356	2.37908	7.52330
5.67	32.1489	2.38118	7.52994
5.68	32.2624	2.38328	7.53658
5.69	32.3761	2.38537	7.54321
5.70	32.4900	2.38747	7.54983
5.71	32.6041	2.38956	7.55645
5.72	32.7184	2.39165	7.56307
5.73	32.8329	2.39374	7.56968
5.74	32.9476	2.39583	7.57628
5.75	33.0625	2.39792	7.58288
5.76	33.1776	2.40000	7.58947
5.77	33.2929	2.40208	7.59605
5.78	33.4084	2.40416	7.60263
5.79	33.5241	2.40624	7.60920
5.80	33.6400	2.40832	7.61577
5.81	33.7561	2.41039	7.62234
5.82	33.8724	2.41247	7.62889
5.83	33.9889	2.41454	7.63544
5.84	34.1056	2.41661	7.64199
5.85	34.2225	2.41868	7.64853
5.86	34.3396	2.42074	7.65506
5.87	34.4569	2.42281	7.66159
5.88	34.5744	2.42487	7.66812
5.89	34.6921	2.42693	7.67463
5.90	34.8100	2.42899	7.68115
5.91	34.9281	2.43105	7.68765
5.92	35.0464	2.43311	7.69415
5.93	35.1649	2.43516	7.70065
5.94	35.2836	2.43721	7.70714
5.95	35.4025	2.43926	7.71362
5.96	35.5216	2.44131	7.72010
5.97	35.6409	2.44336	7.72658
5.98	35.7604	2.44540	7.73305
5.99	35.8801	2.44745	7.73951

J: Squares and Square Roots (continued)

N	N²	√N	√10N
6.00	36.0000	2.44949	7.74597
6.01	36.1201	2.45153	7.75242
6.02	36.2404	2.45357	7.75887
6.03	36.3609	2.45561	7.76531
6.04	36.4816	2.45764	7.77174
6.05	36.6025	2.45967	7.77817
6.06	36.7236	2.46171	7.78460
6.07	36.8449	2.46374	7.79102
6.08	36.9664	2.46577	7.79744
6.09	37.0881	2.46779	7.80385
6.10	37.2100	2.46982	7.81025
6.11	37.3321	2.47184	7.81665
6.12	37.4544	2.47386	7.82304
6.13	37.5769	2.47588	7.82943
6.14	37.6996	2.47790	7.83582
6.15	37.8225	2.47992	7.84219
6.16	37.9456	2.48193	7.84857
6.17	38.0689	2.48395	7.85493
6.18	38.1924	2.48596	7.86130
6.19	38.3161	2.48797	7.86766
6.20	38.4400	2.48998	7.87401
6.21	38.5641	2.49199	7.88036
6.22	38.6884	2.49399	7.88670
6.23	38.8129	2.49600	7.89303
6.24	38.9376	2.49800	7.89937
6.25	39.0625	2.50000	7.90569
6.26	39.1876	2.50200	7.91202
6.27	39.3129	2.50400	7.91833
6.28	39.4384	2.50599	7.92465
6.29	39.5641	2.50799	7.93095
6.30	39.6900	2.50998	7.93725
6.31	39.8161	2.51197	7.94355
6.32	39.9424	2.51396	7.94984
6.33	40.0689	2.51595	7.95613
6.34	40.1956	2.51794	7.96241
6.35	40.3225	2.51992	7.96869
6.36	40.4496	2.52190	7.97496
6.37	40.5769	2.52389	7.98123
6.38	40.7044	2.52587	7.98749
6.39	40.8321	2.52784	7.99375
6.40	40.9600	2.52982	8.00000
6.41	41.0881	2.53180	8.00625
6.42	41.2164	2.53377	8.01249
6.43	41.3449	2.53574	8.01873
6.44	41.4736	2.53772	8.02496
6.45	41.6025	2.53969	8.03119
6.46	41.7316	2.54165	8.03741
6.47	41.8609	2.54362	8.04363
6.48	41.9904	2.54558	8.04984
6.49	42.1201	2.54755	8.05605
6.50	42.2500	2.54951	8.06226
6.51	42.3801	2.55147	8.06846
6.52	42.5104	2.55343	8.07465
6.53	42.6409	2.55539	8.08084
6.54	42.7716	2.55734	8.08703
6.55	42.9025	2.55930	8.09321
6.56	43.0336	2.56125	8.09938
6.57	43.1649	2.56320	8.10555
6.58	43.2964	2.56515	8.11172
6.59	43.4281	2.56710	8.11788
6.60	43.5600	2.56905	8.12404
6.61	43.6921	2.57099	8.13019
6.62	43.8244	2.57294	8.13634
6.63	43.9569	2.57488	8.14248
6.64	44.0896	2.57682	8.14862
6.65	44.2225	2.57876	8.15475
6.66	44.3556	2.58070	8.16088
6.67	44.4889	2.58263	8.16701
6.68	44.6224	2.58457	8.17313
6.69	44.7561	2.58650	8.17924
6.70	44.8900	2.58844	8.18535
6.71	45.0241	2.59037	8.19146
6.72	45.1584	2.59230	8.19756
6.73	45.2929	2.59422	8.20366
6.74	45.4276	2.59615	8.20975
6.75	45.5625	2.59808	8.21584
6.76	45.6976	2.60000	8.22192
6.77	45.8329	2.60192	8.22800
6.78	45.9684	2.60384	8.23408
6.79	46.1041	2.60576	8.24015
6.80	46.2400	2.60768	8.24621
6.81	46.3761	2.60960	8.25227
6.82	46.5124	2.61151	8.25833
6.83	46.6489	2.61343	8.26438
6.84	46.7856	2.61534	8.27043
6.85	46.9225	2.61725	8.27647
6.86	47.0596	2.61916	8.28251
6.87	47.1969	2.62107	8.28855
6.88	47.3344	2.62298	8.29458
6.89	47.4721	2.62488	8.30060
6.90	47.6100	2.62679	8.30662
6.91	47.7481	2.62869	8.31264
6.92	47.8864	2.63059	8.31865
6.93	48.0249	2.63249	8.32466
6.94	48.1636	2.63439	8.33067
6.95	48.3025	2.63629	8.33667
6.96	48.4416	2.63818	8.34266
6.97	48.5809	2.64008	8.34865
6.98	48.7204	2.64197	8.35464
6.99	48.8601	2.64386	8.36062
7.00	49.0000	2.64575	8.36660
7.01	49.1401	2.64764	8.37257
7.02	49.2804	2.64953	8.37854
7.03	49.4209	2.65141	8.38451
7.04	49.5616	2.65330	8.39047
7.05	49.7025	2.65518	8.39643
7.06	49.8436	2.65707	8.40238
7.07	49.9849	2.65895	8.40833
7.08	50.1264	2.66083	8.41427
7.09	50.2681	2.66271	8.42021
7.10	50.4100	2.66458	8.42615
7.11	50.5521	2.66646	8.43208
7.12	50.6944	2.66833	8.43801
7.13	50.8369	2.67021	8.44393
7.14	50.9796	2.67208	8.44985
7.15	51.1225	2.67395	8.45577
7.16	51.2656	2.67582	8.46168
7.17	51.4089	2.67769	8.46759
7.18	51.5524	2.67955	8.47349
7.19	51.6961	2.68142	8.47939
7.20	51.8400	2.68328	8.48528
7.21	51.9841	2.68514	8.49117
7.22	52.1284	2.68701	8.49706
7.23	52.2729	2.68887	8.50294
7.24	52.4176	2.69072	8.50882
7.25	52.5625	2.69258	8.51469
7.26	52.7076	2.69444	8.52056
7.27	52.8529	2.69629	8.52643
7.28	52.9984	2.69815	8.53229
7.29	53.1441	2.70000	8.53815
7.30	53.2900	2.70185	8.54400
7.31	53.4361	2.70370	8.54985
7.32	53.5824	2.70555	8.55570
7.33	53.7289	2.70740	8.56154
7.34	53.8756	2.70924	8.56738
7.35	54.0225	2.71109	8.57321
7.36	54.1696	2.71293	8.57904
7.37	54.3169	2.71477	8.58487
7.38	54.4644	2.71662	8.59069
7.39	54.6121	2.71846	8.59651
7.40	54.7600	2.72029	8.60233
7.41	54.9081	2.72213	8.60814
7.42	55.0564	2.72397	8.61394
7.43	55.2049	2.72580	8.61974
7.44	55.3536	2.72764	8.62554
7.45	55.5025	2.72947	8.63134
7.46	55.6516	2.73130	8.63713
7.47	55.8009	2.73313	8.64292
7.48	55.9504	2.73496	8.64870
7.49	56.1001	2.73679	8.65448
7.50	56.2500	2.73861	8.66025
7.51	56.4001	2.74044	8.66603
7.52	56.5504	2.74226	8.67179
7.53	56.7009	2.74408	8.67756
7.54	56.8516	2.74591	8.68332
7.55	57.0025	2.74773	8.68907
7.56	57.1536	2.74955	8.69483
7.57	57.3049	2.75136	8.70057
7.58	57.4564	2.75318	8.70632
7.59	57.6081	2.75500	8.71206
7.60	57.7600	2.75681	8.71780
7.61	57.9121	2.75862	8.72353
7.62	58.0644	2.76043	8.72926
7.63	58.2169	2.76225	8.73499
7.64	58.3696	2.76405	8.74071
7.65	58.5225	2.76586	8.74643
7.66	58.6756	2.76767	8.75214
7.67	58.8289	2.76948	8.75785
7.68	58.9824	2.77128	8.76356
7.69	59.1361	2.77308	8.76926
7.70	59.2900	2.77489	8.77496
7.71	59.4441	2.77669	8.78066
7.72	59.5984	2.77849	8.78635
7.73	59.7529	2.78029	8.79204
7.74	59.9076	2.78209	8.79773
7.75	60.0625	2.78388	8.80341
7.76	60.2176	2.78568	8.80909
7.77	60.3729	2.78747	8.81476
7.78	60.5284	2.78927	8.82043
7.79	60.6841	2.79106	8.82610
7.80	60.8400	2.79285	8.83176
7.81	60.9961	2.79464	8.83742
7.82	61.1524	2.79643	8.84308
7.83	61.3089	2.79821	8.84873
7.84	61.4656	2.80000	8.85438
7.85	61.6225	2.80179	8.86002
7.86	61.7796	2.80357	8.86566
7.87	61.9369	2.80535	8.87130
7.88	62.0944	2.80713	8.87694
7.89	62.2521	2.80891	8.88257
7.90	62.4100	2.81069	8.88819
7.91	62.5681	2.81247	8.89382
7.92	62.7264	2.81425	8.89944
7.93	62.8849	2.81603	8.90505
7.94	63.0436	2.81780	8.91067
7.95	63.2025	2.81957	8.91628
7.96	63.3616	2.82135	8.92188
7.97	63.5209	2.82312	8.92749
7.98	63.6804	2.82489	8.93308
7.99	63.8401	2.82666	8.93868

J: Squares and Square Roots (continued)

N	N²	√N	√10N
8.00	64.0000	2.82843	8.94427
8.01	64.1601	2.83019	8.94986
8.02	64.3204	2.83196	8.95545
8.03	64.4809	2.83373	8.96103
8.04	64.6416	2.83549	8.96660
8.05	64.8025	2.83725	8.97218
8.06	64.9636	2.83901	8.97775
8.07	65.1249	2.84077	8.98332
8.08	65.2864	2.84253	8.98888
8.09	65.4481	2.84429	8.99444
8.10	65.6100	2.84605	9.00000
8.11	65.7721	2.84781	9.00555
8.12	65.9344	2.84956	9.01110
8.13	66.0969	2.85132	9.01665
8.14	66.2596	2.85307	9.02219
8.15	66.4225	2.85482	9.02774
8.16	66.5856	2.85657	9.03327
8.17	66.7489	2.85832	9.03881
8.18	66.9124	2.86007	9.04434
8.19	67.0761	2.86182	9.04986
8.20	67.2400	2.86356	9.05539
8.21	67.4041	2.86531	9.06091
8.22	67.5684	2.86705	9.06642
8.23	67.7329	2.86880	9.07193
8.24	67.8976	2.87054	9.07744
8.25	68.0625	2.87228	9.08295
8.26	68.2276	2.87402	9.08845
8.27	68.3929	2.87576	9.09395
8.28	68.5584	2.87750	9.09945
8.29	68.7241	2.87924	9.10494
8.30	68.8900	2.88097	9.11043
8.31	69.0561	2.88271	9.11592
8.32	69.2224	2.88444	9.12140
8.33	69.3889	2.88617	9.12688
8.34	69.5556	2.88791	9.13236
8.35	69.7225	2.88964	9.13783
8.36	69.8896	2.89137	9.14330
8.37	70.0569	2.89310	9.14877
8.38	70.2244	2.89482	9.15423
8.39	70.3921	2.89655	9.15969
8.40	70.5600	2.89828	9.16515
8.41	70.7281	2.90000	9.17061
8.42	70.8964	2.90172	9.17606
8.43	71.0649	2.90345	9.18150
8.44	71.2336	2.90517	9.18695
8.45	71.4025	2.90689	9.19239
8.46	71.5716	2.90861	9.19783
8.47	71.7409	2.91033	9.20326
8.48	71.9104	2.91204	9.20869
8.49	72.0801	2.91376	9.21412

N	N²	√N	√10N
8.50	72.2500	2.91548	9.21954
8.51	72.4201	2.91719	9.22497
8.52	72.5904	2.91890	9.23038
8.53	72.7609	2.92062	9.23580
8.54	72.9316	2.92233	9.24121
8.55	73.1025	2.92404	9.24662
8.56	73.2736	2.92575	9.25203
8.57	73.4449	2.92746	9.25743
8.58	73.6164	2.92916	9.26283
8.59	73.7881	2.93087	9.26823
8.60	73.9600	2.93258	9.27362
8.61	74.1321	2.93428	9.27901
8.62	74.3044	2.93598	9.28440
8.63	74.4769	2.93769	9.28978
8.64	74.6496	2.93939	9.29516
8.65	74.8225	2.94109	9.30054
8.66	74.9956	2.94279	9.30591
8.67	75.1689	2.94449	9.31128
8.68	75.3424	2.94618	9.31665
8.69	75.5161	2.94788	9.32202
8.70	75.6900	2.94958	9.32738
8.71	75.8641	2.95127	9.33274
8.72	76.0384	2.95296	9.33809
8.73	76.2129	2.95466	9.34345
8.74	76.3876	2.95635	9.34880
8.75	76.5625	2.95804	9.35414
8.76	76.7376	2.95973	9.35949
8.77	76.9129	2.96142	9.36483
8.78	77.0884	2.96311	9.37017
8.79	77.2641	2.96479	9.37550
8.80	77.4400	2.96648	9.38083
8.81	77.6161	2.96816	9.38616
8.82	77.7924	2.96985	9.39149
8.83	77.9689	2.97153	9.39681
8.84	78.1456	2.97321	9.40213
8.85	78.3225	2.97489	9.40744
8.86	78.4996	2.97658	9.41276
8.87	78.6769	2.97825	9.41807
8.88	78.8544	2.97993	9.42338
8.89	79.0321	2.98161	9.42868
8.90	79.2100	2.98329	9.43398
8.91	79.3881	2.98496	9.43928
8.92	79.5664	2.98664	9.44458
8.93	79.7449	2.98831	9.44987
8.94	79.9236	2.98998	9.45516
8.95	80.1025	2.99166	9.46044
8.96	80.2816	2.99333	9.46573
8.97	80.4609	2.99500	9.47101
8.98	80.6404	2.99666	9.47629
8.99	80.8201	2.99833	9.48156

N	N²	√N	√10N
9.00	81.0000	3.00000	9.48683
9.01	81.1801	3.00167	9.49210
9.02	81.3604	3.00333	9.49737
9.03	81.5409	3.00500	9.50263
9.04	81.7216	3.00666	9.50789
9.05	81.9025	3.00832	9.51315
9.06	82.0836	3.00998	9.51840
9.07	82.2649	3.01164	9.52365
9.08	82.4464	3.01330	9.52890
9.09	82.6281	3.01496	9.53415
9.10	82.8100	3.01662	9.53939
9.11	82.9921	3.01828	9.54463
9.12	83.1744	3.01993	9.54987
9.13	83.3569	3.02159	9.55510
9.14	83.5396	3.02324	9.56033
9.15	83.7225	3.02490	9.56556
9.16	83.9056	3.02655	9.57079
9.17	84.0889	3.02820	9.57601
9.18	84.2724	3.02985	9.58123
9.19	84.4561	3.03150	9.58645
9.20	84.6400	3.03315	9.59166
9.21	84.8241	3.03480	9.59687
9.22	85.0084	3.03645	9.60208
9.23	85.1929	3.03809	9.60729
9.24	85.3776	3.03974	9.61249
9.25	85.5625	3.04138	9.61769
9.26	85.7476	3.04302	9.62289
9.27	85.9329	3.04467	9.62808
9.28	86.1184	3.04631	9.63328
9.29	86.3041	3.04795	9.63846
9.30	86.4900	3.04959	9.64365
9.31	86.6761	3.05123	9.64883
9.32	86.8624	3.05287	9.65401
9.33	87.0489	3.05450	9.65919
9.34	87.2356	3.05614	9.66437
9.35	87.4225	3.05778	9.66954
9.36	87.6096	3.05941	9.67471
9.37	87.7969	3.06105	9.67988
9.38	87.9844	3.06268	9.68504
9.39	88.1721	3.06431	9.69020
9.40	88.3600	3.06594	9.69536
9.41	88.5481	3.06757	9.70052
9.42	88.7364	3.06920	9.70567
9.43	88.9249	3.07083	9.71082
9.44	89.1136	3.07246	9.71597
9.45	89.3025	3.07409	9.72111
9.46	89.4916	3.07571	9.72625
9.47	89.6809	3.07734	9.73139
9.48	89.8704	3.07896	9.73653
9.49	90.0601	3.08058	9.74166

N	N²	√N	√10N
9.50	90.2500	3.08221	9.74679
9.51	90.4401	3.08383	9.75192
9.52	90.6304	3.08545	9.75705
9.53	90.8209	3.08707	9.76217
9.54	91.0116	3.08869	9.76729
9.55	91.2025	3.09031	9.77241
9.56	91.3936	3.09192	9.77753
9.57	91.5849	3.09354	9.78264
9.58	91.7764	3.09516	9.78775
9.59	91.9681	3.09677	9.79285
9.60	92.1600	3.09839	9.79796
9.61	92.3521	3.10000	9.80306
9.62	92.5444	3.10161	9.80816
9.63	92.7369	3.10322	9.81326
9.64	92.9296	3.10483	9.81835
9.65	93.1225	3.10644	9.82344
9.66	93.3156	3.10805	9.82853
9.67	93.5089	3.10966	9.83362
9.68	93.7024	3.11127	9.83870
9.69	93.8961	3.11288	9.84378
9.70	94.0900	3.11448	9.84886
9.71	94.2841	3.11609	9.85393
9.72	94.4784	3.11769	9.85901
9.73	94.6729	3.11929	9.86408
9.74	94.8676	3.12090	9.86914
9.75	95.0625	3.12250	9.87421
9.76	95.2576	3.12410	9.87927
9.77	95.4529	3.12570	9.88433
9.78	95.6484	3.12730	9.88939
9.79	95.8441	3.12890	9.89444
9.80	96.0400	3.13050	9.89949
9.81	96.2361	3.13209	9.90454
9.82	96.4324	3.13369	9.90959
9.83	96.6289	3.13528	9.91464
9.84	96.8256	3.13688	9.91968
9.85	97.0225	3.13847	9.92472
9.86	97.2196	3.14006	9.92975
9.87	97.4169	3.14166	9.93479
9.88	97.6144	3.14325	9.93982
9.89	97.8121	3.14484	9.94485
9.90	98.0100	3.14643	9.94987
9.91	98.2081	3.14802	9.95490
9.92	98.4064	3.14960	9.95992
9.93	98.6049	3.15119	9.96494
9.94	98.8036	3.15278	9.96995
9.95	99.0025	3.15436	9.97497
9.96	99.2016	3.15595	9.97998
9.97	99.4009	3.15753	9.98499
9.98	99.6004	3.15911	9.98999
9.99	99.8001	3.16070	9.99500

Index

Index